Textbook of Microbiology

Textbook of Microbiology

Naveen Kango
M.Sc., Ph.D.
Dept. of Applied Microbiology and Biotechnology
Dr. Hari Singh Gour University
(Central University)
Sagar (M.P.)

TECHSAR™
NEW DELHI

Published and Printed by
Techsar Pvt. Ltd.
4435-36/7, Ansari Road, Daryaganj
New Delhi–110 002 (India)
E-mail: tech_info@techsarworld.com
Website: www.techsarworld.com

ISBN: 978-93-80026-44-2

Reprint 2024

Dedicated to

Aai (my mother)

Preface

Microbiology is a very fascinating branch of biology because here you isolate, watch, grow and manipulate tiny living cells using relatively much simpler methods. A simple medium formulation containing natural components and a Petri dish allows you to play around with microbial life at will. With the advancements in molecular biology, genetic engineering and fermentation technology, microorganisms have gained the centre stage of most biotechnological endeavors. From the very obscure and unwanted form of life, now they are being considered as an asset and in many cases indispensable for our sustainable development. These are now believed to be like small factories which when put in right milieu can churn out wonderful products for us.

This book has been written for the first year undergraduate and postgraduate students of microbiology, industrial microbiology and microbial biotechnology. The book covers all the topics as per the syllabus of B.Sc. Part I Industrial Microbiology and will serve as a ready reference for the young readers. Apart from this particular target audience, the book will also cater to the needs of other degree students of biological sciences and pharmaceutical sciences.

Section-I of the book covers essentials of microbiology including history, scope and milestones in the development of microbiology. This is followed by detailed accounts of characteristics and classification of microorganisms including bacteria, virus, fungi, and actinomycetes. Separate chapters on microscopy, isolation and maintenance of microorganisms, microbial growth and instrumentation will provide a basic idea of these techniques and their application.

Section-II of the book covers biochemistry and microbial genetics including chapters on carbohydrates, proteins, lipids, nucleic acids, gene regulation, translation and transcription along with detailed account of transport of biomolecules.

A lot of planning, effort and time has gone into shaping this volume and I hope that it will serve the purpose. However, I request readers to kindly intimate me their views and suggestions as there is always a possibility of improvement.

At this juncture, I wish to express my thanks to my colleagues and students for providing all the help, support and motivation to take up and complete this task. I express my heartfelt gratitude to my mentors, Prof. S.C. Agrawal and Prof. P.C. Jain for their constant support and encouragement. Special thanks are due to Mr. Sumat Jain, Mr. Hemant Rawat and Mr. Dinesh Mishra for going through the manuscript. I wish to express my sincere gratitude to my father, Shri G.G. Kango, my wife, Rashmi and son Utkarsh, for their constant support, encouragement and patience.

Naveen Kango

Contents

PART II: PRINCIPLES OF BIOCHEMISTRY AND MICROBIAL GENETICS

PART I

Fundamentals of Industrial Microbiology, Tools and Techniques

UNIT I

History and Scope of Microbiology

The history of microbiology or the development of science of microbiology is full of serendipities, diligently made inventions, circumstantial discoveries and efforts. A lot of this has come from efforts made to prevent and cure misery due to dreadful epidemics of that time including smallpox, anthrax, cholera, tuberculosis, leprosy to name a few. Existence of agents invisible to the human eye that spoil decoctions and cause fatal diseases was startling and hard to believe without substantial proof. The course of microbiology as a science is through invention of microscope, disapproval of theory of abiogenesis, germ theory of disease, culturing of microorganisms, vaccination, discovery and mass scale use of antibiotics. Later the role of microorganisms in food, beverage, environmental and agriculture sectors was also realized and worked out. Interestingly, most discoveries and inventions (e.g. antibiotics and microscope) were a culmination of efforts, inclinations and long standing experimentation of individuals rather than teams.

INVENTION OF MICROSCOPE

The development of industrial microbiology can be divided into three distinct phases. A period of ignorance before 18th century, an era of discoveries (1800–1900) including the golden age of microbiology and a period of rapid industrial development (after 1900).

Somewhere in the end of 15th century, Zaccharias Janssen invented a crude compound microscope by combining two lenses. A similar kind of apparatus called *Occiale* was built by the astronomer Galileo in 1610. The term microscope was used for the first time by Fabri in 1625. Before the invention of microscope, microorganisms were invisible and hence apparently non-existent. In the year 1665, Robert Hooke, an Englishman, declared that the smallest unit of life were "Cells". Working with a crude compound microscope he could see the cellular structure of plants and fungi. His lens setup could be compared with that of the today's dissecting microscope. After microscopic examination of cork through a compound microscope he described the regular cubicles as cells in the Journal *Micrographia* (1665).

Dutch merchant Anton van Leeuwenhoek (1632–1723) was the first to observe live microorganisms through magnifying lens (Figure 1.1). On his visit to England in 1668 for cloth merchandise he appreciated drawings of magnifications of cloth and on his return took up lens grinding. While examining a sample of pond water he was surprised to see distinct, well defined organisms. Deeply inquisitive and meticulous, Leeuwenhoek made numerous microscopes from silver and gold and examined almost anything and everything he could including the scum on his teeth. His best single lens tool could magnify ~300–500 folds which allowed him to see algae, protozoa and some larger bacteria and he could draw their exact drawings. He communicated his drawings and observations to the Royal Society of London in 1676. He described a variety of shapes of bacteria (spherical, rod and spiral) and their movements. People were startled and taken aback with Leeuwenhoek's claim that these tiny things were alive and motile. During 1673–1723, he communicated a series of letters to the Royal Society of London describing 'Animalcules' as observed under his indigenous microscope.

In a letter dated 17 September 1683 presented in Philosophical Transactions of the Royal Society of London, Leeuwenhoek described occurrence of microorganisms in his teeth scum and their movement as 'darting through water' making a variety of shapes A,B,G, and so on.

He made more than 250 microscopes using home made lenses mounted in gold, brass or silver. Looking at his fine descriptions and graphic details of shape and movement of microorganisms there is no doubt that he could watch bacteria, fungi and protozoans. In a report dated 16 June 1675, he has described his observations while watching well water as:

"*I discovered, in a tiny drop of water, incredibly many little animalcules, and these of diverse sorts and sizes. They moved with bandings, as an eel, always swimming with its head in front, and never tail first, yet these animalcules swam as well backwards or forwards, though their motion was very slow.*"

For near about 50 years, till his death in 1723, he communicated a series of his findings to Royal Society written in Dutch. These letters were translated and published in English as proceedings of the Royal Society. We should note that Leeuwenhoek was not the first to invent microscope and watch organisms but was certainly the first person to watch and describe fine details of bacteria and make people realize their ubiquitous presence.

DEBATE ON THEORY OF SPONTANEOUS GENERATION OR ABIOGENESIS

Origin of life has perplexed humans since the first human began to reflect on his world. The religious scriptures of ancient civilizations revolved around the sun,

Figure 1.1: (a) Anton van Leeuwenhoek (b) Leeuwenhoek's microscope.

fire, the seasons and the birth of new life and death were mostly construed as miracles. The idea that life can arise spontaneously, i.e. on its own from non-living matter was a common sense interpretation of miraculous life and was discussed for a great deal of time. In spite of observations that all the larger animals and plants reproduced life from previous life, people stuck with the theory of spontaneous generation. Francesco Redi (1626–1697) in 1688, before the discovery of microbial life, stood out to oppose the theory of spontaneous generation. He demonstrated that maggots do not arise spontaneously from decaying meat but from the eggs laid down by flies. He placed a piece of meat in three different jars, one open, one corked securely and the third was covered with fine mesh. No matter how long he left the jars, maggots appeared in the open container only. Redi's experiment was a milestone in disapproving theory of spontaneous generation because of its self-explanatory simplicity, clarity and reproducibility.

In 1745, John Needham, an Englishman, strengthened the theory of spontaneous generation on the basis of his experiments on boiled chicken / corn broth stating that the '*Vital Force*' of life is lost after boiling. Twenty years later an Italian, Spallanzani disapproved the idea saying that microorganism enter in broth after boiling. He showed that the fluid heated in a sealed flask did not develop any growth. Meanwhile, **Laurent Lavoisier**, described importance of oxygen to life and **Spallanzani's** conclusions were criticized on the ground that there was not enough oxygen in the sealed flask to support life.

Rudolf Virchow endorsed the theory of biogenesis claiming that 'living cells can arise only from pre-existing cells (*Omni cellula celluli*) as proposed and challenged the idea of spontaneous generation. Finally, the issue was resolved by the Frenchman in 1859, who was to become the Father of Microbiology in coming years, Louis Pasteur (1822–1895).

He was meticulous in devising conclusive experiments and demonstrated that microorganisms are already present in air and can contaminate sterile solutions. He used S-shaped or goose necked flasks that remained open to the air, but air could only enter by a bent path **(Figure 1.2)**. The flasks were then boiled to destroy contaminating microbes and were then observed for months. The flasks showed no signs of growth, until the sterile broth was touched up into the curved neck where Pasteur suggested that the airborne organisms would have settled. Pasteur claimed that microorganisms do not arise on their own but they are in the air and when the air comes through the goose neck of the flask, microbes settle on the sides of the thin glass necks without reaching the broth. His experiments explained that microorganisms are present in non-living matter, including solids, liquids and air. They can be destroyed by heat, and methods can be devised to block entry of microbes to nutrient environments. These ideas formed the basis of aseptic/pure culture techniques and are recommended as standard laboratory practice.

Louis Pasteur

Later, with the discovery of heat resistant bacterial spores, the occasional spoilage of boiled broths was also clarified. Pasteur continued to discover and establish the basic principles of Microbiology and is credited to lay down foundation for the Science of Microbiology.

FERMENTATION

Fermentation, in 18th century was believed to be a chemical process by leading chemists. In 1837, Schwann was the first person to describe fermentation as a process mediated by yeast. Pasteur confirmed Schwann's views and demonstrated that a living organism, yeast, converts sugars into alcohol. In 1857, Pasteur noticed that sugars were converted into lactic acid by an apparently different microorganism. Pasteur confirmed his findings by isolating these microorganisms and by conducting experimental fermentations. He followed cycles of fermentation and repeatedly isolated microorganisms from ferments and transferred them on sterile media. He

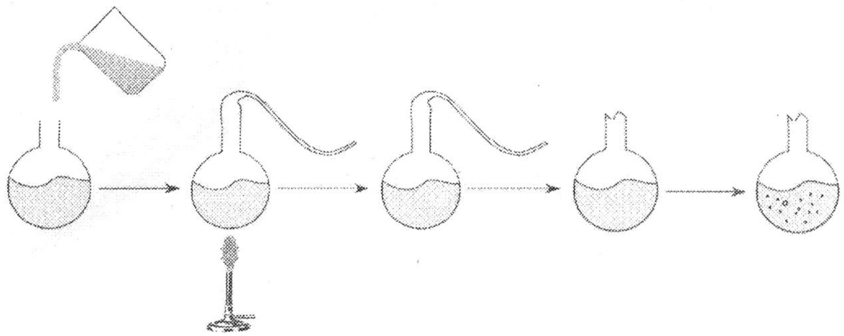

Figure 1.2: Goose necked flask experiment. Pasteur filled a round bottom flask with broth, drew the neck of the flask into S-shape (swan necked flask) and heated it to kill all living forms. Air could enter the flask but entry of microorganisms was prevented due to long S-shape neck. Once the neck was broken, the broth was contaminated with microorganisms present in surrounding air.

noted that good fermentation resulted when proper organisms and conditions were present. During 1857–1877, Pasteur observed a number of important points in fermentation and found that organism associated with production of tartaric acid was a fungus. He also investigated butyric acid fermentation and noted that fermentation can also be conducted in absence of oxygen (anaerobic). He called rod shaped organism associated with this fermentation as *Vibrio*. Pasteur observed that these bacteria were killed upon exposure to oxygen and coined terms, aerobic and anaerobic. He also studied acetic acid fermentation in great detail and termed these bacteria as *Mycoderma* (1862).

In the year 1866, he published a 264 page exclusive memoir on French wine and described that the so called wine-disease (or wine spoilage) was caused due to foreign organisms which altered its chemical and physical properties. Pasteur demonstrated that these spoilage organisms could be killed by heating fruit-juice below boiling point (pasteurization). The same process was later adapted to kill pathogens of milk. Ten years later, in 1876 he published a memoir on beer fermentations.

Pasteur also deciphered the role of yeast in wine production and suggested that the quality of wine could be controlled by altering conditions during wine making. Pasteur found that during making of wine, yeasts convert sugar in alcohol in absence of air (Pasteur Effect).

After the discoveries of Pasteur; there was an explosion of discoveries in the field of microbiology. This period (1857–1914), when most ground breaking discoveries in the field were made, has been referred as '*Golden age of Microbiology*'. Within 60 years (1857–1914), casual organisms of number of diseases were discovered, great strides in understanding microbial metabolism were made and techniques for isolating and characterizing microbes were improvised. Also the role of immune system (vaccination) was also understood to be critical in preventing and controlling diseases.

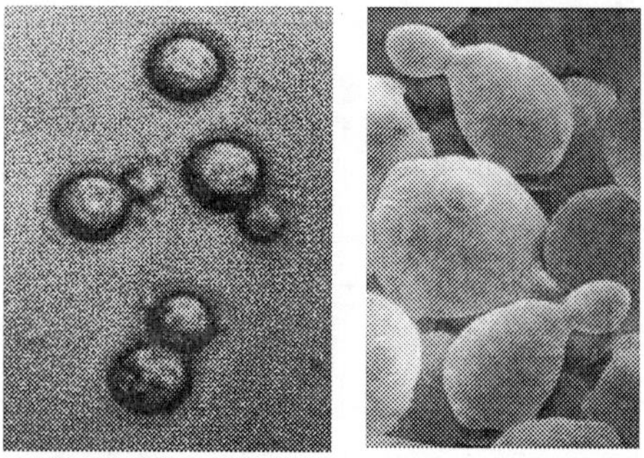

Figure 1.3: Morphology of *Saccharomyces cerevisiae* (a) budding yeasts cells (b) electron microphotograph of yeast.

CELL MEDIATED IMMUNITY

Elie Metchnikoff (full name Ilya Ilyich Mechnikov, 1845–1916) was a Russian microbiologist who established concept of cell mediated immunity (CMI) and explained role of phagocytosis in the immune system. Personal miseries, particularly death of his first and second wife because of tuberculosis and typhoid made him to take up study of microbes with special interest in the immune system. During his study he discovered phagocytosis after experimenting on the larvae of starfish. He proposed that white blood cells ingested and destroyed harmful extraneous bodies such as bacteria. In 1888, Pasteur gave him an appointment at the Pasteur Institute, where he remained for the rest of his life. In 1908, Mechnikov's work on phagocytosis was realized to be path breaking and won him the Nobel Prize. He worked with Émile Roux for development of an ointment that would prevent people from contracting syphilis, a sexually transmitted disease (STD).

Mechnikov also proposed a theory that aging is caused by toxic bacteria in the gut and that lactic acid bacteria could prolong life. Today, these life promoting bacteria are known as probiotics. Following his own theory, he consumed lactic acid bacteria in the form of sour milk every day and lived up to the age of 71 years. He wrote three books: 'Immunity in Infectious Diseases', 'The Nature of Man', and 'The Prolongation of Life: Optimistic Studies'.

The fact that a microorganism is responsible for fermentation and brings about physico-chemical changes in organic material made scientists to explore the possibility of parallel relationships of microorganisms with plants and animals. The concept gave rise to the fact that diseases are caused by live infectious agents rather than being a curse. Joseph Lister, an English surgeon applied this knowledge to medical practices.

In 1841, Ignaz Semmelweis, a young doctor working in maternity ward of Vienna hospital identified that Child Birth Fever or Puerperal Sepsis spreads due to an agent that was transmitted through unhygienic practices of doctors and midwives who went from one to another obstetrical patient. He instituted sanitary measures which said that the doctors should wash their hands with disinfectant and change their dirty aprons. After adopting this practice there was an apparent decline in death due to child birth fever. Lister started concept of aseptic surgery by treating surgical wounds with solution of phenol resulting in low mortalities of surgical patients.

THE GERM THEORY OF DISEASE AND KOCH'S POSTULATES

In 1876, **Robert Koch (1843–1910)**, a German rural physician, during his studies on anthrax (a common disease of both the farmers and their animals at that time), could see a rod shaped bacterium in the blood of a cattle died of anthrax under his microscope. He reasoned that it might be the agent of the disease, and using microbiological techniques he isolated the anthrax bacterium and purified it. Then he inoculated the purified bacteria into healthy animals and observed the development of anthrax. After examination of the inoculated animals he was able to re-isolate the same bacterium from the host. He repeated the isolation, infection and disease cycle until he was sure that he had found the agent of anthrax. Koch became famous because it was an important disease and his techniques could be easily duplicated, tested and used. Koch along with his co-workers collaborated with Pasteur's group and lead to the development of the fundamental techniques of microbiology. These techniques include the pure culture techniques, use of Petri plates, inoculation needles, solid medium, use of agar and gelatin to produce a solid surface, Gram's staining and other staining procedures. In addition, Koch discovered the etiological agents of *cholera* and *tuberculosis*. Together, the studies of Koch and Pasteur established the **Germ Theory** of disease. He won Nobel Prize for his work in the year 1905. His procedure for confirmation of the agent of any disease is referred as **Koch's Postulates** and involves following four steps:

(1) Isolate the suspected agent from a diseased host.
(2) Grow the agent in pure culture.

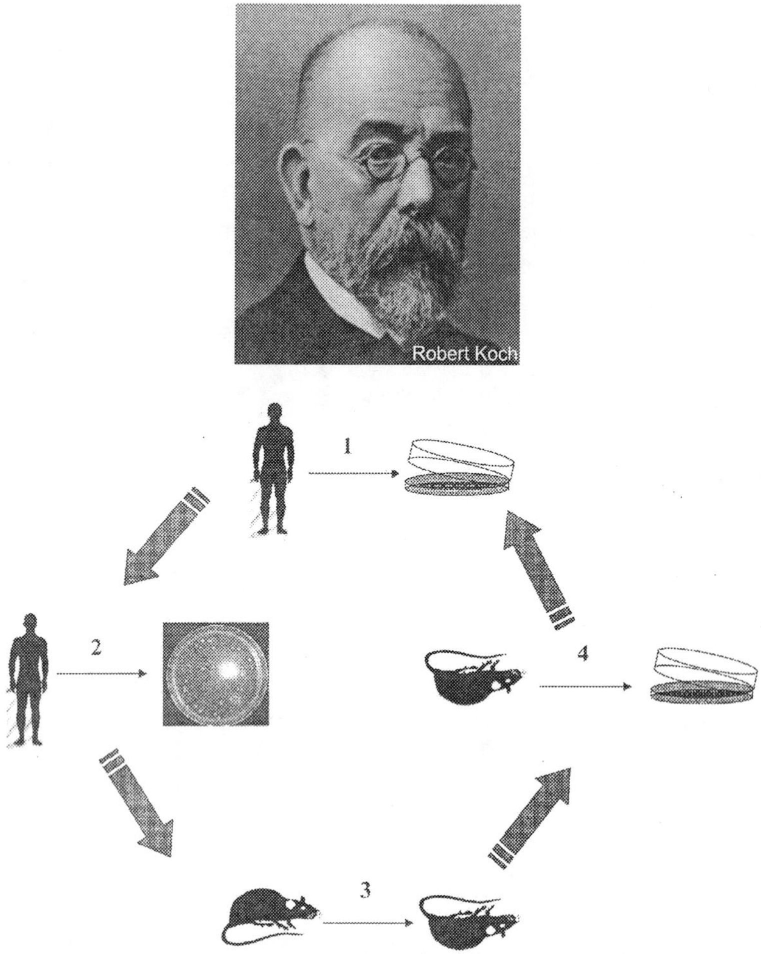

Figure 1.4: Four steps (1–4) for establishing etiology of any disease by Koch's postulates.

(3) Infect a healthy host and show that the organism produces the clinical disease.

(4) Isolate the same organism from the new host.

Even today, we follow these postulates for ascertaining etiology of most diseases excepting few typical diseases where either it is difficult to culture the causative organism *in vitro* or in case of viral diseases, e.g. HIV, which doesn't infect other animals, hence it is not possible to carry out step III.

VACCINATION

Edward Jenner (1749–1823), after training in London and a period as an army surgeon, spent his whole career as a country doctor in his native county of

Gloucestershire in the West of England. His research was based on careful case-studies and clinical observation more than a hundred years before scientists could explain about viruses. On behalf of the success achieved by him in vaccination in 1840, the British government had banned alternative preventive treatments against smallpox. "Vaccination," the word Jenner invented for his treatment (from the Latin *vacca*, a cow), was coined and adopted by Pasteur for immunization against any disease. He observed that milkmaid could not get smallpox because they contact mild infections of cowpox. He inoculated a healthy person with the cowpox material by scratching his arm which developed into a swelling and made the person mildly sick. Once recovered, the volunteer never caught cowpox or smallpox infection. Since then the process of vaccination has been used successfully to prevent many dreadful diseases. Jenner was a diligent student who followed William Harvey's advice, which was very famous among doctors of that time ***"Don't think, try"***. Jenner also contributed papers on angina pectoris, ophthalmia and valvular diseases of the heart but he is better known for his pioneering work in the field of smallpox vaccination.

Edward Jenner

In 1880, Pasteur discovered underlying mechanism of vaccination. He observed that the bacterium causing fowl cholera after repeated culturing in laboratory became avirulent (lost its ability to cause disease) but could trigger an immune response against wild virulent forms. This gave the idea that closely related cowpox virus can induce immunity against both itself and smallpox viruses. Today a variety of vaccines produced from avirulent attenuated microorganisms, killed virulent microbes, immunogenic component of microbe or by recombinant DNA technology are in common use.

DISCOVERY OF ANTIBIOTICS

Alexander Fleming was born in 1881 at Lochfield, Scotland (UK). He worked in Almroth Wright's research team as a research assistant with a strong interest in bacteriology. During the war between Britain and Germany in 1914, Fleming joined the British Royal Army Medical Corps to develop a cure to reduce the number of soldiers dying from infected wounds. He argued that antiseptics were not effective in preventing wounds from becoming infected.

A. Flemming

S. A. Waksman

When World War I was over, Fleming continued working at St. Mary's Hospital. One day in 1928, before tossing some old Petri dishes of culture away, he made an accidental discovery of a blue mold growing on the culture of some harmful kind of bacteria. The mold seemed to be able to kill and resist the growth of bacteria. By virtue of this serendipity, Alexander Fleming, developed the theory of antibiotic action. The fungus was later identified as *Penicillium notatum* and the active principle was called as Penicillin. He published his results in the journal of Experimental Pathology (1929). After a decade two compatriots, Chain and Florey developed the first antibiotic preparation for human use based on Fleming's observations. However, the vast potential of this antibiotic was realized only in 1940 during second world war and then it was produced at large scale and tested clinically. Before this, two antibiotics, Gramicidin and Tyrocidine from a soil bacterium, *Bacillus brevis* were discovered by a French microbiologist, Rene Dubos in 1939. Penicillin being a broad spectrum antibiotic, worked against many kinds of bacteria and was found to be mostly safe for the human body. Finally, in the late 1930s, other scientists found a way to mass-produce penicillin by fermentation and British and American drug companies began to manufacture the drug in large quantities. In 1945, Fleming was awarded the Nobel Prize for Medicine. He exclaimed that, "*Nature makes penicillin; I just found it.*"

Selman Abraham Waksman (1888–1973), a biochemist by profession, is better known for the discovery of Streptomycin (1943). He was born in Erstwhile Russia but moved to USA and performed research in soil bacteriology under Dr. J.G. Lipman at the New Jersey Agricultural Experiment Station. He worked as a Research Fellow at the University of California, Berkeley where he earned his Ph.D. in Biochemistry in 1918. Later he joined as a faculty at Rutgers University

(New Brunswick, NJ, USA) in the Department of Biochemistry and Microbiology. Dr. Waksman coined the term antibiotic and discovered several antibiotics, including actinomycin, clavacin, streptothricin, streptomycin, grisein, neomycin, fradicin, candicidin, candidin, and many other. Out of these, two antibiotics, **Streptomycin** and **Neomycin**, have found extensive application in the treatment of numerous infectious diseases. Streptomycin was the first antibiotic that could be used to cure the tuberculosis. The Nobel Prize in medicine was awarded to Waksman for his discovery of streptomycin, the first antibiotic effective against tuberculosis, and for the development of the methods and techniques that led up to its discovery and the discovery of many other antibiotics in 1952.

Waksman created the *Waksman Foundation for Microbiology* and Waksman Institute of Microbiology in 1951 using his own patent royalties. He has authored several books of varied interests including 'Enzymes' (1926), 'Humus: origin, chemical composition, and importance in nature' (1936, 1938), 'Principles of Soil Microbiology' (1938) and his autobiography 'My Life with the Microbes' (1954).

INDUSTRIAL PRODUCTION OF MICROBIAL PRODUCTS

The start of fermentation industry can be marked in the year 1880 when a plant of lactic acid fermentation was started in Massachusetts, USA. Most fermentation products were realized to have a direct application in food industry (lactic acid, baker's yeast, vinegar), chemical industry (ethanol, butanol, acetone) and textile and detergent industry (amylase, protease). During the first world war (1914–1918) high amount of acetone requirements triggered strong desire of scaling up such fermentations and substantial scientific inputs were made during 1920–1940 resulting in development of large-scale fermentation. These developments marked the emergence of a new multidisciplinary approach from a variety of scientific sectors. The development of chemical industries during first world war also helped development of fermentation industries. Success of microbial processes of citric acid, gluconic acid and acetone and butanol production convinced industrialists about taking up microbial fermentation for economic benefits. The period of second world war saw a similar spurt of interest in antibiotics. This was followed by development of bioprocesses for production of glutamic acid, lysine, biotransformation of steroids to antiinflammatory drugs (prednisoline, cortisone) during 1945–1960. Later on during 1960–1980, state-of-the-art processes for production of several antibiotics, enzymes (rennin), microbial insecticides flavour-enhancers, were introduced.

DISCOVERY OF AGAR-AGAR

Dr. Walther Hesse, a German physician, in 1881 studied many aspects of general public health and bacterial metabolism. His wife Angelina Fannie Hesse used to assist him in the laboratory work and also drew illustrations for his publications. One hot summer when Dr. Hesse had problems with his solidifying agent, gelatin, he asked his wife "Why do your jellies and puddings stay solid in the warm weather?" She explained that she used Agar-Agar, a complex polysaccharide extracted from seaweed *Gelidium*, to keep them solid in hot weather. Dr. Hesse subsequently found that it worked wonderfully in solidifying microbiological media. Since then Agar-Agar has been incorporated in all kinds of microbial culture media specially due to its non-toxicity and inertness to the most microbes, stability at sterilization temperatures, physiologically non-reactivity, high melting (100°C) and low solidifying temperature (45°C). As of today, Agar-Agar is a key ingredient of microbial culture media and is used as solidifying agent for growing microbes as separate entities on solid surface for the study of their colonial characters.

SCOPE OF MICROBIOLOGY

Microbiology is the study of living organisms that cannot be seen by naked eye. To see them, an aid such as magnifying glass or microscope is needed. Such microscopic organisms include bacteria, fungi, algae, protozoa and viruses. This branch of life science deals with identification, classification, form, structure, ecology and metabolism of microorganisms. With the advancement in the knowledge of microorganisms and their successful application to meet our needs, it is very clear that they are very important in human welfare in terms of ecology, production of goods and services.

Study of microorganisms causing human diseases involves etiology and host-pathogen interaction of microbial infection and diseases. Microbial diseases of plants cause severe damage to cash crops and have been responsible for infamous famines in the past. Microorganisms also deteriorate useful materials, e.g. iron pipes, wood, paints, clothes (biodeterioration). Apart from being detrimental there are several beneficial aspects of microorganisms. Microorganisms are involved in curdling of milk, ripening of cheese, production of bread, sofu, jalebi, dosa and numerous other oriental foods and hence microorganisms are indispensable for dairy, baking and other food industries. While they ferment and impart flavours, they also bring about significant losses by spoiling food in shelves and storages. Use of microorganisms for production of commodities such as alcohol, antibiotics, enzymes, organic acids utilizes inherent capability of these cell-factories. They are also claimed to be of utmost importance in rendering services such as bioremediation, biosensors and indicators. A diverse range of microbial activities also include soil fertility (bioinoculants), bio-leaching and recycling of metals and minerals, etc. Looking at the versatility of microorganisms, the research and development in each different area has gained status of a major field in itself.

The quest of improvisation and inventiveness coupled with major advents in genomics and proteomics has lead microbiologists in achieving great success in

Table 1.1 Milestones in the Development of Microbiology

The Development of Basic Techniques in Microbiology	
1664	Robert Hooke coined the term 'cell' to refer compartments of cork and described the fruiting structures of molds using a microscope.
1673	Anton van Leeuwenhoek communicated his observations about shape, size and motility of microorganisms to Royal Society of London.
1872	Ferdinand Cohn reported bacteria and their role in cycling of elements; coined the term '*Bacillus*'.
1872	Oscar Brefeld demonstrated the development of fungal colonies from single spores on gelatin and Joseph Schroeter, a Germen botanist, grew pigmented bacterial colonies on potato slices.
1877	Robert Koch established methods for staining bacteria, micro-photography and visual records of microorganisms on slides.
1881	Koch developed solid media and the methods to obtain pure cultures of bacteria.
1882	Angelina Fannie Hesse and Walther Hesse while working in Koch's laboratory introduced the use of agar in solidifying media.
1884	Hans Christian Gram developed method for differential staining for the identification of bacteria- The Gram's staining technique.
1887	Julius R. Petri reported the use of dishes for growing microorganisms over solid media—The Petri Dish.
1915	M. H. McCrady described a method for quantitative estimation for testing water samples for faecal contamination / potability, called as the most probable number (MPN) test.
1933	Knoll and Ruska developed electron microscope.
Disapproval of Theory of Spontaneous Generation (Abiogenesis)	
1668	Francesco Redi challenged the theory of spontaneous generation.
1745	John Needham reported development of microbes in boiled chick broth and supported the theory of spontaneous generation.
1768	Lazzaro Spallanzani disproved Needham's results and demonstrated that if air is not allowed to enter flasks containing boiled chick broth, no growth occurs.
1859	Louis Pasteur ended the debate on spontaneous generation with experiments in swan-necked or goose necked flasks and theory of abiogenesis / spontaneous generation was disproved.
1877	John Tyndall developed the method for fractional sterilization and reported existence of heat-resistant bacterial spores.
Developments in the Field of Medical Microbiology	
1840–1848	Ignaz Semmelweis demonstrated that washing or disinfecting hands prevented transmission of childbirth fever.
1854	John Snow related Cholera outbreak to drinking of contaminated water
1857	Louis Pasteur established the germ theory of disease
1867	Joseph Lister used Phenol as a disinfectant to treat wounds and disinfection of instruments used for surgery.

1873	G.H.A. Hansen discovered the causal organism of Leprosy—*Mycobacterium leprae* and established that leprosy is a contagious disease and is not inheritable.
1876	Robert Koch and Ferdinand Cohn identified causal organism of Anthrax—*Bacillus anthracis.*
1882	Koch isolated the causal organism of Tuberculosis (TB)—*Mycobacterium tuberculosis.*
1884	Koch proposed postulates to ascertain the causal organism of any disease.
1886	John Brown Buist became the first person who visualized a virus.
1892	Dmitri Ivanowski published the first evidence of the filterability of a pathogenic agent, the virus of tobacco mosaic disease (TMV).
1899	Beijerinck coined the term *contagium vivum fluidum* to refer Ivanowski's discovery.
1899	Friederich Loeffler and Paul Frosch discovered that foot and mouth disease is also caused by a filterable agent.
1915–1917	Frederick Twort and Felix d'Herelle discovered bacterial viruses (bacteriophages).

Treatment and Prevention of Diseases

1796	Edward Jenner used cowpox to immunize against smallpox.
1884	Metchnikoff demonstrated the process of phagocytosis and marked the beginning of immunology (cell mediated immunity).
1885	Paul Ehrlich initiated use of chemicals in the treatment of syphilis.
1886	Theobald Smith and D. E. Salmon developed the treatment of cholera by injecting killed hog cholera microorganisms.
1891	Paul Ehrlich showed role of antibodies in immunity (humoral immunity).
1897	Almwroth Wright and David Sample developed a vaccine against typhoid fever using killed cells of *Salmonella typhi.*
1897	Waldemar Haffkine developed a vaccine against the plague.
1912	Paul Ehrlich announced the discovery of a cure for syphilis—the first specific chemotherapeutic agent for a bacterial disease-Salvarson.
1929	Alexander Fleming published the first paper describing penicillin—the first antibiotic to be discovered.
1938	Max Theiler produced a vaccine against yellow fever by passaging the virus through mice to attenuate it.
1940	Howard Florey and Ernest Chain used extract of penicillin to kill bacteria in animals.
1940	Ernest Chain and E.P. Abraham described a substance from *E. coli* that can inactivate penicillin (antibiotic resistance).
1940	Selman Waksman discovered actinomycin—the first antibiotic from the actinomycetes. Later he discovered tetracycline and streptomycin.
1941	Charles Fletcher made the first administration of penicillin to a human volunteer.
1942	S.A. Waksman suggested the word "antibiotic" to describe the class of compounds produced by one microorganism that inhibit or kill other microorganisms.
1944	Albert Schatz, E. Bugie, and Selman Waksman discovered streptomycin, a very effective drug against tuberculosis.
1967–1979	Smallpox eradication programme by World Health Organization (WHO) started in 1967; Smallpox is declared to be eliminated in 1979.

Developments in Environmental Microbiology

1887	Winogradsky established the concept of autotrophy on the basis of his studies on *Beggiatoa*.
1888	Beijerinck developed enrichment culture technique.
1891	Winogradsky discovered microorganisms responsible for nitrification in soil.
1904	Beijerinck obtained pure culture of sulfur-oxidizing bacterium, *Thiobacillus denitrificans*.
1904	C. J. Koning reported importance of fungi in decomposition of organic matter.
1909	Sigurd Orla-Jensen used physiological characteristics for the classification of bacteria and published a monograph on lactic acid bacteria.
1920	The Society of American Bacteriologists (SAB) presented a report on the characterization and classification of bacterial types which became the basis for Bergey's manual in 1923.
1961	Brian McCarthy and E. T. Bolton described DNA hybridization for use in finding relatedness of two species.
1969	Don Brenner and colleagues established nucleic acid re-association as a parameter for identification of members of Enterobacteriaceae.
1977	Carl Woese established a new form of life, Archaea using rRNA analysis.
1977	Holger Jannasch discovered abundant life in deep sea hydrothermal vents based upon sulfur oxidizing microorganisms.
1982	Karl O. Stetter of Germany isolated hyper-thermophilic microbes able to grow at 105°C and redefined upper temperature limit of living organisms.
1994	Gary Olsen, Carl Woese and Ross Overbeek summarized the state of phylogeny in prokaryotes.

Important Milestones in Molecular Biology

1928	Frederick Griffith discovered transformation in bacteria and established the foundation of molecular genetics.
1941	Beadle and Tatum proposed the one gene – one enzyme hypothesis.
1944	Avery, MacLeod and McCarty gave experimental proof that DNA is the hereditary material.
1946	Lederberg and Tatum discovered another method of gene transfer/ recombination in bacteria-conjugation.
1952	Lederberg and Zinder described transduction in bacteria.
1953	Crick, Wilkins and Watson described the double helical structure of DNA.
1954	Francis Crick, Sydney Brenner and their co-workers proposed the existence of tRNA that mediates transfer of the information from DNA into protein.
1957	Francois Jacob and Elie Wollman demonstrated that *E. coli* has a circular chromosome as genetic material.
1959	Peter Mitchell proposed the chemiosmotic theory of transport.
1959	Arthur Pardee, Francois Jacob, and Jacques Monod showed that α-D-galactosidase enzyme is induced in presence of lactose.
1960	Kornberg demonstrated DNA synthesis in cell-free bacterial extracts and discovered an enzyme complex that catalyzes the synthesis of DNA.
1960	F. Jacob and J. Monod proposed a mechanism of the regulation of gene expression in bacteria via cluster of genes arranged in tandem in the lac operon.

1961	S. Brenner, F. Jacob and M. Meselson showed that ribosomes are the site of protein synthesis. The message from DNA is carried by RNA to direct protein synthesis.
1968	L. Margulis proposed that evolution of cell organelles, mitochondria and chloroplasts is a result of endosymbiosis between primitive bacterial forms.
1970	Temin and Baltimore independently discovered a novel enzyme, reverse transcriptase, a RNA dependent DNA polymerase that uses RNA as template to synthesize DNA. This phenomenon is an exception to the central dogma.

Important Developments in Genetic Engineering

1952	Salvador Luria and Mary Human, and Jean Weigle (independently) unveiled that there is a bacterial system of restriction and modification that restricts bacteriophage. Eventually it led to the discovery of restriction enzymes.
1966	J. Beckwith and E. Signer demonstrated the transfer of the lac region of *E. coli* into another microorganism. This established that the genetic elements can be moved and expressed in other hosts.
1967	Waclaw Szybalski and William Summers developed the technique of DNA-RNA hybridization during their investigations on bacteriophage T7.
1967	Thomas Brock identified *Thermus aquaticus*, a thermophilic bacterium able to grow at 85 °C that helped to establish a separate domain—Archaea. Later Taq DNA polymerase was isolated from it and used in PCR.
1967	Werner Arber showed that bacterial cells have DNA modifying enzymes that alter DNA by addition of methyl groups. Such modified DNA are recognized by the host cell and are not restricted by accompanying nucleases.
1970	Smith and Wilcox described the action of restriction nucleases and purified these enzymes from *Haemophilus influenzae.*
1972	Mertz and Davis characterized the *E. coli* RI restriction enzyme that cuts at a specific site on the DNA leaving the complementary cleaved open ends.
1972	Paul Berg created the first recombinant DNA molecule from viral and bacterial DNA.
1973	Stanley Cohen, Annie Chang, Robert Helling and Herbert Boyer established the method of gene cloning.
1975	A one year moratorium on gene cloning experiments was suggested after discussing issues in gene cloning at the Asilomar conference.
1977	Gilbert and Sanger independently developed methods for determining the sequence of DNA.
1983	Schell and Montagu and Chilton and coworkers produced transgenic plants.
1988	Kary Mullis of Cetus corporation, USA established PCR technology using Taq DNA polymerase.
1992	Complete sequence of one of the chromosome of *Saccharomyces cerevisiae* was determined.
1995	Craig Venter, Hamilton Smith, Claire Fraser, and colleagues at TIGR elucidated the first complete genome sequence of a microorganism, *Haemophilus influenzae*.
2000	The human genome project begun in 1990 is completed; entire human genome is deciphered.

Recent Recognitions for Landmark Findings in Biology
(Actual work was done before the year of award)

2002	Sydney Brenner, John Sulsoton and H. Robert Horvitz were awarded Nobel Prize in Physiology or Medicine of their work on molecular and developmental biology of *C. elegans* that was started in 1974 by Sydney Brenner and it has since been used extensvely as a model organism.
2003	Peter Agre and Roderich Machinon shared Nobel prize in chemistry for giving details of structure and operation of ion channels and aquaporins or water channels in cells.
2004	Aaron Ciechanover, Avram Hershka and Irwin Rose shared Nobel prize in chemistry for providing detailed knowledge of proteosomes and ubiquitin mediated protein degradation.
2006	Roger Kornberg won Nobel Prize in chemistry for describing details of transcription in eukaryotes. Arthur Kornberg, his father received the Nobel Prize in Physiology or Medicine in 1959 for his studies of how genetic information is transferred from one DNA-molecule to another.
	Andrew Z. Fire and Craig C. Mellow for the first time described RNA interference or RNAi in 1998 and shared Nobel Prize in physiology and medicine in 2006.
2007	Mario Capecchi, Martin Evans and Oliver Smithies shared Nobel prize for medicine or physiology for their work on knockout mice which serves as an important animal model.
2008	Barré-Sinoussi discovered AIDS virus in 1983 and recently shared the 2008 Nobel Prize in Physiology or Medicine with LucMontagnier and Harald zur Hausen, who discovered the viral cause of cervical cancer that led to the development of HPV vaccine.
	Osumu Shimomura an organic chemist and marine biologist shared Nobel Prize in Chemistry for the discovery and development of green fluorescent protein (GFP) with two American scientists, Martin Chalfie and Roger Tsien. The GFP is a useful tool for reporting gene expression in molecular biology experiments.

developing desired bugs, running large scale bioprocesses, deciphering genetic and regulatory controls, combating diseases with antibiotics and vaccines, applying biological controls, etc. Further, it promises to expand to newer horizons and hence domain of microbiology continues to be more and more attractive and challenging in coming future.

The field of industrial microbiology encompasses all applications of microbiology that have a high economic impact. The knowledge of microbiology can be used either to generate economic benefits or to avert or avoid economic losses due to biodeterioration, famines, epidemics, etc. Production of goods or commodities such as antibiotics, recombinant vaccines, biofertilizers, organic acids, solvents are some of the chief industrial processes, which involve growth and culture of either bacteria or fungi. On the other hand, biodeterioration of objects of economic importance such as paintings, timber, iron, stored cereals, food, etc. bring about severe economic losses. Thus, we find that the scope of microbiology ranges from the ancient art of wine making and production of oriental foods to production of molecular vaccine using different sorts of microorganism. Looking at the broad area of industrial microbiology and need of large-scale production of microbial products, the nature of subject necessarily becomes interdisciplinary. Apart from microbial physiology, microbial taxonomy, microbial genetics, dairy and food microbiology, agricultural microbiology, it also involves biochemistry, immunology, medical microbiology, bioprocess engineering, microbial biotechnology, fermentation technology, bioenergetics, effluent treatment and disposal.

Along with these disciplines, economics, intellectual properties rights (IPR), patent laws, biosafety regulations, sales, etc. also have direct relevance in successful use of industrial microbiology. Many industries which do not have direct concern with microorganisms have to recruit a microbiologist for reasons such as identification and checking of contamination in raw products, final products, solvents, water, etc. Quite a number of microbiologists share responsibility of quality control, particularly with regards to microbial contamination in a variety of industries, e.g. pharmaceutical industries, mineral water industries, dairy industry, soft drinks and beverage industries to name a few.

A multitude of microbial products such as enzymes, e.g. protease (detergent industry, leather industry, pharmaceutical industry), amylase (starch industry, textile industry, bakery), solvents (ethanol, butanol, acetone, organic acids-acetic acid, citric acid) have known for long, while a number of novel applications are based on use of whole microbial cells. A great deal of explorations, research, modelling and search is based upon the prospected industrial application of microorganisms and economic and environmental benefits thereof.

Most such efforts are focused upon replacement or avoidance of toxic and environment disturbing chemicals by microbial biotechnology based alternatives. For example, use of many enzyme based processes for replacement of toxic chemicals because chemical treatment is harsh and results in release of harmful and toxic by-products in effluent.

Explorations of exotic and extreme habitats have yielded a number of microbial forms that have revolutionized face of biological science as such. The advent of PCR and use of Taq DNA polymerase is one such example. Looking at the current estimate of microbial diversity, what we have seen is merely a tip of the ice-berg.

Industrial microbiology also deals with

(i) Discovering novel metabolites or biologically active molecules from microorganisms.
(ii) Exploring biodiversity for robust novel strains having better productivity than the existing ones.
(iii) Strain Improvement
(iv) Metagenomics

Metagenomics is the study of genetic material (or DNA) recovered directly from soils and other environmental niches. Conventionally, exploration of biodiversity goes along with the isolation and culture of microbes, however, having rapid techniques of DNA isolation, amplification and cloning, metagenomics has been in the centre stage of research. This branch can make use of non-culturable biodiversity by simply isolating their DNA from environment and its cloning in desired host.

Screening of microbial isolates for bioactive molecules is one of the prime thrust area of research and development (R & D) sector of pharmaceutical and biotechnology industries. Here the microbial cultures are studied for elaboration of any of the important metabolites for developing remedies for life-threatening diseases like cancer, AIDS, hepatitis, tuberculosis, etc. For this, isolates are subjected to high-throughput screening. In this regard, a variety of microorganisms from different ecological niches including marine environments, river sediments, alpine forests, and many more are being explored. A large number of microorganisms have also been found to thrive as endophytes which dwell inside the plant tissue and even cells. These are different from pathogens and saprophytes since neither they cause any diseases not they show any signs of tissue deterioration. Recently, there has been an upsurge in isolation and screening of endophytes for bioactive novel molecules.

With the advent in instrumentation research and development in industrial microbiology has seen a lot of change in past few decades. A hefty amount of money goes into the infrastructural set up for detection and assay of microbiological products. State of the art sophisticated instruments are now being used for detection of desired molecules at nano or pico scales. High pressure liquid chromatography (HPLC), gas chromatography (GC), mass spectrophotometer (MS), infrared spectrometer, etc. are now routinely employed for detection, analysis and structure elucidation of microbial products. Further a large array of set-up is also used for purification of these products. The desirability in degree of purity of any microbial product is based on its end use. A high degree of purity (up to homogeneity) is required for products to be administered intravenously (IV) in humans. Primarily, such products should be pyrogen and contamination free.

More recently, MALDI-TOF MS (Matrix-assisted laser desorption / ionization time of flight mass spectrometry) has been used for analysis of protein mixtures produced by microbial cultures. This is a novel technique where a co-precipitate of an UV-light absorbing matrix and a biomolecule is irradiated by a nanosecond laser pulse. Most of the laser energy is absorbed by the matrix, which prevents unwanted fragmentation of the biomolecule. Ionized molecules are accelerated in an electric field and enter the flight tube. During the flight in this tube, different molecules are separated according to their mass to charge ratio and reach the detector at different times. In this way each molecule yields a distinct signal. The method is used for detection and characterization of biomolecules, such as proteins, peptides, oligosaccharides and oligonucleotides, with molecular mass in the range of 0.4 kD and 350 kD. This method is highly sensitive and allows the detection of very low quantities of sample with an accuracy of 0.1- 0.01 %.

CONCEPT OF INDUSTRIAL STRAINS

Industrial microorganisms or industrial strains are supposed to be different from ordinary microorganisms in terms of productivity, genetic stability, robustness, tolerance, growth, etc. Basically, any strain is selected on the basis of relatively high productivity of desired metabolite as compared to other isolates. Such strains can be developed by using artificially induced mutagenesis. Characteristics of an industrial strain include:

(i) Production of high titres of desired metabolite / bioactive molecule.
(ii) Genetically stable over long periods and repeated subculturing and storage.
(iii) Rapid growth and quick product yield – saves operational cost of production process and lessens dangers of contamination.
(iv) Usually have a GRAS (generally regarded as safe) status; should not be pathogenic.
(v) Should allow easy downstream processing.
(vi) Should allow genetic manipulations including induced mutagenesis, recombinant DNA methods, transformation, protoplast fusion etc.
(vii) Should be able to grow in submerged culture conditions.
(viii) Quantification of the cells should be easy for standardization of inoculum and studying growth characteristics.

In addition to above, some extremophiles such as alkalophiles or thermophiles are also preferred. Growth conditions for such organisms are extreme and hence normal mesophilic flora is not able to contaminate the culture.

Usually, it is very difficult to find all these set of conditions in one particular microorganism. However, a microorganism may be directed to undergo mutagenesis and then selected against desired selection pressures or screens. Such mutants if found stable can then be used for the purpose of industrial scale production. Since times immemorial, industrial microbiology has been in practice for production of oriental fermented foods, drinks and condiments. But a truly commercial approach based on microbiology was developed in the 20th century that featured the microbial production of several new products, such as organic acids, solvents, vitamins and enzymes. Acetone fermentation was the first enterprise of industrial microbiology during the first world war (1918) while truly large aerobic fermenters were developed for production of penicillin during the second world war (1944). Later, during the 1950s to the 1980s several new antibiotics and anticancer agents were released by the pharmaceutical industry. The development of recombinant DNA resulted in the birth of biotechnology in the early 1970s. As of now, we are trying to develop sophisticated techniques to find novel therapeutics, enzymes and bioactive molecules by hi-throughput screening and rapid and sensitive detection and assay methods. Exploration of microbial diversity and metagenomics has become a regular feature of R and D endeavors. The diversity and functionality of microbial communities in artificial environments, e.g. those managing to live in industrial effluents under extreme conditions, is of special importance. Apart from finding out novel strains, such strains are also directed to develop into a truly robust industrial strain. Many microbial enzymes are being altered using directed evolution with the help of error prone PCR and other molecular biology techniques. Metabolic engineering is being applied to create novel recombinant strains of microorganisms. Also, marine environment has been recognized as a major resource for the isolation of novel microorganisms. Bioinformatics tools have immensely helped in the heuristics through microbial genomics and proteomics. Advances in proteomics and genomics have allowed us to comprehend genetics of natural product synthesis and the ways bioactive compounds are produced by microorganisms. Many novel derivatives of old compounds have been produced using grouping of the relevant genes and biosynthetic enzymes.

STRAIN IMPROVEMENT

Economy is one of the prime factors for success of any industrial microbial fermentation process. Usually, development of any microbial process starts with isolation of microorganisms and selection of a strain which is a better producer among the natural or native isolates. After identifying an organism that produces the desired valuable product in substantial amounts, it becomes essential, from the point of view of cost-effectiveness, to increase the product yield from fermentation to minimize production costs.

Product yields can be increased by strain development in following manner.

(i) Improving the productivity by hyperproducing strains (ii) Refining the fermentation process and decreasing the cost of individual steps in upstream or downstream processing

(iii) Developing a strain that utilizes a low cost medium for fermentation.

Microbial cells are susceptible to mutagenesis by physical and chemical methods. Their DNA can also be manipulated by recombinant DNA technology to impart desired traits. An industrial microbiologist examines all possible ways for introducing the desirable characters in industrial strain. The techniques and approaches used to genetically alter or modify strains to increase the production of the desired product are called strain improvement.

Traditional methods of strain improvement involve induction of mutation or induced mutagenesis either by growing microbial cells in chemicals (ethidium bromide, NTG) or by exposing them to ultraviolet light. After exposure the daughter cells are selected using an appropriate qualitative or quantitative screen. Further, the strain is characterized for growth, productivity and stability. This method is tedious as it involves screening and selection of a single mutant out of hundreds or thousands. As an alternative, the strain can also be genetically engineered to acquire desired traits. For example, the gene of the desired protein or enzyme can be linked to a very strong promoter so that it is strongly expressed. By manipulating DNA, we can obtain nonrepressible strains which are not repressed by catabolites or products.

Cost-effectiveness of microbial products is the major bottleneck in realization of their desired end use or application. The cost of any microbial product includes all the expenditure incurred towards upstream and downstream processing during production. Improvisation in bioprocess technologies, use of low-value substrates in production, development of rapid purification techniques are some promising interventions in minimizing the cost of microbial products.

SUGGESTED READING

Ainsworth, G.C. 1976. *Introduction to History of Mycology*, Cambridge Univ. Press, London.

Brock, T.D. 1975. *Milestones in Microbiology*, American Society of Mirobiology (ASM), Washington.

Dobell, C. (Ed.) 1932. *Antony van Leeuwenhoek and his 'Little Animals'*, Constable & Co. Ltd., London.

Gardner, E.J. 1972. *Hitory of Biology*, Wiley Eastern Limited, New Delhi, India.

Lechevalier, H.A. and Soltorovsky, M. 1995. *Three Centuries of Microbiology*, McGraw-Hill, NY.

Miller, B.M. and Litsky, W. 1976. *Industrial Microbiology*, McGraw-Hill Book Company, NY.

Roberts, R.M. 1989. *Serendipity: Accidental Discoveries in Science*, John Wiley & Sons, NY.

Doetsch, R.N. 1961. *History of the Microscope. Science* 133 (3456): 946–947.

Waksman, S.A. 1954. *My Life with Microbes*, Simon & Schuster, N.Y.

REVIEW QUESTIONS

1. Discuss important milestones in the development of microbiology.
2. Discuss contributions of following in the development of microbiology:
   a) Anton van Leeuwenhoek b) Alexander Fleming
   c) Selman Waksman d) Robert Koch
3. Discuss importance and scope of microbiology.
4. Define microbiology and discuss its various sub-disciplines.
5. Write short notes on the following:
   a) Discovery of microbial world b) Germ theory of disease
   c) Prevention and cure of diseases d) Pasteur's contribution to microbiology

UNIT II

Microscopy

With the advent of high resolution microscopes modern microbiologists have access to microscope that produces images with high clarity and magnification. The Leeuwenhoek's single lens microscope has been transformed into a high resolution multi-lens combination with magnification upto two thousand times. Further, electron microscopes with magnification upto one lakh times can be used to study fine structure of cells and sub-cellular components. Broadly, microscopes are categorized in two categories: Light (optical) microscopes and Electron microscopes. Light microscopy involves use of optical lenses and light radiations. It can further be categorized as: (1) Bright field microscopy, (2) Dark field microscopy, (3) Phase contrast microscopy, and (4) Fluorescence microscopy. Electron microscopy is of two types (1) Transmission microscopy and (2) Scanning electron microscopy. Light travels as a wave in crests and troughs. The amplitude of the crests and troughs determine the brightness of the light. The number of times complete wave occur per unit time is called as frequency and the distance between two consecutive crests is called as wavelength (λ) of the light. Wavelengths in the range of 400–700 nm make up visible spectrum while the UV region consists of wavelengths ranging from 100–385 nm. Visualizing any object directly by human eye involves incidence and reflection of light in the visual range. Microscopes use day light or light emitted by incandescent bulb. Fluorescent and UV microscope employ UV radiations. Electron microscopes use beam of electrons in place of light and thus object cannot be perceived by our eyes directly. The image produced by electron microscopes is perceived by CRT or X-ray plates.

MAGNIFICATION

The limit of an unaided human eye to visualize any tiny object is 0.1 mm. To see microorganisms, which are much smaller than 0.1 mm, a system of convex lenses is required. When parallel rays pass through the convex lens, they get converged at a point, called as focal length (f) of the lens. The magnification obtainable with a particular lens depends upon its focal length. Therefore, a lens with shorter focal length will have higher magnification power. In a compound microscope the real image is formed by the objective (10X, f = 16 mm; 40X, f = 4 mm; 100X, f = 1.8 mm). The image produced by objective lens falls on the eyepiece lens (10X; 15X) and serves as object for it. The image formed in the eyepiece is perceived by one eye and is the product of magnifications of two lenses. **Figure 2.1** shows a modern compound microscope and its various parts.

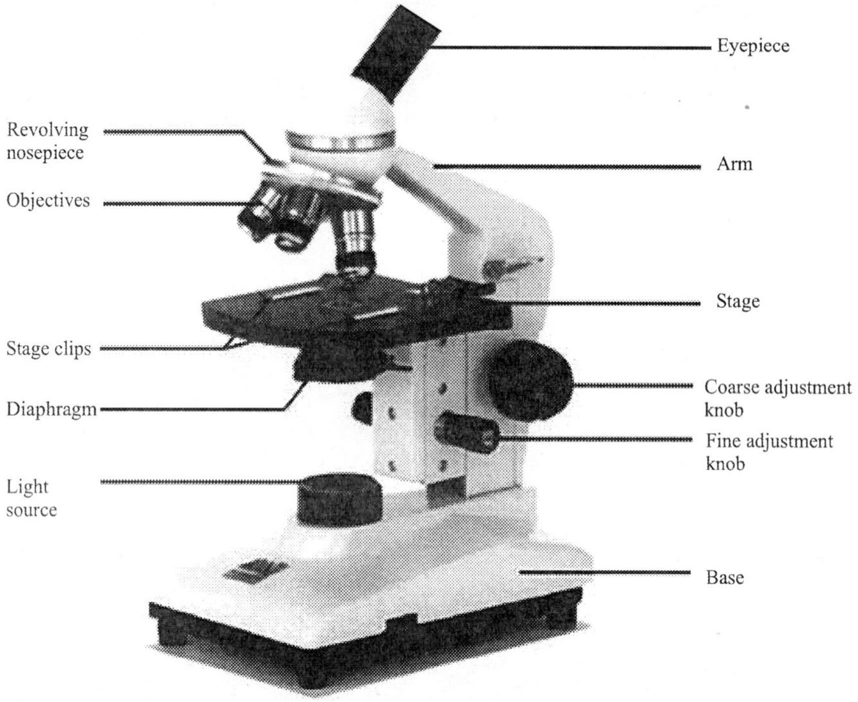

Figure 2.1: Parts of a compound microscope.

RESOLUTION

The ability to see two close objects as two distinct objects is called as resolution. The limit of resolution is the closest distance between two points at which the points still can be distinguished as separate entities. Magnification should be coupled with good resolution to visualize small microorganisms; else magnification alone will produce an inconclusive or blurred image. Resolution is defined as the distance at which two objects can be viewed as separate or distinct objects. The resolution (or limit of resolution) can be calculated as:

$$d = \lambda/2N.A. \quad \text{(i)}$$

or

$$d = 0.5\lambda/ n \sin \theta$$

or

$$d = 0.5 \lambda/ N.A.$$

where λ is wavelength of the light used, n is the refractive index of the medium, and θ is the angle subtended by the optical axis and outermost rays covered by the objective.

For $\lambda = 550$ nm and an immersion oil objective with N.A.=1.3, the maximum resolution will be

$$550 \text{ nm} / (2 \times 1.3) = 211.53 \text{ nm} = 0.211 \text{ µm}$$

The factor n sin θ is referred as numerical aperture (NA). Therefore, the shorter the wavelength (λ) of the light used, the better will be the resolution. Secondly, it depends on θ, the ½ the angle of the cone of light entering the objective (Figure 2.2). Light reaches to the object through a condenser and strikes the specimen. It reaches to the objective lens in form of a light cone. If the cone has a narrow angle (low power, f= 16 mm) it will not spread out much and hence can not separate two very close points, thus resulting in low resolution. If the cone of light has a wide angle (high power, f= 1.8 mm) two points are widely separated as two distinct entities.

Another crucial factor in the resolution is refractive index (n) of the medium. The medium, through which light passes, deviates the path of the light as per the refractive index and imparts change in NA. The refractive index of air is 1.0 and sin θ = 1 and numerical aperture (NA) is therefore 1.0 **(Figure 2.2a and b)**. Numerical aperture can be enhanced by changing the medium with suitable material that has higher refractive index such as immersion oil or cedar wood oil (refractive index = 1.56).

Table 2.1: Relationship between optical parameters (refractive index and aperture angle) and resolution

Optical parameter	Air	Canada balsam	Immersion oil
n (refractive index)	1.0	1.0	1.56
θ (½ aperture angle)	32°	48°	58°
N.A. (n sin θ)	0.6	0.8	1.3
Resolution (for white light)	0.458 µm	0.343 µm	0.211 µm

UNITS IN MICROSCOPY

Microorganisms and their structural components are measured in micrometers (µm) or micron (µ), nanometers (nm) and angstrom (Å). 1 µm or 1 µ is equal to 10^{-6} meters. The prefix micro indicates that the unit following it should be divided

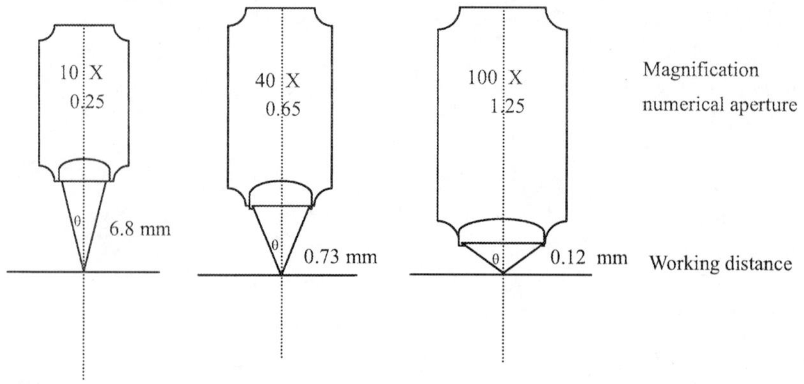

Figure 2.2(a): Numerical aperture and working distance of the 10X, 40X, and 100X (oil immersion) objectives of a compound microscope.

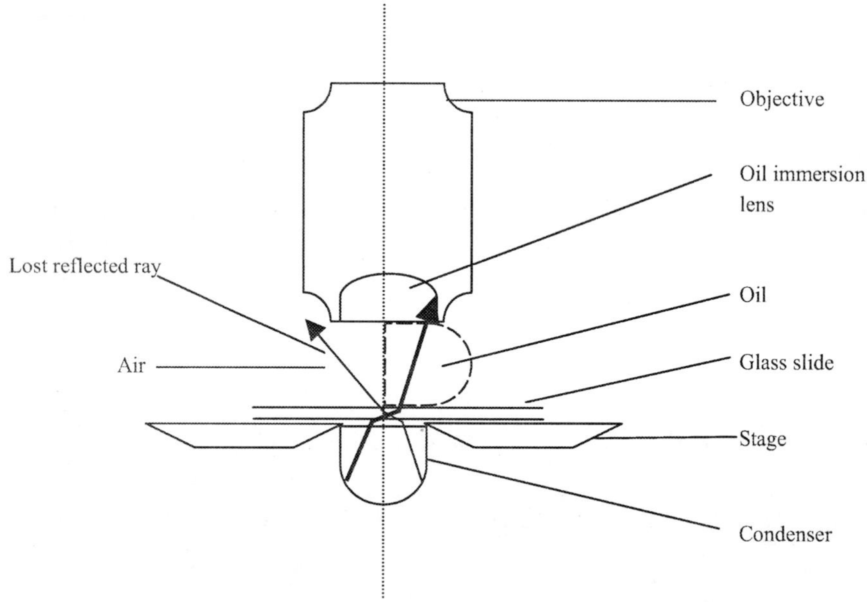

Figure 2.2(b): Ray diagram showing effect of immersion oil on cone of light (numerical aperture).

by one million (10^{-6}). A nanometer (millimicron) is equal to 10^{-9} meters. An angstrom (Å) is equal to 10^{-10} meters. The choice of microscopic unit is grossly dictated by the specimen size. However, all these above mentioned units are used in scientific literature and hence one should be acquainted with all these.

LIGHT MICROSCOPES

Simple Microscope

The microscope invented by Anton van Leeuwenhoek was a simple microscope having a single lens to see through. Today, these kinds of microscopes are not a very common sight and are only used for dissection purposes (also referred as Dissecting Microscopes).

Compound Microscope

A microscope with multiple lenses is referred as compound microscope. The invention of first compound microscope is credited to the efforts of Zaccharias Janssen. It uses visible light as source of illumination through a combination of finely ground lenses and results in a focused and magnified image. The light rays coming from a light source (illuminator) are converged or condensed through a condenser and are directed to fall upon the object (specimen). Further, these rays enter into objective and finally produce a magnified image by ocular lens in the eyepiece **(Figure 2.3)**. The resulting magnification of specimen can be calculated by multiplying the objective lens magnification power by the ocular lens magnification power. Most compound microscopes used in microbiological laboratories have 2–3 objective lenses including 10X (low power), 40X (high power)

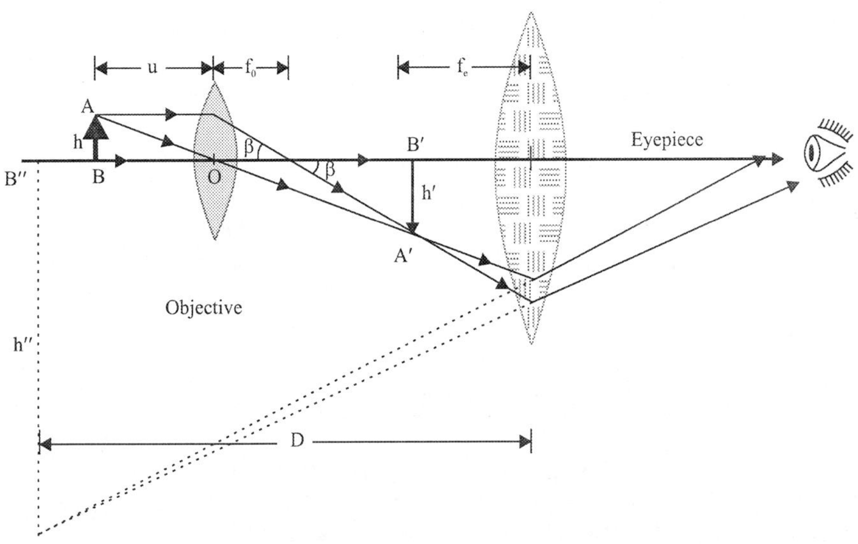

Figure 2.3: Ray diagram of a compound microscope showing formation of virtual, inverted and magnified image of the specimen by combination of objective and eyepiece lenses. The objective forms a real, inverted, magnified image of the object (A′B′). This image serves as the object for the second lens, the eyepiece, which forms a virtual, magnified and inverted image (A″B″) for the eye. (u: Distance of specimen from objective, f_0: focal length of objective lens, f_e: focal length of eyepiece lens, D: Distance of specimen from eyepiece, h″: length of magnified image).

and 100X (oil immersion). Most ocular lenses have magnifying power of 10X or 15X. As mentioned earlier, the resulting magnifying power will be obtained by multiplying the magnification powers of objective and ocular lenses. For an ocular lens of 10X, magnification produced by low power objective will be 10X × 10X = 100X and similarly 400X and 1000X for high power and oil immersion objectives, respectively.

Next to magnification a very important parameter in microscopy is resolution (resolving power). It is defined as the ability of the microscopic lens setup to distinguish between two close points. General principle in case of microscopes is that shorter the wavelength of illuminating light, the greater will be the resolution. The wavelength of commonly used white light is long and can effect resolution up to 2 nm only. Considering this and other practical constraints, the limit of magnification of the best compound microscope can be upto 2000X only.

For a sharp and clear image in compound microscope the specimen should sharply contrast with the medium (air). In order to attain such contrast, the refractive index of the specimen should be changed from that of its own medium. The refractive index is defined as the measure of relative velocity with which light passes through any material. The refractive index of any specimen can be changed by staining it. In normal course, light rays move in a straight line through a single medium. After staining, the incident light has to pass through two different media, i.e. air and the stained specimen. Due to difference in refractive indices the rays adapt a bending and refracting pattern at the edges of specimen and hence the contrast between medium and specimen is increased. The rays that scatter away from the specimen are spread out resulting in separation of close points, i.e. better resolution.

Objectives with high magnification (100X) have small lenses (small diameter) which are not sufficient enough to gather refracted rays. Light rays emerging out from the slide are refracted as they enter the air and the high power lens being small is not able to capture them resulting in a fuzzy image with low resolution. Thus, to maintain the direction of light rays at 100X, immersion oil is put between glass slide and objective lens. Immersion oil has similar refractive index to that of the glass and becomes part of optical path **(Figure 2.3)** and does not allow the light rays emerging from specimen to get away.

IMPROVISATIONS IN COMPOUND LIGHT MICROSCOPES

Bright Field Microscope

In normal course the field of a compound light microscope is brightly illuminated by focused natural sunlight or white light (electrical illuminator) resulting in a bright field image of the stained specimen. Such commonly used microscopes are called as bright field microscopes. However, certain improvisations in conventional light microscope produce special images that are used to draw important scientific conclusions.

Dark Field Microscope

Under some circumstances it may not be desirable to stain the specimen, e.g. while examining live microorganisms. Specimens that are not visible in ordinary light microscope or cannot be stained by standard methods or are deformed beyond limit by stain cannot be examined using ordinary light microscopes. Dark field microscopes have a special dark field condenser containing an opaque disc which blocks the light that would enter the objective directly and only light reflected by the specimen enters the objective lens. Since the background light is restricted, the specimen appears light against a black background (dark field). This kind of microscopic examination is used for viewing live, unstained microorganisms suspended in liquid.

Phase Contrast Microscope

Internal structures of living microorganisms can be examined under a phase contrast microscope without fixing or staining the specimen. A phase contrast microscope uses a special condenser containing an annular diaphragm with phase shifting element or annular stop and phase plate. This creates a diffraction phase by allowing a ring of light to pass through the condenser. The diffracted and undiffracted rays are then adjusted into phase with each other to produce image in the ocular lens. The first phase contrast microscope was devised by Fredericke Zernike in 1933.

The velocity of light rays passing through a specimen is altered by differences in the thickness and physical nature of parts of the specimen. The diffracted rays, therefore, travel differently and the phase differences are observed as different degrees of brightness in the eyepiece **(Figure 2.4)**. The internal structures of the specimen can be seen in sharp contrast with each other. Differential interference contrast microscope is an improvised version of phase contrast microscope. It is also based on differences in refractive index of diffracted light but uses two beams of light emerging through prisms. Prisms split light beams and provide contrasting colours to the specimen producing a high resolution coloured image. Essentially, PCM is useful for the study of microbial motility, determining the shape of living cells, and detecting bacterial components such as endospores and inclusion bodies that contain poly-β-hydroxybutyrate (PHB), poly metaphosphate and sulphur.

Fluorescence Microscope

Fluorescence is a phenomenon in which fluorescent substance absorbs short wavelengths of light (e.g. UV light) and emits a light of longer wavelength. Some organisms fluoresce naturally under UV light due to presence of intracellular fluorescent compounds, e.g. as in *Pseudomonas aeruginosa*. Otherwise also, the specimen can be stained with fluorescent dyes containing fluorochromes and examined under fluorescent microscope equipped with special light filters, viz. exciter filters and barrier filters **(Figure 2.5 a)**. The exciter filter allows only blue light to pass through and fall on the object while the barrier filter allows reflected

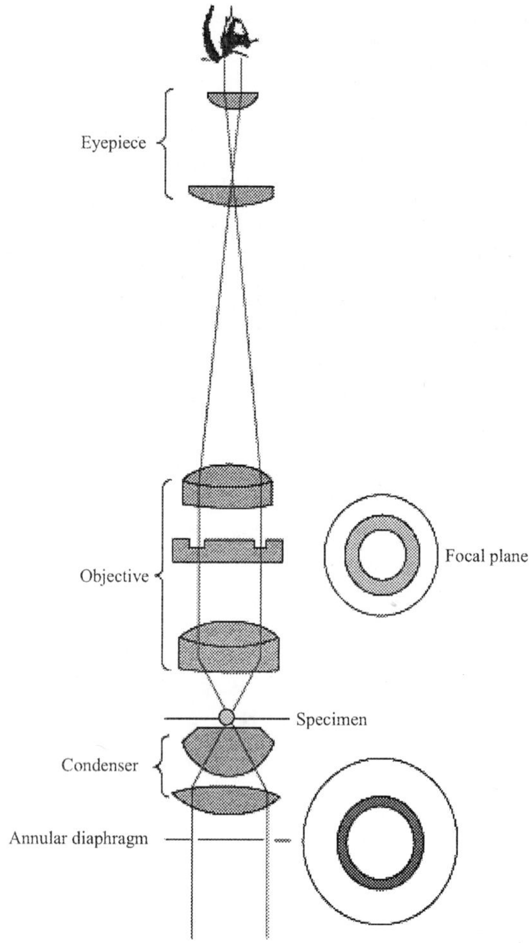

Figure 2.4: Diagram showing working of phase contrast microscope.

fluorescent light excepting blue. The specimen appears as a luminescent bright object against a dark background. Fluorescent microscopy is also used effectively for diagnostic or identification purposes. For example, flourescein isothiocyanate (FITC) selectively stains *Bacillus anthracis* that appear green while auramine 'O' is selective for *Mycobacterium tuberculosis* that appear yellow when fluoresce. The most important use of this technique is made in immunodiagnostics and is referred as immunofluorescence. In this technique, the fluorescent group is attached to homologous antibodies raised against the pathogen. Suspected clinical sample such as sputum, blood, etc. is mixed with labeled antibody on a slide. If the sample contains the pathogen, fluorescent antibodies bind firmly to the antigenic determinants of the pathogen making it fluoresce under UV light **(Figure 2.5b)**. Fluorescent treponemal antibody absorption (FTA-ABS) is a good example of

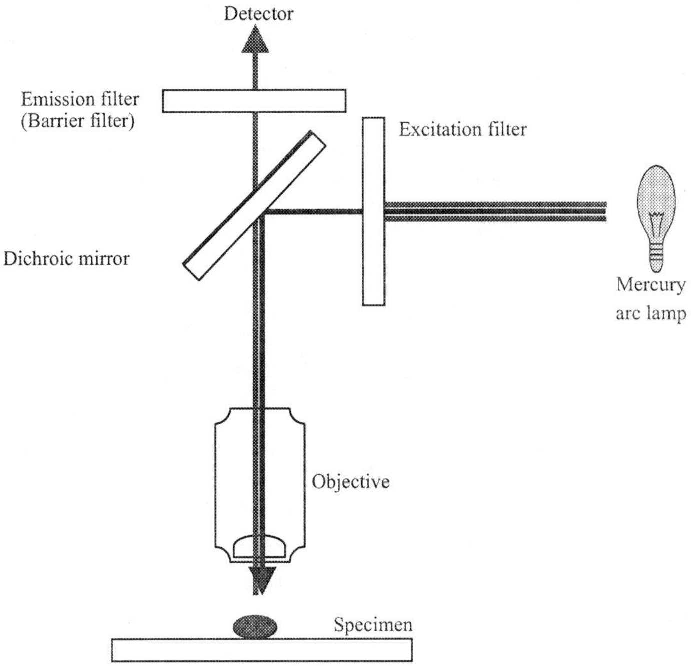

Figure 2.5(a): Line diagram of working of a fluorescent microscope.

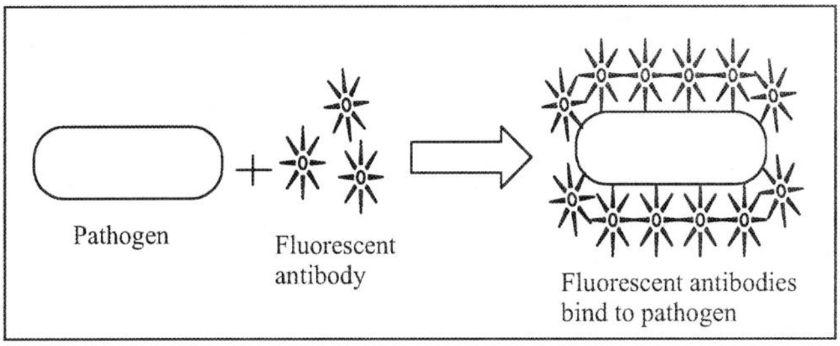

Figure 2.5(b): Fluorescent treponemal antibody absorption (FTA-ABS) for diagnosis of syphilis.

diagnosis of treponemal infection (Syphilis). When excitation light is transmitted from underneath or below the specimen, the phenomenon is called as transmitted fluorescence while if the excitation light is made to fall upon the specimen through objective lens, the phenomenon is referred as epi-fluorescence.

Confocal Microscopy

Confocal microscopy is a technique applied to enhance the contrast of photomicrograph for developing three-dimensional images by using a spatial pinhole to eliminate out-of-focus light or flare in specimens that are thicker than the focal plane. It was described and patented by Marvin Minsky in 1957. A confocal microscope uses point illumination and a pinhole in an optically conjugate plane in front of the detector to eliminate out-of-focus information. Thus, only the light within the focal plane can be detected, so the undesired flare of light is avoided and image quality is enhanced than that of normal optical images. Since a single point is illuminated at a time in confocal microscopy, 2D or 3D images are developed by scanning over a regular raster on the object.

MICROMETRY

A compound microscope is a combination of lenses resulting in formation of a virtual and magnified image of the microorganism or any specimen. For the purpose of characterization and identification of any species it is important to measure the dimensions micrometrically. **Figure 2.6** shows a comparison of dimensions of a variety of microorganisms. In this technique, firstly, the magnification of the lenses is calculated using a standard stage micrometer. For this an ocular micrometer (OM) having 100 equal spaced divisions without any standard unit is placed in the eyepiece and stage micrometer (SM) is placed on the stage. The value of ocular micrometer is calibrated with stage micrometer by focusing the scale of stage micrometer (having divisions spaced at 10 μm) through objective and superimposing its graduations with those of ocular micrometer **(Figure 2.7)**. The value of one division of ocular micrometer (OM) is calculated as:

1 division of O.M. = (No. of divisions of S.M. × 10 μm) / No. of divisions of O.M.

Stage micrometer is removed after calibration of ocular micrometer, and the slide of specimen is placed on the stage and examined. The dimensions of desired microscopic structures are measured using calibrated ocular micrometer. In case of fungi, length and breadth of spores, sporangiophores and other structures of taxonomic interest are recorded and compared with their taxonomic or monographic descriptions for identification.

Camera Lucida

This is an accessory instrument that allows us to draw the exact drawings of microscopic images seen through a microscope. The principle of camera lucida is based on simple optics. The image formed by the microscope in the eyepiece is reflected on the paper by a prism and reflecting mirror. The observer then moves the pencil around the desired structure of the microorganism visible against the white background of paper and draws the exact figure of the microorganism. There are three main parts of a camera lucida—the attachment ring, the prism and the mirror. The attachment ring attaches the camera lucida on the body tube around eyepiece. The prism is placed above the eyepiece when the instrument is attached with the microscope. The image of the specimen is reflected by the prism horizontally onto the plain mirror. The mirror is set at an angle of 45° with reference

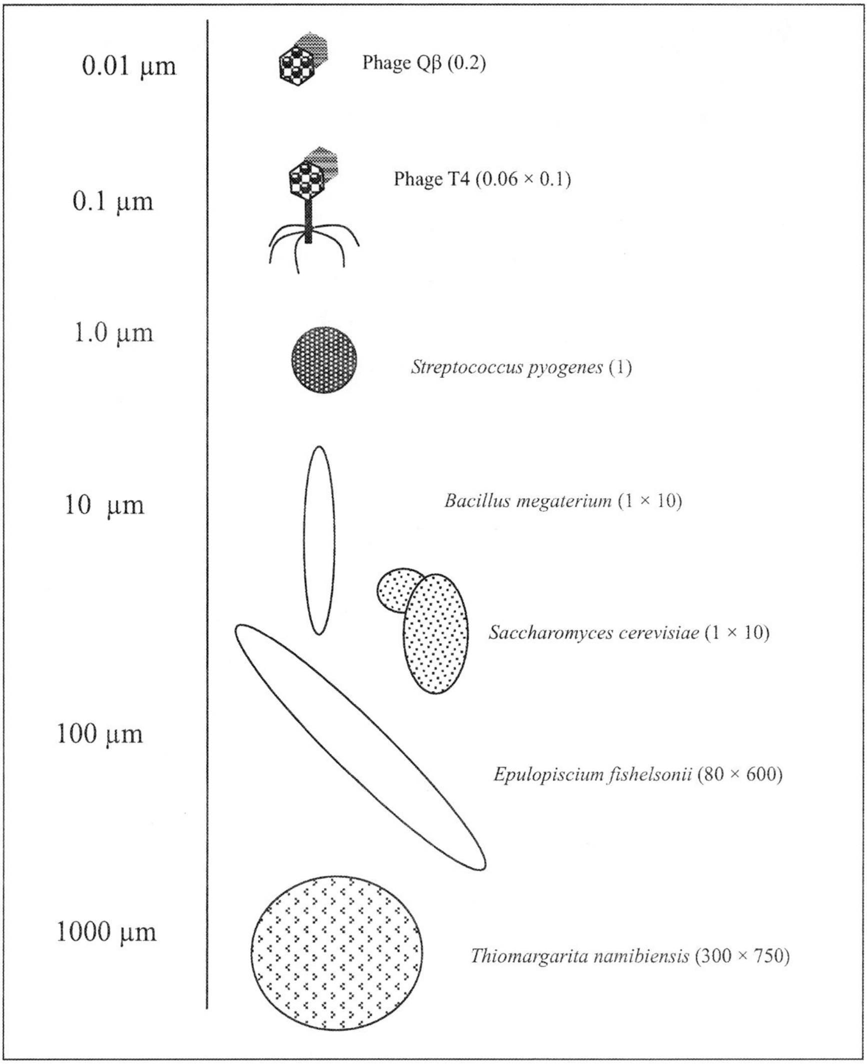

Figure 2.6: Range of size of various microbial forms.

to the prism and the plain paper and allows the light to pass from paper to prism and also from prism to paper **(Figure 2.8)**. A scale for indicating the size of the specimen can also be drawn using a stage micrometer. A camera lucida allows us to draw exact drawings of the microscopic structures as per our desire and need. This may not be the case with the camera where the entire visible field is photographed. The drawing made using a camera lucida can be later arranged using simple tracing and the structures can be arranged in a meaningful way, e.g. all the types of spores can be arranged according to size to describe the range of the taxon.

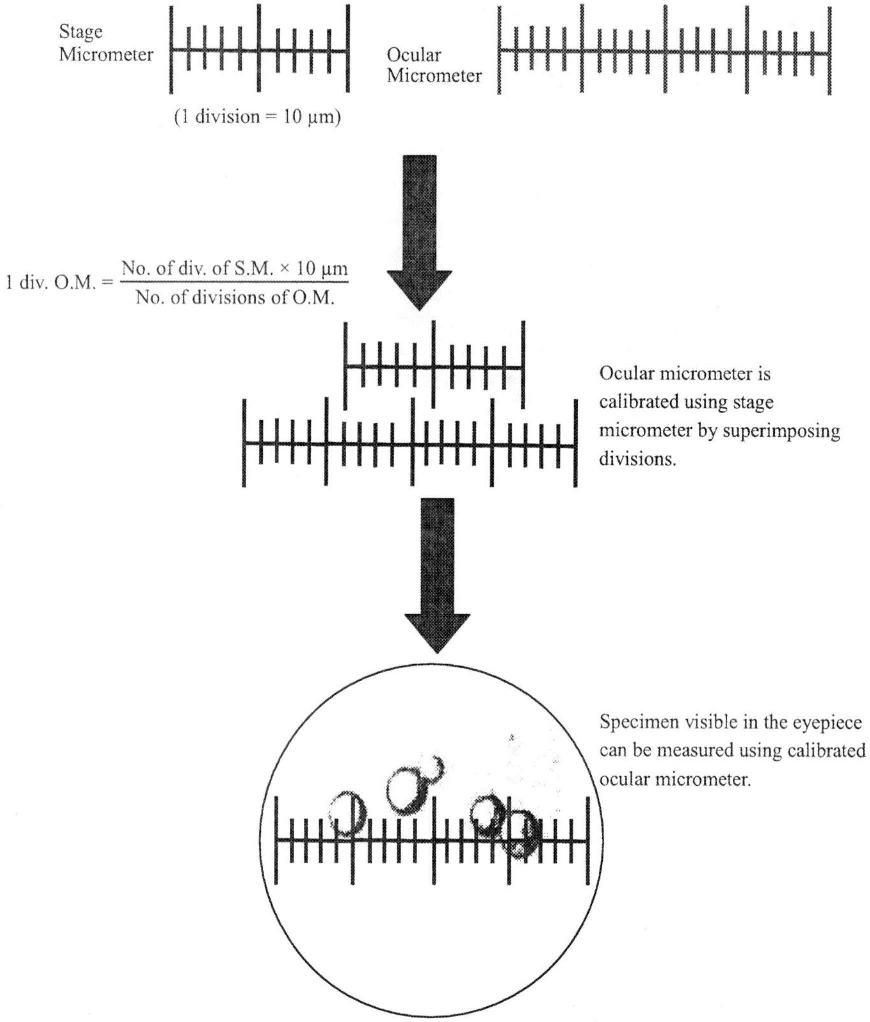

Figure 2.7: Calibration of ocular micrometer using stage micrometer.

ELECTRON MICROSCOPY

Looking at the equation (i), it is very clear that shorter the wavelength, higher is the resolution achieved. The electron microscope uses beam of electrons and strong magnetic fields in place of light and glass lenses, respectively. The wavelength of electron beam is only 0.05 Å or 0.5 nm whereas the white light used in light microscopes has wavelength in the range of 400–700 nm. The use of such shorter wavelengths in electron microscope results in magnifications up to 400,000 X. This also enables an electron microscope to achieve high resolution of objects (10

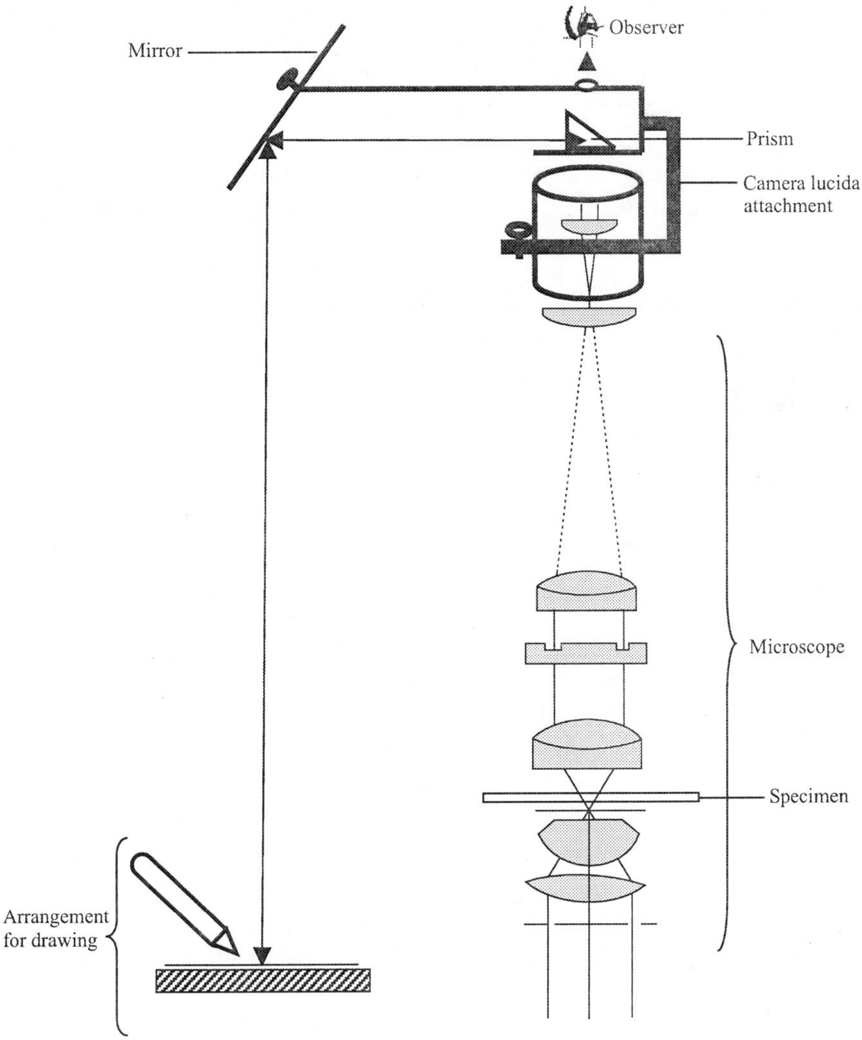

Figure 2.8: Schematic showing working of a camera lucida.

Å) which is almost 100 times than that of a light microscope. The magnified image can be viewed on a fluorescent screen (CRT- cathode ray tube or TV) or recorded on a photographic plate.

The specimen to be studied is prepared as a very thin dry film on small screens and is put in the instrument at a point (equivalent to stage of a light microscope) between the objective and condenser. The resulting magnified image can be viewed on a fluorescent screen through an airtight window or can be printed on a photographic plate with the help of an inbuilt camera.

Several techniques such as staining, shadow casting, freeze fracturing, etc. when used with electron microscope make it a highly useful instrument for study of finest details of cellular structure. *Shadow casting* involves deposition of a thin layer of metal (e.g. gold, platinum) at an oblique angle on the organism so that the organism produces a shadow on the uncoated side. This produces an image that provides topographical information, e.g. rough surface or echinulations of spore surface. Compounds that do not allow electrons to pass through (electron dense materials) such as phosphotungstic acid can be used to stain the outline of the object (*negative staining*). The deposition of such electron dense materials on the cell surface highlights its clefts and crevices and helps to derive fine details of the cell surface and appendages (flagella, pili, etc.)

Visualizing intracellular structures requires sectioning of the specimen cells. The cells are fixed in a block and then *ultrathin sections* (60 nm) are cut with ultramicrotome using fine glass knives. The resulting ribbon of thin slices is then examined and as one can imagine, different slices reveal different level and angle of intracellular structure. The contrast among the intracellular structures can be enhanced by using stains, e.g. uranium salts. Thin sectioning of specimen under frozen condition is called as *freeze etching*. Carbon replicas of the structures thus exposed reveal internal structures of the cell. Chemical fixation and staining procedures that may produce distortions or artifacts are avoided. Intracellular components of a cell can also be localized using immunochemical techniques. Thin sections can be treated with antibody labeled with electron dense substance such as Ferritin. The combination of such labeled antibody with specific cell antigens produces an image that highlights the ferritin in sharp contrast. Similarly, enzymes and other molecules can also be localized using homologous antibodies raised against a particular enzyme or molecule. The antibodies are labeled with colloidal gold or electron dense materials for providing a sharp contrast to the enzyme-antibody complex in electron microscopic image.

In transmission electron microscope (TEM) electron beams traveling through magnetic condenser lens fall upon specimen and create an image having high resolution at the final aperture **(Figure 2.9 a)**. In scanning electron microscopy (SEM) the specimen is scanned or rapidly moved over by a narrow beam of electrons resulting in release of secondary electrons and other radiations from the specimen surface **(Figure 2.9 b)**. The intensity of these secondary electrons varies with the shape and chemical composition of the specimen surface. These secondary electrons and radiations are collected by a detector which generates an electronic signal. These signals are then accumulated and scanned to produce an image at the cathode ray tube (similar to that of television). In this case the resolution is not as good as TEM but the pictures reveal three-dimensional topography of the specimen along with the depth of the field.

In electron microscope, specimens are examined in a chamber that is under a very high vacuum and hence these can not be examined in living state. Due to low

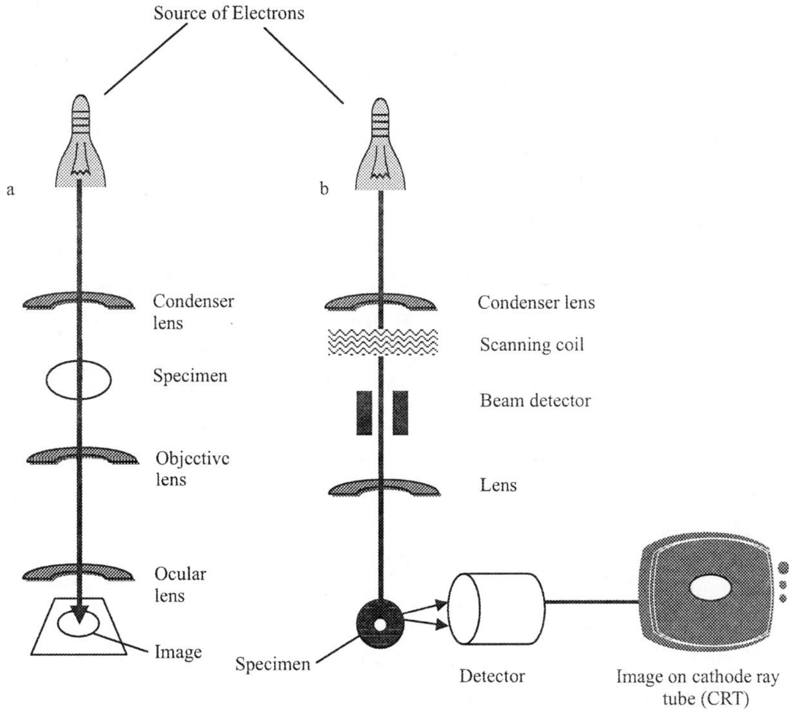

Figure 2.9: Diagram showing working of
(a) Transmission electron and (b) Scanning electron microscope.

penetration power of the electron beams, the specimen is to be prepared as a thin film and has to undergo processing, e.g. drying, freezing, staining, etc. which can lead to misleading morphological alterations, distortions or artifacts in it.

Scanning Tunnelling Microscopy

Scanning tunnelling microscopy (STM) is specifically meant for the study of the three-dimensional structure of the proteins and nucleic acids. A normal SEM picture of the macromolecule will lack desired vertical resolution while direct hit of the electrons during TEM distorts these molecules. In STM, images of the molecules are obtained only by scanning its surface under ultra-high vacuum. By using STM, we can determine intramolecular distances in materials with surfaces that can conduct electrons. Gerd Binning and Heinrich Rohrer were awarded Nobel prize in Physics in 1986 for inventing *scanning tunneling microscope.*

Immunoelectron Microscopy

This microscopy technique utilizes the power of an electron microscope to analyze cellular structural and functional details based on antigen and antibody reactions. Immunoglobulins (Igs) are labeled with electron dense materials, e.g. ferritin, phosphotungstic acid, colloidal gold, etc. Such antibodies when used for detection bind with their specific antigen and thus the antigen can be detected based on dense staining. With this method we can also study and visualize the movement or translocation of molecules in the cell and biosynthetic pathways.

STAINING TECHNIQUES

Several chromogenic synthetic and natural organic compounds (dyes) are used as stains to impart colour to the microscopic specimens. This enhances the visibility of the cells due to enhanced sharpness, clarity and contrast. Further, specific combinations of stains can be used to group or distinguish similar or different species of microorganisms (differential or selective staining). The staining procedure in most cases (especially bacteria) involves preparation of smear, fixation and application of stain and finally washing (to remove excess of stain).

The dyes are complex chemical compounds and may be classified on the basis of their chemical composition (thiazine, oxazine and triphenylmethane) or nature (acidic, basic or neutral). The selection of a dye is based upon the charge of the cell component to be stained, e.g. acidic dyes are used to stain basic cell components. Staining may involve ion exchange mechanism between the dye and the cell component. Methylene blue chloride (MB^+Cl^-, a blue coloured dye) ionizes as MB^+ and chloride ion (Cl^-). The exchange of ions during staining of a bacterial cell which is negatively charged and surrounded with sodium ions can be represented as:

$$(\text{Bacterial cell}^-)(Na^+) + MB^+Cl^- \rightarrow (\text{Bacterial cell}^-)(MB^+) + Na^+Cl^-$$

Simple staining involves use of single stain imparting colour of the dye to the microorganism by flooding it with the stain on a microscopic slide for a short duration and then washing off the excess stain with water. Cells pick up stain uniformly and hence the visibility and contrast is enhanced. Intracellular granules, e.g. polyhydroxy butyrate (PHB) may also become distinctly visible. Bacterial cells are stained using dyes such as methylene blue, malachite green, safranin, etc. after fixation while the fungal specimens are stained with lactophenol cotton blue. Sometimes, contaminants, if any, can be pointed out on the basis of a distinct staining pattern or intensity.

Bacteria exhibit a striking difference in their ability to retain crystal violet. The differential behaviour is attributed to the differences in their cell wall structure. In 1884, Christian Gram established the method of differential staining which formed the basis of bacterial classification into two groups, viz. Gram positive and Gram negative. Gram staining involves application of dyes and reagents in the following sequence: Crystal violet (violet colour stain), iodine solution (mordant), alcohol (decolourizing agent) and safranin (red coloured counterstain). Bacteria that retain violet colour after the whole procedure are referred as Gram positive. Here crystal violet binds firmly to their cell wall and resists decolourization with alcohol. In bacteria that appear red (Gram negative), crystal violet gets washed away with the decolourizing agent and hence they take up the counter stain colour.

The most plausible explanation of this differential behaviour lies in the chemistry and structure of cell wall. Gram-positive bacteria have a thick peptidogylcan cell wall while Gram-negative bacteria have thin layers of peptidoglycan and lipids. During the treatment with alcohol, the lipids of Gram-negative cell wall get washed away leading to increased porosity through which CV-Iodine complex escapes and the cells get decolourized. When treated with safranin they get stained to appear red. On the other hand, the porosity of thick peptidoglycan cell wall of Gram-positive bacteria is reduced due to dehydration by alcohol and CV-Iodine complex is retained with it providing a deep violet colour to the cells.

The behaviour can also be explained on the basis of structure and porosity of the cell walls. Gram-positive cell walls are made up of highly cross-linked peptidoglycan and bear very small pores whereas Gram-negative bacteria have very thin and less cross-linked peptidoglycan with larger pores. In Gram-positive bacteria the pore size is reduced after the alcohol treatment and CV-I complex is entrapped. Due to less peptidoglycan, there is not much diminution in the pore size of Gram-negative cell wall after the alcohol treatment and the CV-I is able to escape with decolourizing agent. Also, the fact that protoplasts of Gram-positive bacteria do not retain CV-I complex supports the idea that the retention of this stain depends upon their characteristic cell wall.

Old Gram-positive cells exhibit Gram variable behaviour, i.e. they many lose CV and get stained red with safranin. This alteration is attributed to changes in the cell wall with age and environment. The cell wall of Archaebacteria have entirely different structure and composition and some members of the group are Gram positive while some are Gram negative. The walls of Gram-negative archaebacteria are thinner than Gram-positive archaebacteria. Thus, the Gram behaviour is an important parameter for bacterial classification and identification. Gram staining does not apply to other microbial groups, e.g. protozoa and fungi, however, yeasts, unicellular ascomycetous fungi, are always Gram positive.

Staining in which the acidic dyes are repelled by the cell and thus background appears coloured while the specimen is light is called as negative staining, e.g. staining with Nigrosine. Usually, positive staining is used in which stain binds and imparts colour to the specimen. Acid fast staining or Zeihl-Nelson staining is also a differential staining techinique for differentiating acid fast bacteria from non-acid fast bacteria. Also bacterial appendages (flagella), capsules, endospores and intracellular deposits of starch, glycogen, poly-hydroxybutyrate, and lipids can be specially stained and visualized under light microscope.

SUGGESTED READING

Bradbury, S. 1984. *An Introduction to Optical Microscope*. Oxford Univ. Press, London.

Costerton, J.W. 1979. *The role of electron microscopy in the elucidation of bacterial structure and function. Annual Review of Microbiology* 33: 459–479.

Dunlap, D.D. and Bustamante, C. 1989. *Images of single stranded nucleic acids by scanning tunneling microscopy. Nature* 41: 204–206.

Goodhew, P.J. and Cartwright, L.E. 1975. *Electron Microscopy and Analysis*. Crane, Rusack & Co., Inc., N.Y.

Wischnitzer, S. 1981. *Introduction to Electron Microscopy*. Pergamon Press, N.Y.

Douglas B. Murphy 2001. *Fundamentals of Light Microscopy and Electronic Imaging*, Wiley-Liss Publishers.

REVIEW QUESTIONS

1. Draw a ray diagram of working of a compound microscope.
2. Discuss following in detail:
   a) Numerical aperture b) Magnification
   c) Resolution c) Oil immersion objective
3. Discuss the relationship among various optical parameters in microscopy.
4. Compare and contrast between:
   a) Light microscope and electron microscope
   b) Bright field and phase contrast microscope (PCM)
   c) Scanning electron microscope (SEM) and Transmission electron microscope (TEM)
5. Discuss the working and applications of fluorescent microscope.
6. Write notes on aplications of
   a) Micrometry
   b) Camera Lucida

UNIT III

General Characteristics and Classification of Bacteria

Microbiology is the study of living organisms of microscopic size which cannot be seen without any aid. These include bacteria, fungi, algae, protozoa and much smaller *obligate intracellular parasites*, viruses. Microbiology involves study of their form, structure, reproduction, physiology, metabolism, and classification. More precisely, microbes are an important object of study since many of these are involved in causing diseases in animals and plants, food spoilage and mineral recycling in nature (decomposers). More recently, these have been thought of as tiny factories, when put in a conducive atmosphere, churn out commercially important commodities and deliver various services, e.g. bioremediation of xenobiotics, biocontrol, etc.

Bacteria are small unicellular, prokaryotic microorganisms that normally measure in the range of 0.5–1.0 μm in diameter. Due to their smaller size, the surface area: volume ratio is very high which helps them to acquire nourishment from their surroundings via their surface. There is hardly any circulatory mechanism or cytoplasmic streaming in bacterial cells. We will discuss detailed characteristics of this heterogeneous and diverse group in this chapter.

Prokaryotes are classified as either archaea or eubacteria based on evolutionary studies of their 16S rRNA. In addition to this difference in their RNAs; the two domains also differ in the chemistry of their cell walls. Bacteria have murein and archaea have pseudomurein.

MORPHOLOGICAL FEATURES OF BACTERIA: SIZE, SHAPE AND ARRANGEMENTS

Morphological characteristics of bacteria involve their size, shape, structure, appendages and cell organization. Although dominated by spherical (cocci) and rod-like (bacilli) shape, several other shapes such as comma (vibrio), spiral and club, etc. are found. Moreover, often these cells are seen in arrangements such as regular or irregular clusters, chains, branched filaments, etc. The morphology of bacteria is important since many a times it is a determinative parameter of a taxonomic group, i.e. genus. Development of electron microscope has enabled us to see through the bacterial cell. With the help of molecular biology techniques, more precise comparisons of bacterial species are possible now.

Bacteria have a definite shape due to a characteristic cell wall. Due to the cell wall, these were initially believed to be plants but later on the basis of genetic and cellular makeup were placed in the kingdom Protista. The kingdom has been divided into two groups—Prokaryotes and Eukaryotes. Bacteria and blue-green algae belong to prokaryotes, i.e. those having an incipient nucleoid or naked DNA while fungi, algae, and protozoa are eukaryotes with a well-defined nucleus covered with a nuclear membrane.

Majority of bacterial cells are either spherical (cocci, sing. Coccus; *kokkos* meaning berry), straight rods (bacilli, sing. Bacillus; *baculus* meaning rod) or spiral rods (spirila, sing. spirilium). However, many other shapes such as comma, curved, trichomes, branched along with some pleomorphic forms are also observed. Small rod with a single twist that appear like a comma is called as vibrioid while the longer ones with more than one twist appear like a helix are called as spirillium (e.g. *Spirillium*). Spirochaetes (*spiro*-coil and *chaete*-hair) are flexible spiral bacteria.

Next to their shape, bacteria also exhibit characteristic arrangement or organization of cells as well. Many times the arrangement is indicative of a particular genus or species. Apart from single cells, the bacterial rods (bacilli) occur in pairs (diplobacilli) or chain of rods (streptobacilli). Some rods join to form long trichomes. The trichomes are also kind of chains but the cells are more stringently attached to each other with larger contact area at the ends as in *Beggiatoa* and *Saprospira* sp. Another arrangement is similar to that of palisade tissue of leaves in which cells lie side by side as in *Corynebacterium diphtheriae*. Members of Actinomycetes (acti-ray, mykes-fungi, also called as *ray fungi*) form long branched, multinucleate filaments similar to fungal hyphae that collectively give an appearance like mycelia.

However, many bacteria not conforming to aforesaid shapes are also observed. These include pear-shaped (*Pasteuria*), lobed spheres (*Sulfolobus*), rods with straight blunt ends (*Bacillus anthracis*), disks (*Caryophanon*), etc. Mycoplasma or (PPLO) are cell wall deficient bacteria and hence are pleomorphic, i.e. without any rigid or well-defined shape. They are seen as round or oval bodies and are filterable due to pliabitity.

CELL WALL

The bacterial cell acquires its shape from a rigid cell wall surrounding the cytoplasm. With the help of this wall, bacteria are able to survive osmotic turgor pressure in hypotonic environments and retain shape under pressure and other kind of physical stress conditions. Among eubacteria, Gram-positive bacteria have thicker (20–80 nm) cell wall than Gram negative bacteria (10–15 nm). Archaebacterial cell walls have much different chemical composition. Cell wall performs several important functions for the bacterial cell and its removal (formation of protoplast) causes disturbances in cell and physiology.

Components of the Cell Wall

Peptidoglycan: The bacterial cell wall is made up of peptidoglycan which is a porous, insoluble polymer of sugar cross-linked with peptides (peptido: peptides; glycan: sugar). Exclusively found in prokaryotes, the peptidoglycan forms a firm and rigid sac or bag-like structure around the cytoplasmic membrane. These are the heteropolymers of substituted acetylated amino sugars (N-acetyl glucosamine: NAG, and N-acetyl muramic acid: NAM, also called as amino sugars) and short peptides having L-alanine, D-alanine, D-glutamate and a diamino acid (L-meso diaminopimelic acid and L-lysine). Across the different bacterial cell wall, amino acids vary uniquely and hence over 100 chemotypes of peptidoglycan are known. The peptidoglycan of archaebacteria is different from that of eubacteria with a conspicuous absence of N-acetylmuramic acid (replaced with N-acetyltalosaminouronic acid or pseudopeptidoglycan or pseudomurein).

The amino sugars form glycan composed of alternating residues of NAG and NAM linked with β-1, 4 glycosidic bond with each strand containing 10–65 disaccharide units. The peptide chain is attached at carboxyl group of lactic acid moiety of NAM by a peptide bond. The peptide sequence shows a great diversity, however, the most common sequence is L-ala—D-glu—*meso* DAP—D-ala, present in most Gram negative and some Gram-positive bacteria (**Figure 3.1 a** and **b**). However, in Gram +ve bacteria, meso DAP is replaced by L-lysine. Many adjacent peptide chains may get cross-linked by formation of a peptide between terminal alanine residue of one and the α-amino group of *meso*-diaminopimelic acid of the other. Such cross-linking of peptide chains gives peptidoglycan a lattice or mesh character good enough to enclose protoplasm. Peptidoglycans can also differ in terms of cross-linking position of peptides. The cross-linking depends upon the availability and vicinity of free carboxyl and amino groups of amino acids. In the most common (also called as direct type) L-ala—D-glu—*meso*DAP—D-ala peptide, cross-linking occurs between the terminal D-alanine and meso-diaminopimelic acid. On the basis of position of cross-link among peptides, the peptidoglycans are classified as Group A (at third amino acid) and Group B (cross-linked at second amino acid). All di-amino acids (L-diaminopimelic acid, L-diaminobutyric acid, L-lysine and L-ornithine) always participate in the cross-linking. In Group A peptidoglycans the third amino acid residue of the peptide chain binds either directly to the terminal D-alanine residue of an adjacent chain (mDAP direct type linkage) or through a peptide bridge of variable length and composition (**Figure 3.2 a** and **b**).

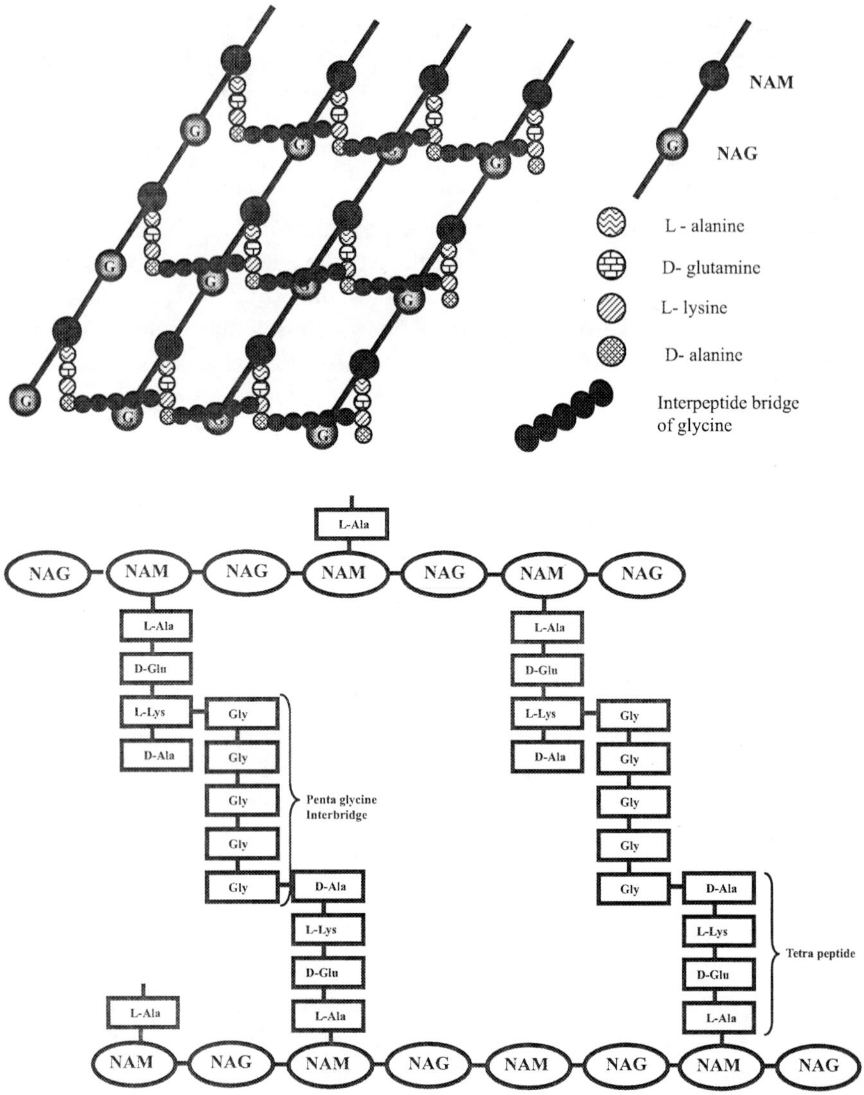

Figure 3.1 (a): Structure of Gram-positive cell wall (e.g. *Staphylococcus aureus*). The NAM-NAG polymer is cross-linked by tetrapeptides of amino acids and interbridge of pentaglycines.

GRAM STAINING OR GRAM REACTION

Bacteria are frequently divided into two groups on the basis of their reaction to staining devised by Christian Gram (1884). The differential reaction to the staining procedure is because of differences in the structure of their cell walls. In this, the heat fixed cells are stained with crystal violet followed by iodine which forms

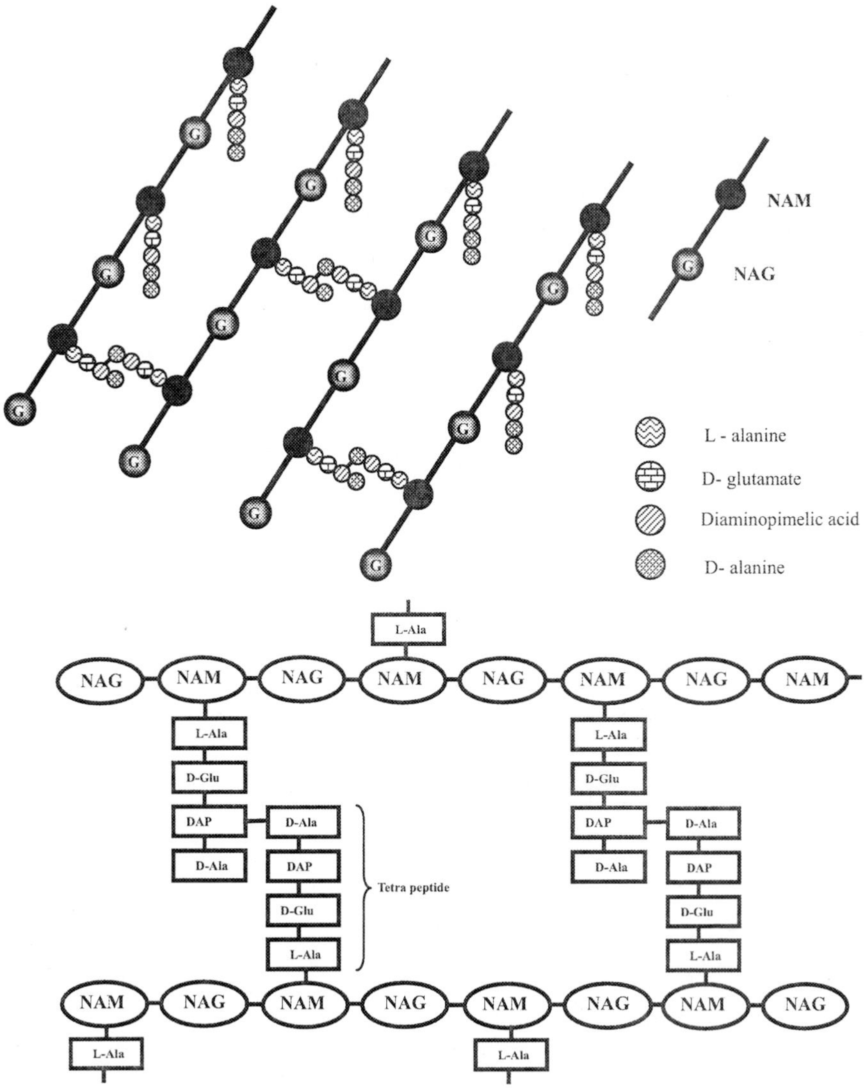

Figure 3.1 (b): Gram-negative peptidoglycan (e.g. cell wall of *E. coli*). The site of action of lysozyme is the bond that links G to M.

complex with the stain in the cell wall. Alcohol is used to decolourize crystal violet-iodine complex out of thin walled Gram-negative bacteria but thick walled Gram-positive cell retain the CV-I complex. Finally, a counter stain, safranin, is added to colour decolourized cell walls. Gram +ve bacteria appear violet while Gram-negative bacteria appear pink colour.

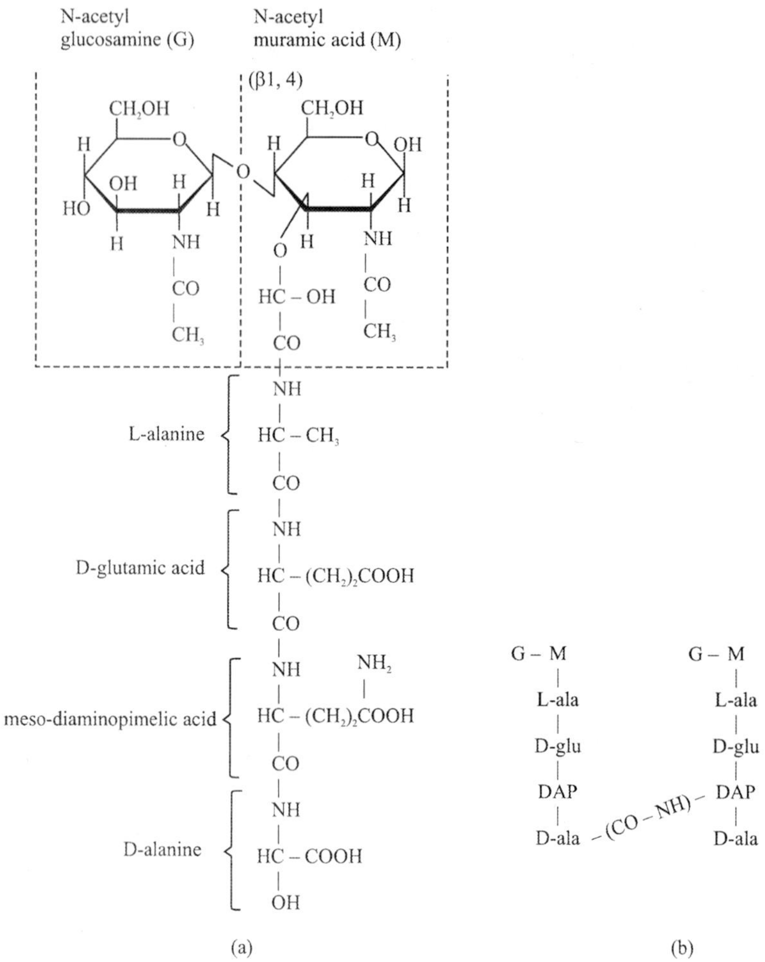

Figure 3.2: Chemical structure of peptidoglycan. (a) The glycan backbone is polymer of two amino sugars, N-acetylglucosamine (G) and N-acetylmuramic acid (M). A tetrapeptide side chain consisting of L-ala-D-glu-DAP-D-ala is attached to N-acetylmuramic acid. (b) Adjacent tetrapeptide side chains are linked to one another by an interpeptide bond between DAP on one chain and D-ala on the other.

Gram-negative Cell Walls

The location of the peptidoglycan in Gram-negative bacteria was disclosed by W. Weidel and his co workers in *E. coli.* They found it to occur in the innermost layer of the multilayered wall and that it could be isolated as a thin sac after removal of the other cell wall components. These peptidoglycans display a much lower degree of cross-linkage. Many a times it is only a thin monomolecular layer. It is

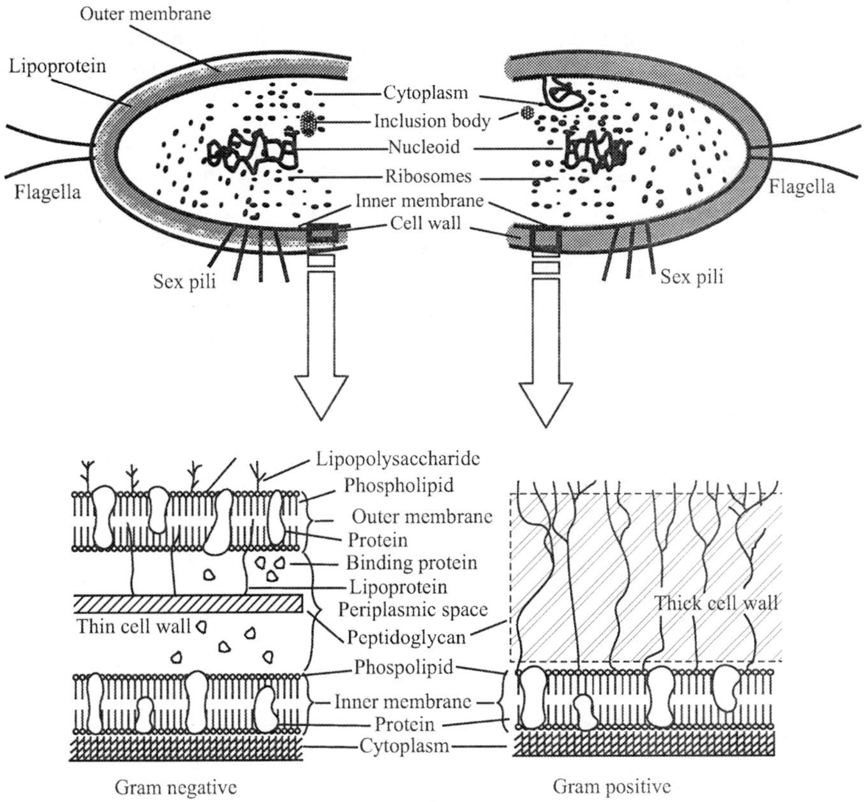

Figure 3.3: Cell wall of Gram (–ve) and Gram (+ve) bacteria.

superimposed with an outer membrane which is similar to cell membrane in composition. It is made of phospholipids, proteins and lipopolysaccharides (LPS).

Lipopolysaccharides (LPS) contain different unusual sugars, e.g. rhamnose, galactose, heptose, mannose, etc. The outer side of the layer is completely dominated by the presence of LPS which are complex molecules with molecular weights over 10,000 and vary widely in composition throughout the Gram-negative groups. LPS is made up of three distinct regions: lipid A (hydrophobic), the R core region and the O side chain (hydrophilic). Lipid A consists of the hydrophobic region and carries six or seven saturated fatty acid residues attached to a phosphorylated glucosamine dimer. The R core is a short chain oligosaccharide made up of 2-keto-3-deoxyoctonoic acid (KDO) and heptose is attached to the 6 C of one glucosamine residue in lipid A. This R core in turn bears the hydrophilic O side chains, again composed of many repeating tetra or pentasaccharides. The peptidoglycan layer of the cell wall also bears small lipoproteins that form anchorage between outer membrane. The outer layer of Gram-negative cell wall serves many purposes including: (i) forms the outer limit of the periplasm, the region between the two layers that contains digestive enzymes; (ii) makes the outer surface highly negatively charged to evade phagocytosis and complement action; and (iii) act as a permeability barrier. Thus, Gram negative bacteria can evade many harmful destructive agents, e.g. intestinal enzymes, antibiotics, lysozymes, bile salts are evaded by this outer membrane. This layer also facilitates diffusion of nutrients through embedded proteins called as porins, e.g. OmpF (outer membrane porin) in *E. coli* which forms a pore of 1.2 nm. The appearance of ompF is highly regulated with regards to temperature. It gets repressed at 37°C indicating that at this physiological temperature within the eukaryotic host the bacteria does not need these pores due to availability of nutrients in large quantity. Some typical specific channels are also observed, e.g. inducible *maltoporin,* that allows entrance of maltose in the bacterial cell. Porins, Omp C, Omp D, Omp E help in transferring substances like sugar, amino acid, inorganic ions, vit-B_{12}, iron chelates, etc. Porins also contain attachment site for bacteriophage. The characteristic make up of the Gram negative outer membrane and its impermeability to hydrophobic molecules accounts for resistance of Gram-negative bacteria to dyes like methylene blue and brilliant green. This is used in formulation of some selective media for Gram-negative bacteria. The region between the cell membrane and the outer membrane is designated as Perisplasm. This space is marked by presence of a number of proteins.

Gram-positive Cell Walls

Gram positive bacterial cell walls are dominated by the presence of peptidoglycan (40–90%; 10–80 nm). The peptidoglycan of Gram positive bacteria contains lysine (diamino acid) attached by pentaglycine cross link which form lattice structure of peptidoglycan. This peptidoglycan layer is covalently linked with other macromolecular cell wall constituents, mainly, polysaccharides and teichoic acids (polyphosphate acids). Teichoic acids are water soluble compounds containing ribitol or glycerol residues linked through phosphodiester linkages. The cell wall teichoic acids lie at the external surface and make the major antigenic determinants of the Gram-positive bacteria. *Lactobacillus* produces polyglycerol teichoic acid while *Bacillus subtilis* and *B. aureus* produce polyribitol teichoic acid. These are also found attached with cell membrane (membrane teichoic acids). The cell wall of most Gram-positive bacteria do not contain proteins and lipids (except *Mycobacteria, Corynebacteia* and *Nocardia* that contain long chain fatty acids) and are characteristically dominated by the presence of teichoic acid. Most teichoic acids contain large amounts of D-alanine as a substituent attached to glycerol or ribitol. Other common substituents include glucose, galactose, N-acetyl muramic acid and glucosamine. Apart from being the antigenic determinants, the major function of the teichoic acids is believed to be in transport of ions.

L FORMS

Penicillin inhibits the last step of cell wall formation, i.e. cross linking between peptides and results in a weak cell wall making the cell susceptible to lysis. The L forms are typical cell wall deficient bacterial cells in which the synthesis of peptidoglycan doesn't occur but growth continues. These cells may or may not revert to have cell walls. These can be produced on osmotically balanced penicillin containing medium. L forms of Gram-negative bacterium lacking peptidoglycan layer can also be obtained, however, their outer layer remains intact.

Capsules and Extracellular polysaccharides

Many prokaryotes synthesize organic polymers that get deposited outside the cell wall as an amorphous or slimy layer termed as capsule. Capsule is defined as an outer covering that remains attached to the cell wall and can be visualized by negative staining. Negative staining of the capsule is done by India ink. Capsules are important for the cell with respect to its adherence to surfaces, virus reception, interaction with cells, resistance to phagocytosis and antimicrobial agents, etc. However, capsules are construed to be non-essential for cell survival or metabolism since its removal does not affect the cell.

Many bacteria produce extracellular polysaccharides (EPS) synthesized from sugar-nucleotide precursors by formation of a polysaccharide chain (e.g. UDP-glucose, UDP-galactose, UDP-mannose, etc.). These sugars are synthesized by the cell through intermediary metabolism. These can be either homopolymeric e.g. EPS of *Acetobacter xylinum* is cellulose or it may be complex heteropolysaccharides in others. Dextrans and levans make a separate class of EPS, since their production depends upon presence of exogenous sucrose as substrate. These are synthesized directly by addition of glycosyl units to sucrose molecule. In formation of levan, there is successive addition of fructosyl residues to the fructose residue of acceptor sucrose molecule. The levan molecule has a terminal glucose unit and a chain of α-fructosyl moieties linked by 2, 6 glycosidic bonds. In formation of dextran, there is successive addition of α-glucosyl residues by transglycosylation to the glucose residue of acceptor sucrose molecule and thus dextrans have a terminal α-fructosyl residue with a chain of glucosyl moieties. Several kinds of EPS mainly composed of repeating disaccharides units have been noticed among bacteria.

Appendages: Flagella and Pili

Appendages of bacterial cell originate from the cell membrane and can go up to 10 times of the diameter of the cell. These are made up of low molecular weight (17,000–40,000) specific protein monomers, flagellins and pilins. The protein monomers are arranged in helical fashion and seem to coil around a hollow core. Size of flagella ranges in between 12–18 nm whereas that of pili ranges between 4–10 nm. Sheathed flagella are covered by an extruding cell membrane. Some bacteria show mixed flagellation with one polar sheathed flagella and peritrichous unsheathed flagella. Flagella of *Salmonella typhimurium* is made up of flagellin subunits (40 kDa) and is 14 nm in diameter.

The type I pilus of *E. coli* is 7 nm in diameter and the molecular weight of constituent pilin is 17 kDa. Its main function is to adhere to the surfaces. Pilin subunits of sex pili are synthesized by sex plasmids that determine conjugative abilities of bacteria. The sex pili help in conjugation by attachment in donor and recipient strains.

Deep down, the flagellar apparatus has a special basal structure consisting of an L-shaped hook and a basal body inserted in the cell envelope. The basal body consists of a small shaft inserted into two pairs of rings. The outer pair (L, P) is situated at the outer membrane and act as bush during the movement. The inner pair (S, M) is located in the cell membrane just below the peptidoglycan (**Figure 3.4**). In Gram positive bacteria only the lower pair of S and M ring is present since there is no need of support to the rod as it passes through the thick peptidoglycan. This also implies that only lower rings can suffice the functional requirements of the flagellum. The movement of flagella originates from the movement of flagellar motor which in turn is governed by transmembrane protonmotive force. The S and M rings rotate relative to each other and impart motion to flagellar shaft either in clockwise or anticlockwise rotation. The abbreviations L, P, S and M denote lipopolysaccharide, peptidoglycan, supermembrane and membrane rings, respectively.

Bacteria motile with flagella show movement towards the selected organic compounds that can serve as nutrients (attractants), e.g. maltose can attract *E. coli*. Many of such chemoreceptors are known for *E. coli* (N-acetylglucosamine,

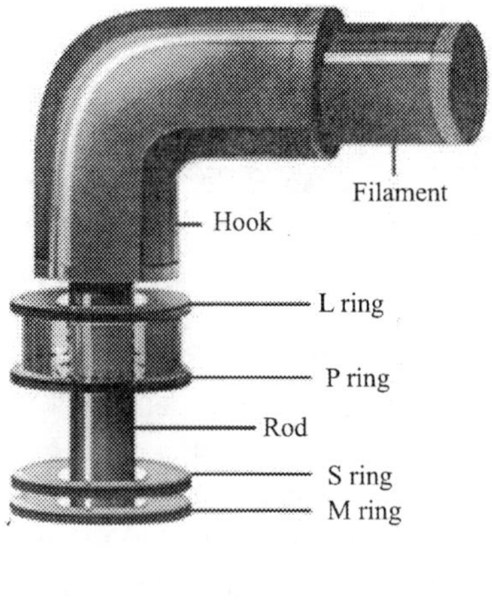

Figure 3.4: Schematic showing basal structure of a flagellar apparatus.

mannitol, fructose, serine, aspartate, etc.). The movement of strict aerobes towards molecular oxygen is called as aerotactic movement. Similarly, photosynthetic bacteria respond to light intensity (phototaxis). Interestingly, studies on phototaxis in lights of different wavelengths have shown that the action spectrum of phototaxis corresponds to that of photosynthesis.

Inclusion Bodies

Often the cytoplasm of some bacteria contains some concentrated deposits of metachromatic granules or volutin made up of polyphosphates that can be observed by light microscopes. Poly-β-hydroxybutyrate (PHB) granules are the reserve polymer of many aerobic bacteria, e.g. *Rhizobium*. PHB granules can be stained with Carbol fuschin or Nile blue stains. Glycogen granules are also seen in some bacteria. These can be stained dark blue with iodine. Globules of elemental sulphur may accumulate in bacteria growing in hydrogen sulphide rich environment.

Spores

These are thick walled, highly refractile resistant bodies produced inside the bacterial cell by many genera including *Bacillus, Clostridium*, etc. The endospores are produced by cells reaching decline phase of their growth. Spores are produced as resistant bodies that can survive under extreme conditions. All endospores show presence of dipicolinic acid (DPA) in combination with calcium. The DPA-calcium complex is believed to play a role in heat resistance of spores. Some genera, e.g. *Methylosinus* produce exospores by terminal budding. These exospores are also resistant to extreme conditions but can be differentiated from endospores by marked absence of DPA. Actinomycetes or ray fungi produce spores contained in a vesicle, sporangium or naked spores, conidia. Some bacteria, e.g. *Azotobacter*, produce spore like structures called cysts from vegetative cells.

Gas Vacuoles and Vesicles

Presence of distinct gas vacuoles can be seen as refractile bodies under light or phase contrast illumination in a wide variety of prokaryotes including archaebacteria, cyanobacteria, green bacteria, etc. These are gas filled structures and provide buoyancy to the aquatic bacteria. These are made up of smaller cylindrical gas vesicles (75 nm in diameter) surrounded by a thin layer of protein. The composition and amount of gases in the vesicles depend upon gases in the surrounding medium.

Chlorosomes

The antenna pigments of photosynthetic green bacteria are housed in special distinct vesicles arranged immediately below the cell membrane. These are the site of photosynthesis in green bacteria. The location and structure of photosynthetic apparatus varies in green, purple and cyanobacteria.

MODES OF NUTRITION IN BACTERIA

Both heterotrophic and autotrophic modes of nutrition are observed in bacteria. The heterotrophic bacteria make a large chunk of bacterial genera and need a diverse range of nutritional requirements including simple to highly complex sources of carbon and nitrogen. Growth media are formulated on the basis of simple or complex requirements. Generally, these media contain rich organic sources of nitrogen (beef extract, peptone, yeast extract) and carbon sources (cellulose and soluble starch) of plant or animal origin. An example of commonly used growth medium, Nutrient agar contains following: Beef extract: 3 g; Peptone: 5 g and Agar: 15 g per litre of water. It is a general purpose medium used for cultivation of heterotrophic bacteria. Agar serves as non-nutritive solidifying agent. Some heterotrophs also require vitamins and other specific growth factors. Among heterotrophs, apart from saprophytes, some bacteria are obligate parasites which propagate only in association with hosts, e.g. rickettsias, chlamydias, *Treponema pallidum, Mycobacterium leprae*, etc. Often these include causative agents of animal or human diseases. Autotrophic bacteria need simple media containing mostly mineral salts without any carbon source. Such media contain a specific amount of standard salts and are examples of *chemically defined media.*

If the particular requirement of a specific group of bacteria is known, media that specifically enhance their growth can be formulated. Such media are called as *selective media.* Attributes such as ability to grow on a specific carbon source and resistance towards an antibiotic or chemical (dye) are used to formulate such media. For example, *Neisseria gonorrhoeae* can grow on media containing antibiotics while most bacteria are inhibited. Therefore, the media serves to enrich the population of *N. gonorrhoeae* while it suppresses the growth of other sensitive bacteria. Apart from supporting selective growth, media can also be used to differentiate two or more bacteria on the basis of cultural characteristics exhibited on a specifically designed media. *Differential media* help in differentiating bacteria on the basis of colonial characters produced on a medium that contains a special ingredient as per the physiological properties of the bacteria to be differentiated. For example, media containing blood can be used to differentiate haemolytic or non-haemolytic bacteria. The conventional media preparation involves weighing of medium ingredients followed by its dissolution in appropriate amount of water. The medium is then autoclaved at 121°C and 15 lbs per sq. inch pressure for 15 minutes. Thermolabile or heat sensitive ingredients (e.g. glucose, antibiotics) are filter sterilized separately. If the bacteria are to be cultured on solid surface, agar is added (15–20 gm per thousand ml or per litre of media). Once inoculated, the flasks or plates are incubated at required temperature. Most bacteria are incubated at 37°C which is the body temperature that suits to human pathogens (e.g. *Salmonella*) or commensals (e.g. *E. coli*). Thermophilic bacteria need higher incubation temperatures similar to their warm ecological niche (e.g. hot springs). Conversely, psychrophilic bacteria need lower temperatures in the range of

0–20°C for their optimum growth. The bacterial growth pattern follows a typical *growth curve* including four distinct phases: lag phase, log phase, stationary and decline or death phase (Unit VI; Figure 6.1). The lag phase is marked by slow growth, representing the adaptation of cells to the new substrate and growth conditions. In the log phase, bacterial cells undergo rapid proliferation. This phase is also referred as exponential growth phase as bacteria multiply in geometric progression. Stationary phase represents the time period during which the magnitude of growth and death is almost equal and hence no net increase is witnessed in viable count. Finally, decline phase marked by old cells and dwindled media ingredients appears.

Some bacteria are fastidious regarding their typical gaseous requirements. *Strict anaerobic* bacteria need total absence of oxygen. They lack energy generating mechanisms that require oxygen as terminal electron acceptor. These are unable to tolerate even very low levels of oxygen and hence specific arrangements (anaerobic chambers) are to be made to avoid oxygen. *Facultative anaerobes* are the class of bacteria that can manage to grow in the presence of oxygen but do not need oxygen for their growth. *Microaerophilic* bacteria need low levels of oxygen and are unable to tolerate its high levels. Most routinely cultured bacteria are *aerobic* and need oxygen as terminal electron acceptor for operation of energy generating process, respiration. Conversely, the same oxygen is toxic for strict anaerobes and induces inactivation of their enzymes by oxidizing the essential reduced group of amino acids (e.g. Nitrogenase inactivation). Oxygen produces derivatives such as superoxide, hydrogen peroxide and hydroxyl radicals which are highly reactive (oxidants) and damage important molecules of bacterial cells. Aerobic bacteria have mechanisms to mitigate or avoid damage due to such radicals. These have enzymes, viz. catalase, peroxidase and superoxide dismutase that help eliminate these radicals. The pH of the medium also needs to be defined since most bacteria require specific pH range for their optimum growth. Different bacteria exhibit wide range of optimum pH for their growth. Alkalophiles or acidophiles need pH as high as 10.0–12.0 or as low as 2.0–3.0, respectively. The initial pH of the medium is shifted by physiological activities of bacterial cells (e.g. acid production) and hence needs to be maintained by incorporating buffer solution. It is obvious that any phototrophic bacteria will need illumination as a primary requirement for its growth (e.g. Cyanobacteria). *Barophiles* isolated from deep sea or earth need high pressure for their growth. Therefore, the cultures should be maintained under pressure by appropriate mechanism.

IDENTIFICATION OF BACTERIA

After isolation of bacteria, the purity of the culture is ensured and morphological characters are seen by physical examination under microscope. The morphology of the microorganism sometimes also depends on culture medium, culture conditions, age of culture and subculturing.

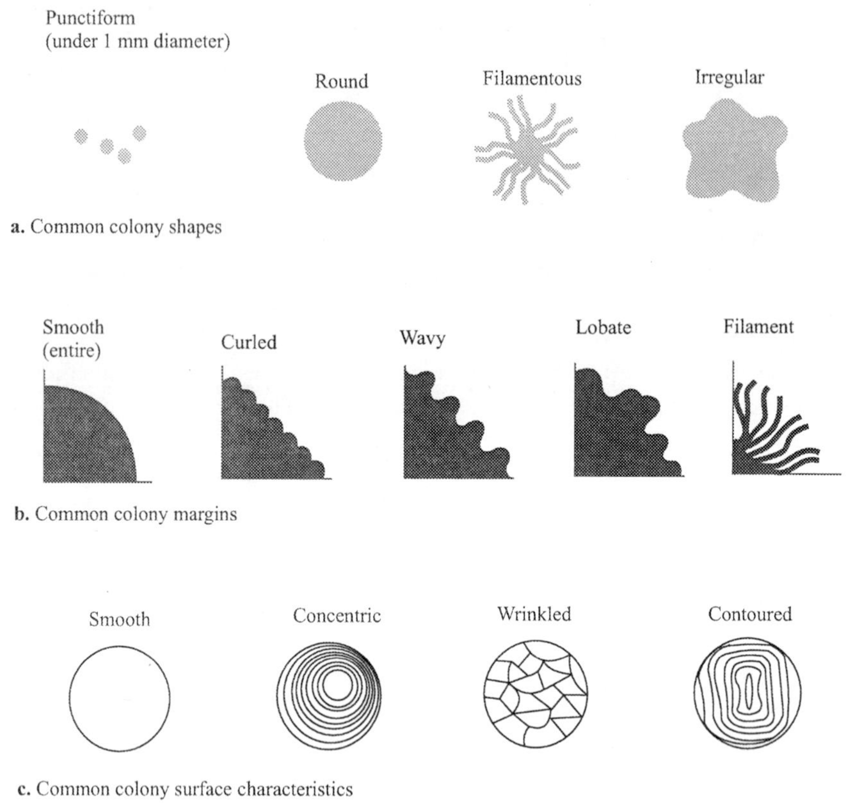

Figure 3.5: Shapes, margins and surface charateristics of bacterial colonies.

Visual observations of the culture growing on solid medium in a Petri plate on a specific media are noted: colour, pigmentation, elevation, margin, surface (smooth, wavy, rough, granular, slimy, and powdery **Figure 3.5**). In a liquid medium the amount of growth (profuse or scanty), turbidity, surface growth (formation of a mat or pellicle), and colour imparted to liquid medium and odour should also be recorded.

Morphology of Bacteria

The important morphological characteristics, viz. shape, size, arrangement, motility, and presence of flagella, spores, and capsules are recorded by microscopic examination. Although dominated by sphericals (cocci) and rods (bacilli), bacteria present a wide range of shapes including filamentous, curved, comma or spiral. Also great variations with regard to length and breadth of the bacteria belonging to different species can be seen. The sides of the bacterium may be straight, convex, concave, or irregular with straight, rounded or tapering ends (**Figure 3.6**). Often,

Bacterial Morphology		Example
	Straight rod	*Escherichia coli*
	Endospore forming rod	*Bacillus*
	Club-shaped rod	*Corynebacterium*
	Branched or filamentous rod	*Actinomyces*
	Comma-shaped	*Vibrio*
	Spiral	*Spirochaeta*
	Coccus	*Micrococcus*

Figure 3.6: Morphological types (shapes) of bacteria.

the bacterial cells are found to be uniquely arranged in chains (strepto), tetrads or packets of eight (sarcina), and irregular packets or bundles (staphylo-). This organization is a typical characteristic of the bacterial genera and is based on the plane of division by which bacteria divide to reproduce (**Figure 3.7**).

A variety of flagellar arrangements are seen in bacteria (**Figure 3.8**). They may have one (monotrichate) or more than one flagella arranged as clusters (lophotrichate), at the two ends (amphitrichate) or all around (peritrichate) or non-flagellate (atrichate). The spores may be seen outside or inside (endospores) the cells. These may be oval, spherical or ellipsoidal with size generally smaller than bacterium. They may be positioned equatorially, terminally or subterminally in bacterium. To study these characteristics the simple light microscopy is not sufficient and one has to improvise on slide preparation (hanging drop, wet mount), staining technique (differential, spore, capsule, negative staining), illumination (dark field or phase contrast). Electron microscopy is the ultimate technique for ascertaining fine structure of bacterial cell, its appendages and intracellular

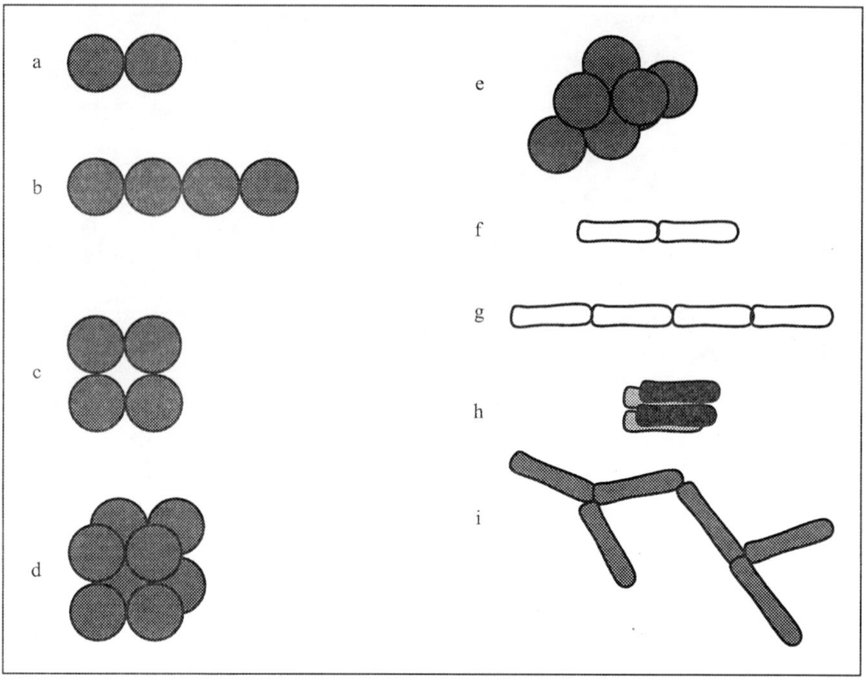

Figure. 3.7: Organization of bacteria in groups (a) diplococci (b) streptococci (c) tetrad (d) sarcina (e) staphylococcus (f) diplobacillus (g) streptobacillus (h) pallisade (i) filamentous.

structures. Bacteria can also be characterized for their resistance or sensitivity towards antibiotics (drug resistance), bacteriocins, temperature relationship and ability to grow in absence or presence of oxygen.

Based on these observations, bacteria can be preliminary identified as belonging to a typical morphological group, e.g. a culture may be identified as Gram positive endospore forming rod.

BIOCHEMICAL CHARACTERISTICS OF BACTERIA

To identify any bacterial culture in terms of genus and species, it is important to study its biochemical characteristics that relate to its physiology or presence of a specific metabolic pathway or enzyme in the organism. Some standard tests **(biochemical tests)** performed for identification of enteric bacteria are listed below:

1. Methyl red test: Bacteria ferment sugar and change the pH of the medium to acidic. This test is aimed to examine the ability of a bacterium to produce and maintain stable acid end products from sugar (glucose) fermentation. Addition of a pH sensitive indicator, e.g. methyl red brings about a colour change implying fermentation of sugar. Few drops of 0.04% solution of methyl red are added to the

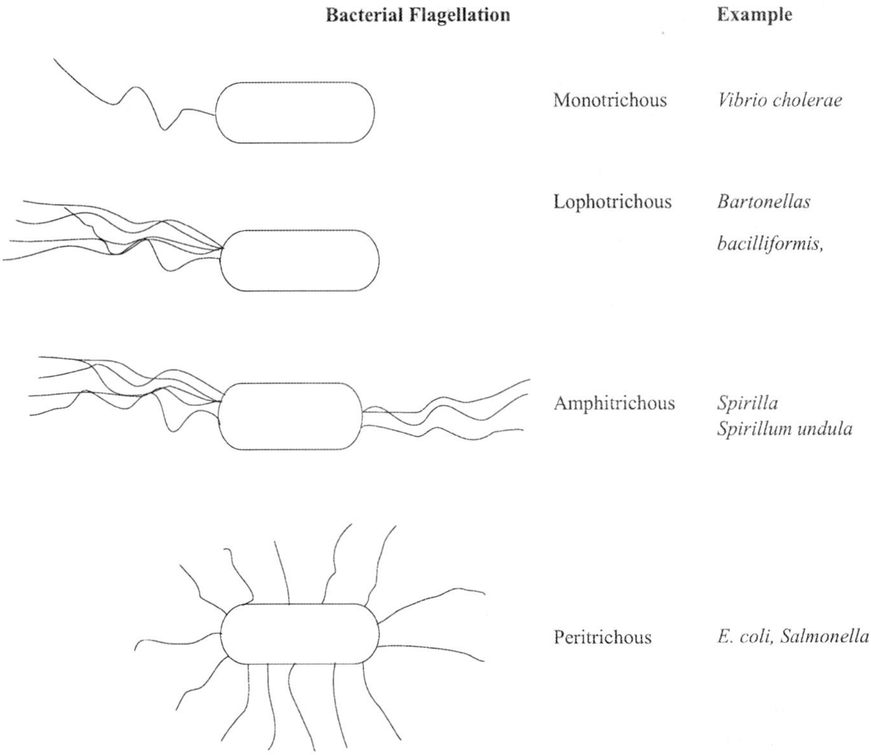

Figure 3.8: Kinds of flagellation seen in bacteria.

culture grown in **glucose phosphate broth** (buffered glucose medium). Appearance of red colour indicates a positive result or acid production while no change in original colour (yellow) implies a negative result. The test is used to differentiate between bacterial genera, viz. *E. coli* (MR +ve) from *Enterobacter* (MR –ve); *Yersinia* sp. (MR +ve) from other Gram negative non-enteric bacteria (MR –ve).

2. Indole production: This test demonstrates the production of indole from tryptophan by the action of tryptophanase enzyme. A peptone broth (serves as a source of tryptophan) when inoculated with a bacterial culture producing indole, gives red colour on addition of Kovac's reagent (Paradimethylaminobenzaldehyde: 10 g, Isoamyl alcohol: 150 ml; Conc. HCl: 50 ml). This test is specially used for the differentiation among genera and species, viz. *E. coli* (+) from *Enterobacter* (–) and *Proteus mirabilis* (–) from other *Proteus* species.

3. Voges-Proskauer test (VP): This test confirms production of acetyl methylcarbinol ($CH_3.CHOH.CO.CH_3$) from pyruvic acid, as an intermediary metabolite in its conversion to 2, 3-butylene glycol. In alkaline conditions the acetyl methyl carbinol produced in the medium is oxidized to form diacetyl ($CH_3.CO.CO.CH_3$). This diacetyl reacts with peptone to give red colour. This test is performed by addition of 0.6 ml of a 5% solution of α-naphthol in ethanol and 0.2 ml of 40% KOH to 1 ml of glucose-phosphate culture broth of the organism. The positive result is indicated by development of pink colour that darkens to magenta in 30 minutes time. The test is used to differentiate between bacterial genera e.g. *Enterobacter* (VP positive) from *E. coli* (VP negative). Opposite results are usually obtained for M.R. and V.P. tests.

4. Citrate utilization: This test proves the ability of the bacterium to utilize citrate as the sole source of carbon with resulting alkalinity. Bromothymol blue is green when acidic (pH- 6.8 or low) and blue when alkaline (pH- 7.6 and higher). This dye is used as a pH indicator in the medium containing citrate as the sole source of carbon. Development of colony on Koser's citrate or Simmon's citrate agar plate indicates that the test organism is able to utilize citrate as a source of carbon. The growth is accompanied with a rise in pH to change the colour of the medium from green to deep blue.

These four tests are routinely performed for identification of enteric bacteria (enterobacteriaceae) and together are referred as **IMViC** tests.

5. Gas production: The production of gas (CO_2) as a result of fermentation of sugars may be confirmed on the basis of cracks produced in slants or stabs by the escaping gas. The gas may be collected in Durham's tubes as done in MPN test.

6. Nitrate reduction: This test is performed to test the ability of bacteria to reduce nitrate present in the culture broth (1% KNO_3) to nitrite or to N_2. Nitrate broth is inoculated with bacteria and incubated at 37°C for 4–5 days. The reagent is freshly prepared by mixing equal volumes of sulphanilic acid and α-naphthylamine in 5N acetic acid just before use. 0.1 ml of this reagent is added to the culture broth. Appearance of red colour indicates a positive result confirming the presence of nitrate reductase. Broths showing a negative reaction should be further tested by adding a pinch of zinc. The appearance of a red colour indicates that nitrate is still present and hence has not been reduced by the organism. If there is no change in the colour of solution, then it can be inferred that the organism has reduced the nitrate through nitrite to nitrogen gas and the result is positive. This test is used in identification of Enterobacteriaceae members which usually give a positive result.

7. Ammonia production: Production of ammonia can be confirmed by production of brown colour on addition of a few drops of Nessler's reagent to the bacterial culture grown on peptone water.

8. Urease test: Urea is reduced to ammonia in presence of urease enzyme. This test employs Christensen's urease agar medium in form of a slant. The appearance of purple-pink colour indicates formation of ammonia.

9. Production of hydrogen sulphide (H_2S): Several bacteria liberate hydrogen sulphide from sulphur containing amino acids. Production of H_2S can be detected by incorporating lead acetate in the nutrient medium which turns black or brown due to formation of lead sulphide (PbS).

Table 3.1 Features used to assist the identification of enteric bacteria

S.No	Species	MR	VP	Lac.	Glu.	Suc.	Oxidase	Indole	Citrate	H_2S	Gas	Urease
1.	*Escherichia coli*	+	–	+	+	D	–	+	–	–	+/–	–
2.	*Shigella sp.*	+	–	–	+	–	–	D	–	–	–/+	–
3.	*Salmonella typhi*	+	–	–	+	–	–	–	–	+	–	–
4.	*Salmonella paratyphi*	+	–	–	+	–	–	–	–	–	+	–
5.	Other *Salmonella sp.*	+	–	–	+	–	–	–	D	+/–	D	–
6.	*Citrobacter freundii*			Late	+	D	–	–/+	+	D	+	D
7.	*Klebsiella pneumoniae*	–	+	+	+	+	–	–/+	+	–	+	Slow
8.	*Enterobacter sp.*			+	+	D	–	–	+/–	–	+	–
9.	*Serratia marcesens*	D		D	+	+	–	–	+	–	D	D
10.	*Proteus vulgaris*	+	–	–	+	+	–	+	D	+	D	+
11.	*Proteus mirabilis*	–	+	–	+	D	–	–	+/–	+	+	+
12.	*Morganella morganii*			–	+	–	–	+	–	–	D	+
13.	*Providencia sp.*	+	–	–	+	D	–	+	+	–	D	D
14.	*Yersinia enterocolitica*			–	+	+	–	D	–	–	–	Slow
15.	*Vibrio cholerae*	+	–	24 hrs	+	+	+	+	D	–	–	–
16.	*Vibrio parahaemolyticus*			–	+	–	+	+	D	–	–	–
17.	*Pseudomonas aeruginosa*	–	–	–	D	–	+	–	+	–	–	D
18.	*Alcaligenes sp.*			–	–	–	+	–	D	–	–	–

Abbreviations: MR: Methyl red test (acid production) ; VP: Voges-Proskauer (acetyle methyl carbinol production); Lac- Lactose utilization; Glu- Glucose utilization; Suc: Sucrose utilization

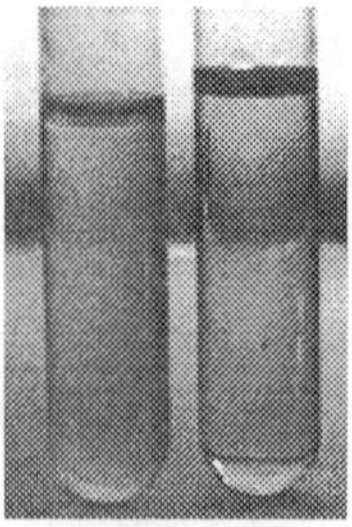

Indole production

Methyl red test

Voges-Proskauer test

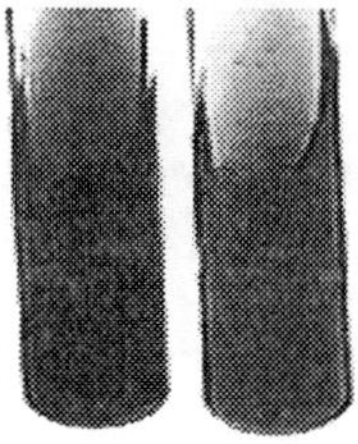

Citrate test

Growth of different bacteria on Triple Sugar Iron Agar (TSIA)

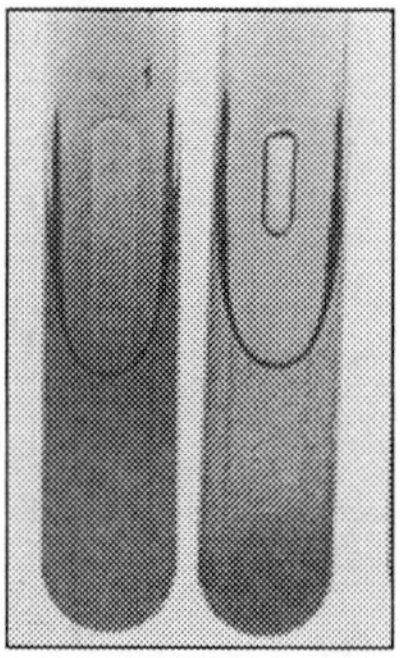

MPN: Most probable number test for potability of water

Figure 3.9: Biochemical testing for identification of bacteria.

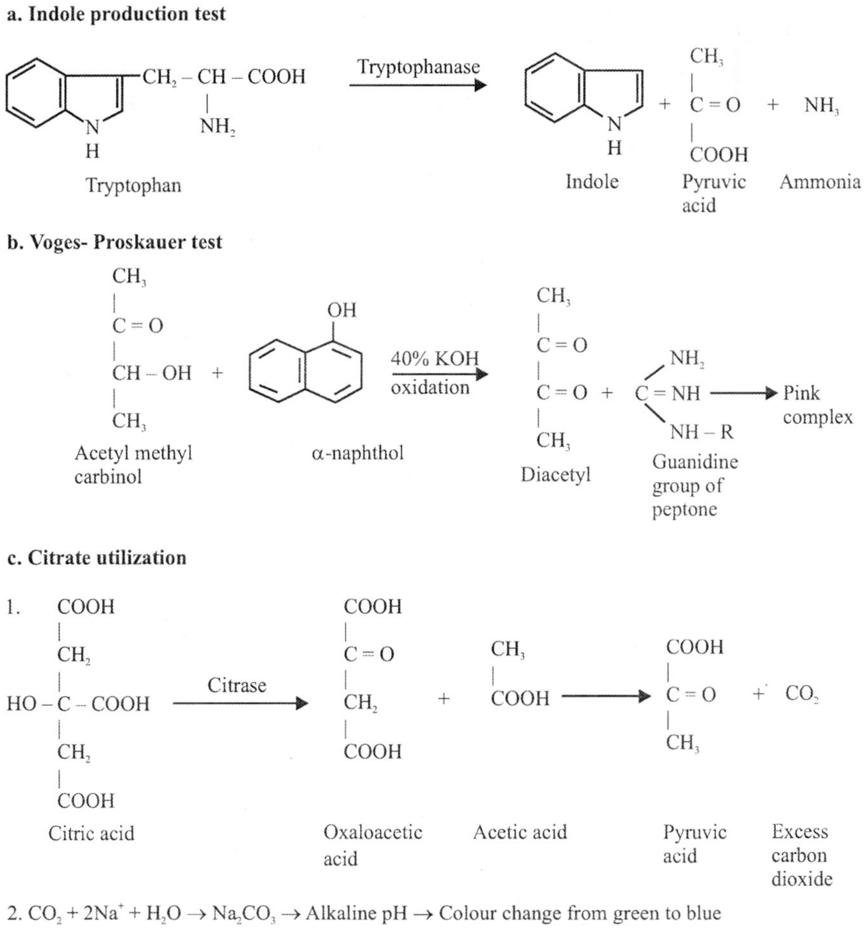

Figure 3.10: Chemical reactions of some biochemical tests

10. Methylene blue reduction test (MBRT): Bacteria reduce aqueous methylene blue and change it into colourless (leuco) form. This test is specially used for grading milk on the basis of number of bacterial contaminants. Few drops of methylene blue are added to fresh milk, mixed and incubated at 37°C. More the number of bacteria, sooner the blue colour disappears indicating poor milk quality.

11. Catalase production: Catalase is a common enzyme found in most bacteria. It catalyses the breakdown of hydrogen peroxide (H_2O_2) with the release of free oxygen. Bacterial cells of the test colony are mixed with few drops of 3% H_2O_2 and formation of bubbles indicates a positive reaction. Catalase is found in most aerobic and facultatively anaerobic bacteria excepting the genus *Streptococcus* while all the anaerobes are catalase negative.

12. Litmus milk test: This test determines the effect of bacteria on milk having litmus indicator. Milk is a complete food and contains lactose, glucose, casein, fats, minerals and vitamins, thus offering complete nutrition to most bacteria. Litmus milk is inoculated with bacterial culture and incubated at 37°C for 48 hrs or more. Depending upon the bacteria, several kinds of changes can be seen in the litmus milk. Clotting, peptonisation or saponification may occur with or without production of gas. Production of pink colour indicates that small amount of acid is produced from glucose but lactose remains unutilized while appearance of dark pink colour indicates substantial acid production and utilization of both sugars. Clotting of milk and pink colour confirms precipitation of casein. Alkaline pH turns the dye blue and indicates liberation of amines or ammonia while the clearing of solution is indicative of proteolytic activity of bacteria.

13. Oxidase test: This test is performed to determine the presence of cytochrome oxidase enzymes which catalyze oxidation by molecular oxygen. The test employs 1% solution of tetramethyl-p-phenylenediamine or filter paper strip impregnated with this reagent. This reagent serves as an alternate substrate for the cytochrome oxidase reaction and when oxidised becomes purple. The result should be noted quickly within 10–30 seconds. This test is useful in differentiation of enteric bacteria (oxidase negative) and *Pseudomonas* (oxidase positive). Also it should be remembered that an oxidase positive organism will be catalase positive as well and most obligate anaerobe are oxidase negative.

14. Oxidation-Fermentation (OF test): This test determines the ability of any bacterium to carry out oxidative or fermentative metabolism of a carbohydrate. It indicates that the test bacterium has the necessary enzymes for the aerobic breakdown of glucose (oxidation) and/or for the fermentation (anaerobic breakdown) of glucose. Two stabs of OF medium (green colour) are inoculated for each kind of test organism. One tube is overlaid with sterile paraffin oil to eliminate all oxygen (anaerobic condition to trigger fermentation). Both the inoculated tubes are incubated at 37°C for 48 hours.

Results can be interpreted as:

Tube 1 (Aerobic)	**Tube 2 (Anaerobic)**	**Inference**
Yellow	Green	Oxidation
Yellow	Yellow	Fermentation
Green	Green	Glucose remains unutilized

The test is used to differentiate among *Micrococcus* sp. (oxidative) from *Staphylococcus* sp. (fermentative). Members of enterobacteriaceae usually have fermentative abilities.

15. Triple Sugar Iron Agar (TSIA): This test uses a composite media (TSIA) that contains glucose, lactose, and sucrose. It determines the ability of an organism to attack a specific carbohydrate incorporated into a basal growth medium, with or without the production of gas, along with production of hydrogen sulphide (indicated by presence of a black precipitate). TSIA slants are inoculated by streaking the slant surface and stabbing the butt with needle and by incubating at 37°C for 24 hours. Results can be interpreted as follows:

Slant colour	Butt colour	Inference
Red	Yellow	Only glucose is fermented
Yellow	Yellow	Glucose, lactose and sucrose fermented
Red	Red	Sugars remains unutilized

A complete yellow slant with cracks and black coloured patches indicate that the organism has utilized all the three sugars and produced gas and hydrogen sulphide as well.

Serotyping

Serotyping of bacteria based on their reactions with different sera is a technique to identify and compare bacteria carrying similar antigens. This kind of grouping is referred as serotyping and is a kind of immunological identification.

ACTINOMYCETES

Actinomycetes are a heterogeneous group of Gram positive eubacteria present in a wide variety of environments, either as dormant spores or as actively growing colonies. These bacteria exhibit a mycelial growth habit and characteristic chemical and structural properties. They include some of the most common soil inhabitants and play an important role in humus formation, decomposition and recycling of organic matter. They are also a well known source of life-saving antibiotics. S. A. Waksman (1940) revealed that some peculiar soil bacteria produce actinomycin. Later the significance of this discovery was realized and he was honored with Nobel Prize. Extensive investigations in soil microbiology concerning the group actinomycetes were carried on over many years by Dr. Waksman and his associates in the Department of Microbiology of Rutgers University and the New Jersey Agricultural Experiment Station. Since then several members of actinomycetes, particularly those belonging to genus *Streptomyces* have been found to produce antibiotics. Some Actinobacteria inhibit plants and animals, including a few pathogens, such as *Mycobacterium, Corynebacterium, Nocardia* and *Rhodococcus*. Actinomycetes form branching filaments which resemble the mycelia of moulds among which they were originally classified as *ray (actino) fungi (mycetes)*. Mostly these are aerobic but for few exceptions like *Actinomyces israelii*, which can grow under anaerobic conditions. Unlike the Firmicutes, the other main group of Gram-positive bacteria, they have DNA with a high GC-content and some Actinomycetes species produce external spores. Members of this group are well known for production of secondary metabolites and bioactive molecules including antibiotics and enzymes having high pharmacological and commercial value.

The life cycle of actinomycetes has two distinct phases (i) vegetative mycelial phase and (ii) sporulation phase. Sporulation starts under the adverse or starving conditions. These two stages can be seen at the edge and the centre, respectively, of an actively growing colony (**Figure 3.11**). Both substrate mycelium (*completely embedded in the substrate*) and aerial mycelium may be witnessed in an actinomycetes colony. Spores can be seen at the tips of the mature and older aerial hyphae. Substrate mycelium is slender, long and septate. Members of this heterogeneous group exhibit a wide array of structures including short filaments to highly branched, curved mycelia.

They also show a great diversity in their cell wall and their cell wall composition has been used as a basis of taxonomy of these bacteria. More than 60 types of mureins have been found in members of actinomycetes. Motility is limited to spores of few genera producing zoospores. Major groups of actinomycetes are listed in **Table 3.2.**

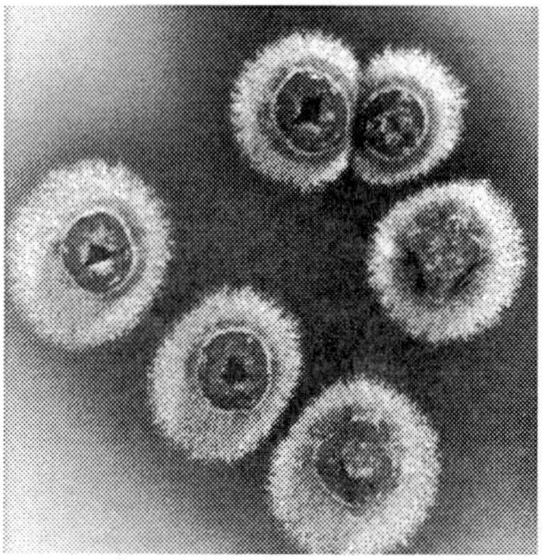

Figure 3.11: Typical morphology of actinomycetes (*Streptomyces*).

Actinobacteria

This group consists of unicellular, strictly aerobic non-filamentous cocci and rods. *Micrococci* occur as tetrads and are brightly pigmented due to accumulation of large amounts of red, yellow and pink caretenoids. These are particularly present in dust particles and human skin. *Arthrobacter* makes a large part of chemoheterotrophic bacterial population of soil involved in mineralization. These cells exhibit a sharp variation in shape from coccus to rod and from rod to coccus during their life cycle (rod-coccus morphogenesis). A distinctive V-shape can also be seen due to peculiar post-fission movement of daughter cells, called as snapping

Table 3.2 Major groups of actinomycetes.

Group	Genera
• Actinobacteria	*Micrococcus* *Arthrobacter* *Cellulomonas* *Actinomycetes*
• Nocardioform	*Corynebacterium* *Mycobacterium* *Rhodococcus* *Nocardia*
• Dermatophilus	*Dermatophilus* *Geodermatophilus* *Frankia*
• Streptomycetes	*Streptomyces*
• Actinoplanetes	*Micromonospora*

fission. Member of genus *Cellulomonas* have a distinct ability to utilize cellulose as source of carbon. These mostly motile facultative anaerobes require biotin and thiamin for growth. *Actinomyces* can be isolated from mouth and throat of human and animals. These predominantly mycelial aeroduric anaerobes require higher partial pressure of CO_2 for growth and produce formate, succinate and lactate as products of anaerobic fermentation.

Nocardioform

The genus Nocardia is named after Edmond Nocard (1850–1903), a French bacteriologist. Collectively, they are called as nocardiforms. The main cell wall carbohydrate of nocardias is arabinogalacatan (a polymer of arabinose and galactose) bound by phosphodiester bonds to acetyl or glycolyl moieties of the murein. Some distinctive lipids (hydroxyl fatty acids) of this group, called as mycolic acids, are found esterified to arabinogalactans. Presence of large amounts of lipids reduces permeability of staining reagents into the cell wall. Thus, it is necessary to heat cells up to 100°C for permeation of stain. Once stained, its decolourization is again very difficult. This makes the principle of acid fast staining during staining. Due to mycolic acid on the cell surface some colonies of nocardioform show waxy appearance. Morphologically the members of this group are diverse. *Corynebacterium* are club shaped pleomorphic rods while *Mycobacterium* gives a fragmented mycelial appearance. *Rhodococcus* show rod-cocci morphogenesis while *Nocardia* are typically mycelial. *Corynebacterium diptheriae*, *Mycobacterium tuberculosis* and *M. leprae* are human pathogens causing dreaded diphtheria, TB and leprosy, respectively.

Dermatophilus

The typical members of this group *Dermatophilus*, *Geodermatophilus* and *Frankia* have characteristic sporangial structures (*sporangium*). *Dermatophilus* and *Geodermatophilus* develop substrate mycelia into sclerotia while intercalary swellings of *Frankia* develop into a distinct sporangium containing spores. *Frankia* infects root nodules of a number of non-leguminous plants belonging to casurinales, myricales, cucurbitales, etc. where they live as endophytes and fix nitrogen.

Streptomycetes

Members of this group have typical mycelial structure and the aerial mycelium terminates into non-motile spores. These bacteria predominantly inhabit humus and soil and produce volatiles responsible for earthy smell of damp soil. In fact, the smell of the moist earth is largely a result of production of volatile substances such as geosmin by the members of genus *Streptomyces*. These are strict aerobes and have specific nutritional preferences. They can hydrolyze a number of polymers including chitin, lignocellulose, latex, etc. and hence play a significant role in mineralization and humification. The group has several famous antibiotic producers, e.g. *Streptomyces griseus. Chainia* species of this group produces distinct sclerotia containing a chemical L-2, 3-diaminopropionic acid.

Actinoplanetes

This group is typically identified on the basis of a sporangium arising from the substrate mycelium. The sporangium is formed due to coiling and fragmentation of sporangial mycelium in the sporangial wall. The group includes genus *Actinoplanes* and *Micromonospora.* Although aerobic, some strictly anaerobic *Micromonospora* have been isolated from the termite gut.

MYCOPLASMA

Mycoplasma belong to the mollicutes group of eubacteria which have a common characteristic feature – they lack cell wall. This results in their pleomorphic shape and resistance to antibiotics that inhibit cell wall synthesis such as penicillin. These are highly sensitive to osmotic pressures and tetracyclins. These bacteria can pass through common bacterial filters due to their pleomorphic shape. All the members of this group are known to be parasitic and have fastidious growth factor requirements including vitamin, sterol, fatty acid, purines, pyrimidines, etc. The media used for culturing mycoplasma thus includes blood serum to furnish sterols and fatty acids. Mycoplasmas have a small genome of 0.58–1.38 megabases, thus have few biosynthetic capabilities and obligatory dependence on host. Their optimum growth temperature is often the temperature of their host (37°C in case of humans).

They form small colonies on solid media having a typical fried-egg appearance (**Figure 3.12**). The cell membrane of the mollicutes is surrounded with ample of

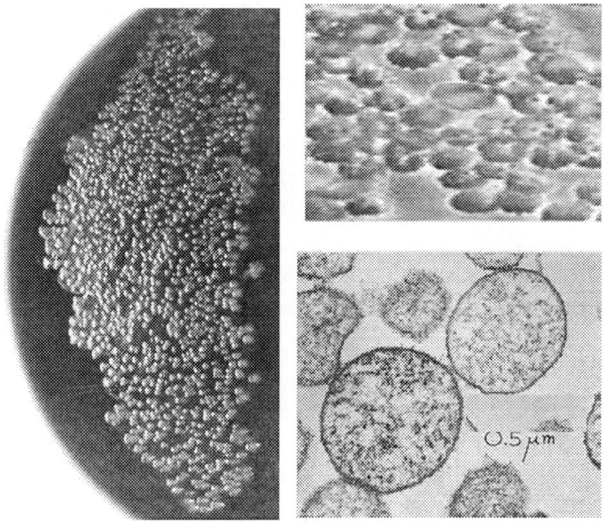

Figure 3.12: Colonial and morphological features of Mycoplasma.

polysaccharides comprised of glucosamine, galactosamine and N-acetyl glucosamine. The thick polysaccharide layer probably compensates for the missing cell wall. Earlier the group was believed to have originated as stable L-forms. But recently, their origin based on similarity index of 16S rRNA and gene content has been confirmed and they are thought to be related to *Clostridia* and *Lactobacillus* which belong to firmicutes.

Although they lack cell wall, their shape is known to contribute to their nutrition. *M. pneumoniae* cells through their 'tip-structure', protruding from the coccoid cell body adhere to host cells. *M. pneumoniae* cells are of small size and pleomorphic, but with a rough shape in longitudinal cross-section resemble that of a round-bottomed flask. A few of these show unidirectional gliding ability due to polarization of the cell content as leading protruding anterior end and lagging rod-like structure. Exceptionally, *Mycoplasma* protein synthesis features a stop codon UGA as a codon for tryptophan.

Nocard and Roux (1898) reported first isolation of the causative agent of a dreaded disease of cattles, contagious bovine pleuropneumonia (CBPP) on medium containing pulmonary fluid. Later it was known to be caused by *Mycoplasma mycoides.* Today, more than 60 species of mycoplasma are known. They are stained by certain polychrome stain, e.g. Giemsa and Costaneda's rickettasia stains.

Mycoplasmas are usually found as parasites of animal mucous membranes of respiratory tracts and synovial membranes of joints resulting in pneumonia or arthritis, respectively. These are microaerophilic or facultative anaerobes and can ferment arginine or carbohydrates or both. Their fermentation product is predominantly lactate. Other members of mollicutes include *Ureoplasma*,

Acholeplasma, *Anaeroplasma* and *Spiroplasma*. Mycoplasma are capable of reproduction by following means: (a) Binary fission, (b) Budding and (c) Elementary bodies.

CYANOBACTERIA

Cyanobacteria (or blue-green algae) are **autotrophic photosynthetic eubacteria** that derive their energy from photosynthesis. Characteristically they have chlorophyll a, phycocyanin (blue) and phycoerythrin (red) pigments. The name cyanobacteria is given due to blue-green appearance of some of the members, however, a great range in colour from brown to olive green is also observed.

Mostly these are found in aquatic environments or water bodies including oceans, ponds, lakes, and also damp places like caves, moist soils, etc. Usually, cyanobacteria exist mostly as colonies and filaments and rarely as single cells also.

The cyanobacteria exhibit subtle forms of movements, e.g. *Oscillatoria sp.* rotate in a screw-like manner whereas the gelatinous forms show gliding movement in a mucoid slime they produce. Blue-green algae produce gelatinous capsules, being lighter than water it keeps them afloat on the surface of the water where cyanobacteria can procure sunlight for photosynthesis. Cyanobacteria are known to reproduce by fission only.

As compared to other bacteria, cyanobacteria show considerable level of complexity and evolution in their organization. Their chlorophyll is arranged into thylakoids which are invaginations of the cell membrane. This is in contrast with other photosynthetic bacteria where chlorophyll and pigments are dispersed in the cell. The biochemical nature of photosynthetic mechanism in the cyanobacteria is nearly identical to that of the algae and the higher plants. Their photosynthetic pigments include chlorophyll *a* and the accessory pigment beta-carotene, although they lack chlorophyll *b*. The glucose produced by the cyanobacteria in the process of photosynthesis is stored in their own form of starch, which is similar to animal glycogen. Interestingly, from evolutionary point of view these resemblances make cyanobacteria as incipient eukaryotic chloroplast.

As a group, cyanobacteria are a heterogeneous pool of Gram-negative eubacteria characterized by their capability to carry out oxygenic photosynthesis. They exhibit a great deal of structural diversity along with genetic heterogeneity as represented by the varied range of G + C values (35 to 71 per cent) and genome sizes (MW of 1.6×10^9 to 8.6×10^9 Dalton).

All members are photoautotrophic without any requirement of growth factors; however, some marine cyanobacteria may require vitamin B_{12}. Sequestration of CO_2 is achieved through the Calvin-Benson cycle as in higher plants and results in accumulation of glycogen as intracellular storage material.

Most cyanobacteria are *obligate phototrophs*, being completely incapable of growth in dark while some chemotrophic members can manage to grow in the dark as well. They have limited range of utilizable sources of carbon and show absence of TCA cycle. Cyanobacteria lack a key enzyme of this cycle, α-ketoglutarate dehydrogenase, a metabolic peculiarity that they share with many obligately anaerobic chemoheterotrophs and aerobic chemoautotrophs. Respiratory metabolism occurs through the oxidative pentose phosphate cycle, most reactions of which are in large part common to those of the Calvin-Benson cycle. The obligate photoautotrophic characteristic of many cyanobacteria are due to the absence of the specific permeases that limit the uptake of exogenous sugars by the cell.

The metabolic machinery of the oxidative pentose phosphate cycle is present in all cyanobacteria and is utilized for the generation of ATP in the dark through endogenous respiration of stored glycogen. Cyanobacteria are ubiquitous and they have been isolated from oceans, fresh water bodies, rocks, soils, crop fields, etc. Most are found in fresh water, while others are marine, some occur in damp soil, or even over temporarily moistened rocks and deserts. Some members occur as endo-symbionts in lichens, plants, protists and sponges and carry out photosynthesis part for the host.

Cyanobacteria include unicellular and colonial species. Colonies may form filaments, sheets or even hollow balls. Some filamentous cyanobacteria show the ability to differentiate into several different cell types: vegetative cells (photosynthetic cells that are formed under favourable growing conditions), akinetes (the climate-resistant spores); and thick-walled *heterocysts* (fix nitrogen through nitrogenase enzyme). It is important to note that these are the only examples where both phenomena of photosynthesis and nitogen-fixation occur in one single organism. Motility in cyanobacteria is seen in the form of motile filaments, called hormogonia. These filaments move away from the main biomass to bud and form new colonies elsewhere. The cells in a hormogonium are often thinner with tapered cells on either end of the filament. The breaking of hormogonium from parent thallus is facilitated by a weaker cell, necridium.

CLASSIFICATION OF CYANOBACTERIA

On the basis of morphology, the cyanobacteria were traditionally classified into five sections, (I–V). However, the first three—Chroococcales, Pleurocapsales, and Oscillatoriales are not supported by more recent phylogenetic studies. Rest two—Nostocales and Stigonematales—are monophyletic, and include all the heterocystous cyanobacteria. The members of chroococales are unicellular and usually aggregated in colonies. The classic taxonomic criterion has been the cell morphology and the plane of cell division. In pleurocapsales, the cells have the ability to form internal spores (baeocytes). The rest of the sections include filamentous species. In oscillatorialles, the cells are uniseriately arranged and do not form specialized cells such as akinetes and heterocysts. In nostocales and stigonematales the cells have the ability to develop heterocysts in certain conditions. Stigonematales, unlike nostocales include species with truly branched trichome. Most genera included in the phylum or division cyanobacteria are discussed in the Bergey's manual of determinative bacteriology (Group 11).

Nitrogen Fixation

Cyanobacteria are extremely significant organisms for the health and growth of many plants. They are one of very few groups of organisms that can fix inert atmospheric nitrogen into an organic form, such as **nitrate** or ammonia. These fix atmospheric nitrogen needed by plants for their growth.

Nitrification cannot occur in the presence of oxygen, so nitrogen is fixed in specialized cells called **heterocysts**. These cells have an especially thickened wall that maintain an anaerobic environment.

The symbiotic relationship between blue-green algae and plants has been used to great benefit in the cultivation of rice, where the fern, *Azolla* is distributed among the rice paddies. The fern houses colonies of the cyanobacterium *Anabaena* in its leaves, where it fixes nitrogen. Ferns therefore supply an inexpensive natural fertilizer and nitrogen source for the rice plants when they die at the end of the season. Cyanobacteria also form symbiotic relationships with many fungi, forming complex symbiotic "organisms" known as lichens.

Photosynthesis

Cyanobacteria have an elaborate and highly organized system of internal membranes which house the photosynthetic apparatus. Photosynthesis in cyanobacteria generally uses water as an electron donor and produces oxygen as a by-product, though some may also use hydrogen sulphide as seen in other photosynthetic bacteria. Carbon dioxide (CO_2) is reduced to form carbohydrates via the Calvin cycle. In most forms the photosynthetic machinery is embedded into folds of the cell membrane called thylakoids. The large amounts of oxygen in the atmosphere are considered to have been first created by the activities of ancient cyanobacteria. Due to their ability to fix nitrogen in aerobic conditions they are often found as symbionts with a number of other groups of organisms such as fungi (lichens), corals, pteridophytes (*Azolla*), angiosperms (*Gunnera*), etc.

Cyanobacteria are the only group of organisms that are able to reduce nitrogen and carbon in aerobic conditions. The water-oxidizing photosynthesis is accomplished by coupling the activity of photosystem (PS) II and I (Z -scheme). In anaerobic conditions, they are also able to use only PS I — cyclic photophosphorylation—with electron donors other than water (hydrogen sulphide, thiosulphate, or even molecular hydrogen) just like purple photosynthetic bacteria. Furthermore, they share an archaeal property, the ability to reduce elemental sulphur by anaerobic respiration in the dark.

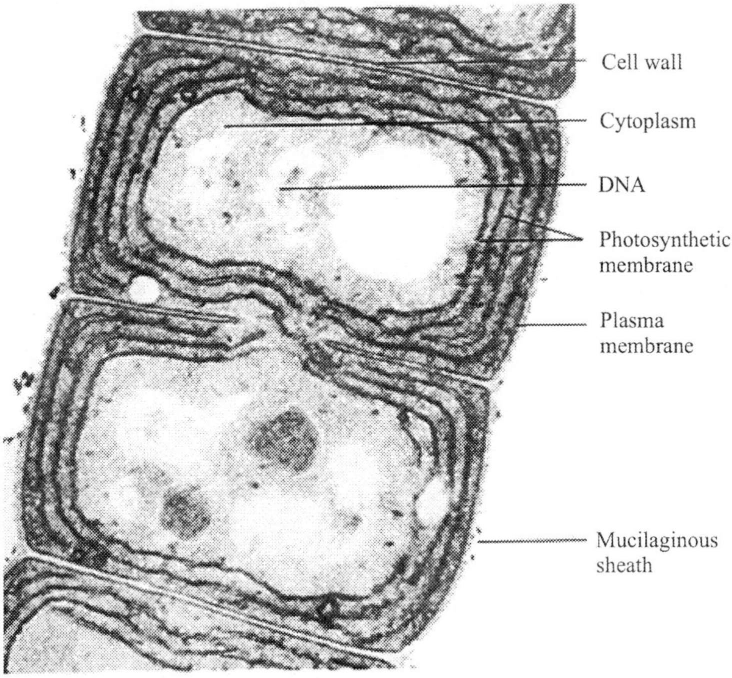

Figure 3.13: A typical cyanobacterial cell.

Attached to thylakoid membrane are phycobilisomes that act as light harvesting antennae for the photosystems. The phycobilisome components (phycobiliproteins) are responsible for the blue-green pigmentation of most cyanobacteria. The variations are mainly due to carotenoids and phycoerythrins which give the cells the red-brownish colouration. In some cyanobacteria, the colour of light influences the composition of phycobilisomes. In green light, the cells accumulate more phycoerythrin, whereas in red light they produce more phycocyanin. Thus, the bacteria appear green in red light and red in green light. This process is known as complementary chromatic adaptation and is a way for the cells to maximize the use of available light for photosynthesis.

A few genera, however, lack phycobilisomes and have chlorophyll b instead (*Prochloron*, *Prochlorococcus*, *Prochlorothrix*). These were originally grouped together as the Prochlorophytes or Chloroxybacteria, but appear to have developed in several different lines of cyanobacteria. For this reason, they are now considered as part of the cyanobacterial group.

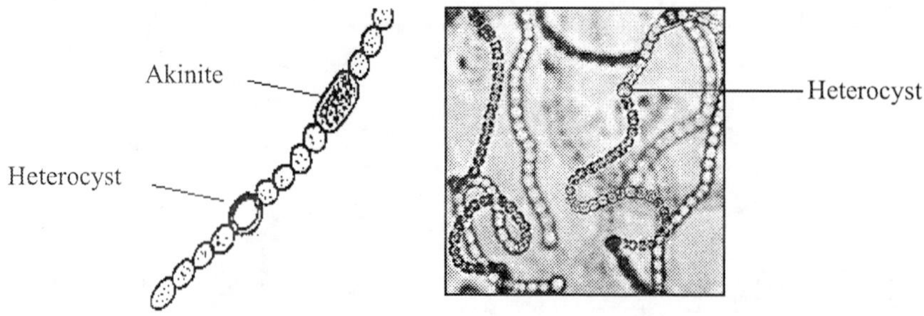

***Anabaena* sp.**

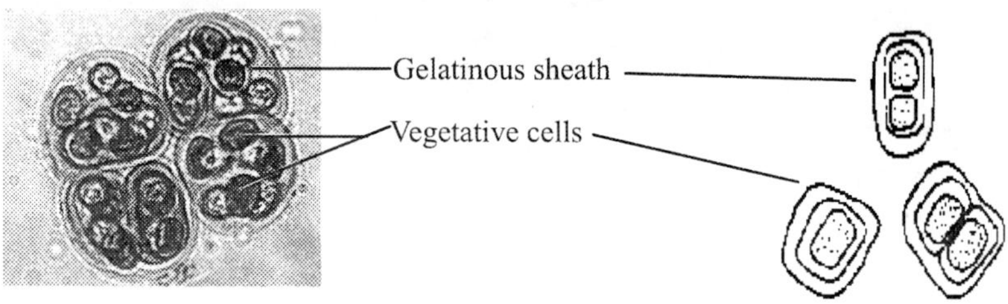

***Gloeocapsa* sp.**

Figure 3.14: Morphological features of *Anabaena* and *Gloeocapsa* sp.

CLASSIFICATION OF BACTERIA

The concept of "species" as applied to higher living organisms (plants and animals) reproducing sexually cannot be used for bacteria. Species is defined in phylogenetic or evolutionary terms meaning similar groups which may inbreed to reproduce similar offsprings. Most microorganisms are **haploid (n)** and reproduce by asexual methods. The bacterial progeny is produced by amitotic division or fission of bacterial cells and therefore freely evolves divergently. Bacteria also carry out reproduction by some *sexual means or parasexual* recombination or by other mechanisms of recombination (e.g. transduction). However, it is difficult to assess the effect of these processes since their frequency and recombining genome may vary all together.

Furthermore, the presence of extrachromosomal elements (plasmids such as F factor) which initiate chromosomal transfer and rearrangements also lead to ambiguity. Thus, the working and practicable definition of a *bacterial species* is 'a group of strains that show a high degree of phenotypic similarity and differ from related groups with respect to many independent physiological or biochemical characteristics'. As of now, with the advent of technique of *in vitro* DNA

Table 3.3 Definition of some of the terms of common use in microbial taxonomy

Species	Group of individuals that are generally similar in characteristics and are adapted to particular ecological niche and are capable of fertile interbreeding (non applicable for bacteria).
Isolate	Each primary culture isolated from natural source.
Strain	Line/population of bacteria descendent from single ancestral bacterium found in natural habitat.
Clones	Young colonies consist exclusively of the progeny of single cell. They do not include demonstrable mutant cell.
Bacterial species	Basic taxonomic group of collection of strain having similar characteristics.
Type strain	Type strain is the strain that is designated to be the permanent reference specimen for the strains included in the species.

amplification (PCR) and sequence alignment, the genotypic relatedness has become a prime identification parameter apart from complete phenotypic description.

Bacteria are also named on the basis of the Linnean binomial nomenclature scheme of genus and species as per the International Code of Nomenclature of Bacteria. Scientific names of all taxa (categories) or those words used to refer to any taxonomic group are either of Latin origin or are Latinized. According to the bionomial system of nomenclature, each distinct bacterial species is provided a name consisting of two words, e.g. *Escherichia coli.*

The first word is the name of the genus and is always written with first alphabet as a capital letter. The second word is a specific epithet and is not capitalized. A set of taxonomic categories (taxons) is employed to group related organisms at various levels of similarity (taxonomical hierarchy). A bacterial species is defined as a single kind of bacterium, while a bacterial genus includes a group of species all of which bear sufficient resemblance to one another to be considered closely related and are easily distinguishable from the members of other genera.

Most bacterial genera are sharply defined, e.g. the genus *Bacillus, Escherichia, and Vibrio* are clearly distinguished from each other. The genus *Bacillus* includes Gram positive, endospore forming rods, *Escherichia* are Gram negative, facultative, non-spore forming rods and *Vibrio* are comma shaped, Gram-negative bacteria. However, it is difficult to define boundaries of microbial species and hence some terms such as *strain or clone* having specific indication are used. A strain is a culture of a given species while a clone is a strain derived from a single cell. A type species is a culture that is thoroughly studied and easily identifiable as a species. The name of a type species generally conveys the characteristics of the group (see **Table 3.3**).

Bacterial taxonomy deserves special attention due to its highly complex systematics. David Hendriks Bergey was the first person to present a comprehensive system of bacterial systematics in his book Bergey's manual of determinative bacteriology. This manual was updated in coming years by collective efforts of several authors from all over the world and was called as Bergey's manual of Systematic bacteriology. Today, the manual comprises four volumes and is considered to be the most comprehensive treatise on bacterial classification. It includes all the known forms of bacteria and is a suitable guide for identification.

In this manual, Bacteria are placed in kingdom Procaryotae which is further subdivided into four divisions. No further taxonomic grouping is provided beyond this but the bacteria are categorized into groups or sections. Various sections are dealt separately as per the available expertise and literature. Criteria used for identification of bacteria include shape, size, staining reactions, motility, spore production, GC content, oxygen, pH and temperature requirement and more importantly physiological abilities as determined by the biochemical tests. Apart from this, pathogenicity, mode of nutrition and immunological characteristics are also noted. Most modern procedures involve 16s rRNA sequencing and preparation of dendrogram based on biological data resources. Using this, exact position of the bacterial strain in question is determined on the phylogenetic tree.

All microorganisms are named and cited as per the rules of binomial nomenclature and thus have Latin (or Latinized) a genus and species. Briefly, the first letter of the genus is capitalized, remainder is written in lower case. Both genus and species should be printed in italics, or if not possible, these should be underlined, e.g. *Escherichia coli*. For any subsequent citation of the same microorganism, genus is abbreviated, e.g. *E. coli*. Nomenclatures of microorganisms have very interesting genesis and criteria. Sometimes its their shape (*Vibrio*- comma shaped; *Spirillium*), organization (*Staphylococcus* – A bundle of spheres, *Streptococcus* – a chain of spheres), food habit (*Hemophilus* – blood loving, *Saccharomyces* – sugar fungus), or physiological character (*Thermomyces* – high temperature fungi). Many a times the genus is named after the discoverer, e.g. the genus of the fungus *Saksenia vesiformis* and *Escherichia coli* are named after their discoverer S.B. Saksena and Theodor Escherisch, respectively.

Characters Used in Bacterial Identification

The classification of bacteria is based upon the results of microscopic, colonial, genetic, biochemical and physiological observations. Some specialized techniques of strain identification such as serotyping, bacteriophage typing and genotypic data including the measurement of guanine and cytosine (G + C) content of DNA, DNA-DNA homology, rRNA sequencing are also commonly used nowadays.

Colonial Characteristics

Bacteria show typical growth characteristics on solid media under appropriate cultural conditions and the *colony morphology* can be used in presumptive identification. The colonies can be varying in size and diameter, in outline (circular, wavy, rhizoid), elevation (flat, raised, convex) and opacity (transparent, opaque, and translucent). The colour of the colony or the changes imparted to surrounding medium are important tools in the tentative identification of the bacteria. For example, colonies of streptococci on blood agar medium are small, beadlike and have an opalescent grey colour with smooth or slightly rough edges.

Bacterial Morphology

Microscopic examination of bacterial cells to learn their shape, size and staining characteristics is the next step. Based on this, a very clear primary information regarding their possible group is obtained.

Primarily based on obvious differences in morphological, cultural and colony characteristics bacteria can only be pooled differently. However, further, biochemical, serological and specific tests are often required to be performed before bacteria are conclusively identified.

Nutritional and Biochemical Characteristics of Bacteria

Bacteria show a diverse range of nutritional characteristics ranging from obligate intracellular parasitic mode to saprotrophy to autotrophy. Similarly, the *in vitro* culture of different bacteria requires entirely different media and conditions. Some common saprotrophs or commensals can grow easily on general purpose medium, nutrient agar, while the fastidious may require blood agar medium. Also, the range of growth conditions needed for the growth of bacteria may vary sharply. For example, provision of light is a must for photoautotrophic bacteria. Aerobic bacteria must have access to air while strict anaerobes cannot tolerate even traces of air. Also the biochemical needs of bacteria vary and the ability or inability of an organism to grow on a medium containing a selective inhibitory factor (antibiotics, low pH, bile salts, etc.) can also be an important criterion. The growth and multiplication of bacteria is the consequence of active metabolic pathways and utilization of a substrate. Production of certain metabolic end products indicates presence or absence of specific enzymes, metabolic ability or biochemical pathway.

Serological Examination

Most bacteria trigger an immune response. Serological examination (antigen-antibody reaction) is used as a very sensitive and specific method for diagnosis of diseases (Widal test, VDRL test). This involves the interaction between a specific antigen or toxin and antisera. For example, the final identification of a typhoid bacteria (*Salmonella typhi*) is done using Widal test. Bacterial species and types can be identified by specific antigen antibody reactions. Antigens are substances that induce the production of homologous antibodies in a foreign host. Bacterial cell wall, toxins, appendages, etc. act as antigens. Some surface antigens that have been extensively studied for identification of enteric bacteria (e.g. *S. typhi*) include O antigens (lipopolysaccharides in the outer membrane), K antigens (capsular polysaccharides) and H antigens (flagellar proteins).

The test includes production of antibodies in an animal host and testing of the antiserum by either the agglutination or precipitation test. In the agglutination test, a drop of the culture of a particular bacterium is mixed on a slide with the anti-serum of an infected individual. If related, the antigen and antibody react visibly to produce agglutination or precipitation. If clumping occurs, the test bacterium is considered to be the same or closely related to the bacterium used as the antigen. A range of such antigen-antibody reactions is available including simple slide test to most sensitive ELISA (enzyme linked immunosorbent assay).

Phagetyping

This method is used for type-classification of bacteria by testing the susceptibility of the bacterial culture to lysis by each of a set of type specific lytic bacteriophages. The phage type of the bacterial culture is identified based on its pattern of susceptibility to the different phage strains. Phage typing allows further subdivision of a serological entity, e.g. *S. typhi* serotypes can be further classified into more than 100 different phage types.

Numerical or Adansonian Taxonomy

Michael Adanson proposed a system of classification of microorganisms on the basis of evolutionary affinities and the arrangement of organisms into groups on the basis of inherited and stable structural and physiological characters. This method of classification of bacteria is known as *Adansonian or Numerical taxonomy*. This arrangement is also known as the Natural or the Phylogenetic classification. It is based on the assumption that when each phenotypic character is given equal weightage in terms of numbers, then it should be possible to express the taxonomic relatedness between organisms by numbers. The determination of similarity coefficient and matching coefficient for two bacterial strains, both characterized with respect to different characters A, B, C, D, etc. can be determined as follows:

Number of characters positive in both strains = A

Number of characters positive in strain 1 and negative in strain 2 = B

Number of characters negative in strain 1 and positive in strain 2 = C

Number of characters negative in both strains = D

Similarity coefficient (Sj) = A / A+B+C

Matching coefficient (Ss) = A+D / A+B+C+D

The similarity coefficient does not take into account the characters negative for both organisms while the matching coefficient includes both positive and negative characters. After similarity coefficient and matching coefficient are determined for organisms under study, the data are arranged in similarity matrix. Such matrices can be reordered to bring similar strains close to one another. Such data are then transposed over a dendrogram which act as the basis for determining taxonomic relatedness. The vertical dotted lines in a dendrogram indicate similarity levels which might be considered for recognizing two different taxonomic ranks such as a genus and a species.

Earlier, the Adansonian approach appeared impractical because of the magnitude of numerical operations involved. This difficulty has now been overcome by the use of computers which can be programmed to compute the data and determine the degree of similarity. Although numerical taxonomy does not show any phylogenetic relationship, yet, it provides a stable basis for the construction of taxonomic groupings.

Use of Software in Bacterial Identification

Bacterial characteristics are studied by conventional methods including morphological, biochemical and serological tests. However, since a great volume of the data is generated, the problem arises in comparing differences and similarities. To overcome this, computers equipped with specialized software are used for estimating the differentiating power of the tests as well as for estimating the overall similarity with known organisms. The speed and memory of computer allows the rapid comparison of characteristics in as many as species in the identification process by comparing the properties of unidentified organisms with those in the computer hard disk. The use of computer minimizes the possibility of error in identification due to infrequent occurrence of an organism or the occurrence of a more common organism with, superficial resemblance to other organisms. Several computer programmes for bacterial identification have now been developed and are commonly used in bacteriological laboratories.

Genetic Methods in Taxonomy

The central dogma (DNA $\rightarrow$ mRNA $\rightarrow$ Protein) of molecular biology states that phenotypic characters of any organism are direct expressions of their genomes and hence the overall phenotype is a result of the DNA base sequence. Hence, the best method of comparing two organisms should be on the basis of parity in their genetic material. The total guanine and cytosine content of DNA (G+C %) of a given organism is one such criteria.

DNA contains four bases namely, adenine (A), guanine (G), thymine (T) and cytosine (C) and in a double stranded DNA, A pairs with T and G pairs with C. The relative percentage of guanine and cytosine, [(G+C / A+T+G+C) x 100] varies widely with different bacteria. The composition of chromosomal DNA is a fixed property of each cell and is independent of age and other environmental influences.

The homology between two DNA or two RNA sequences is also being used as conclusive criteria for comparison and relatedness. The homology can be found out using hybridization or sequence alignment tools. The basis of this test is that the degree of hybridization would be an indication of the degree of relatedness or homology.

DNA base % composition is an extremely useful character which complements the phenotypic characterization of microorganisms with similar genotype. Members usually possess similar base composition (within 1–2%) in certain groups of bacteria within the tribes such as *Escherichiae* or genera *Pseudomonas* and *Rhizobium*.

The base composition can be determined either by subjecting the purified DNA to increasing temperature and determining the increase in hypochromicity due to denaturation or by centrifugation of the DNA in cesium chloride density gradients. The basis of the first method, i.e. the melting point method, is that when double-stranded DNA is subjected to increasing temperature, the two DNA strands separate at a characteristic temperature. The melting temperature depends upon the G+C content of the DNA. Higher the G+C content, higher will be the melting point.

The mean temperature at which thermal denaturation of DNA occurs is called the melting point (T_m) and this is determined by noting the change in optical density of DNA solution at 260 nm during the heating period. Based on the melting point or T_m value, the GC can be calculated as below:

$$(G + C)\% = Tm \times 63.54/0.47$$

Ribosomal RNA Sequencing and Cataloguing

The prokaryotic relatedness can be ascertained by analyzing the relatedness in rRNA sequences of two microorganisms. The rRNA molecules are of choice because they are universally present in all organisms, have a constant function and do not get altered over longer periods of time. The 16 S rRNA is most suitable for the purpose as it is sufficiently large and easy to handle with a high degree of precision. On the other hand, 5S is too small while 23 S molecules are too large for the purpose. For this, the 16 S rRNA of an organism is digested with ribonuclease T1 and the resulting nucleotides are sequenced using automated sequencer machines. The rRNA sequences of several thousand species of bacteria and eukaryotes have been generated and compared by alignment tools to find out their relatedness.

Chemotaxonomy

Chemotaxonomy is based on identification of any chemical or biochemical profile of biomolecules in bacteria. Microbial cell walls also differ considerably in their chemical composition and hence protein, lipid and carbohydrate profile can reveal difference or relatedness between any two bacteria. Ester or ether linkages in lipid molecules and amino acid sequences of enzyme or proteins are useful in chemotaxonomy profile.

Analysis of cell components such as carbohydrates, fatty acids, hydroxyacids, organic acids, amino acids and amines can also be done by gas chromatography. Similarly, characteristic metabolic products of bacteria may also be analyzed by gas chromatography and used in identification. For example, gas chromatographic methods have been used routinely as aids in the identification of anaerobic bacteria through analysis of end products such as alcohols, fatty acids and other volatile compounds.

BERGEY'S MANUAL OF DETERMINATIVE BACTERIOLOGY

The treatise called as *Bergey's Manual of Determinative Bacteriology* was initiated by David Bergey, Professor of bacteriology and co-workers at the university of Pennsylvania, USA in 1923. The manual gained status of an international reference after the establishment of an independent Bergey's manual Trust in 1936. Since then the Bergey's Manual of Determinative Bacteriology has been edited nine times and the present volume is the 9th edition. It is edited by John G. Holt and published by Williams & Wilkins, Baltimore, USA in the year 1994.

The very first edition of Bergey's Manual of Determinative Bacteriology took its shape under the the auspices of Society of American bacteriologists or SAB (now American Society for Microbiology or ASM). The society appoined an editorial Board having David H. Bergey as chairman and Francis C. Harrison, Robert S. Breed, Bernard W. Hammer and Frank M. Huntoon as members. This Board brought the first manual into print in 1923. The Board under the chairmanship of Dr Bergey published its second edition in 1925 and a third edition in 1930. Subsequent editions of the manual were published by the trust which is a non-profit organization and its income is used for the purpose of preparing, editing and publishing revisions and successive editions of the Bergey's manual.

The Bergey's manual of determinative bacteriology can be referred as a compact, utilitarian aid for bacterial identification derived out from the larger and voluminous Bergey's manual of systematic bacteriology. The 9th edition of the manual categorizes bacteria into 35 groups, which are comparable with the "Sections" described in the 1st edition of Bergey's manual of Systematic bacteriology (1984–1989, Table 3.4). These groups do not represent taxons or systematic position but attempt to group bacteria into easily recognized phenotypic groups for identification and diagnostic purposes. The book was compiled by abstracting the phenotypic information contained in the four volumes of Bergey's Manual of Systematic bacteriology. This manual includes all the published bacterial genera published through the January 1992 issue of the International Journal of Systematic Bacteriology (ISJB). It carries abstract and general descriptions of bacteria including morphological features (shape and size, Gram reaction, etc.), motility and flagella, oxygen relationships, nutrition and metabolism type, carbon and energy sources, habitat and ecology. The manual includes several keys or tables to genera and species, genus descriptions, synonyms, and other nomenclatural information for help in identification.

Table 3.4 Bergey's Manual of Systematic Bacteriology, Ist Edition, Eds. John G. Holt *et al.*, published by Williams & Wilkins, Baltimore, USA in Four volumes over the period of six years (1984–1989).

Volume 1 (1984)	Gram-negative Bacteria of general, Medical or industrial importance
Volume 2 (1986)	Gram-positive Bacteria other than Actinomycetes
Volume 3 (1989)	Archaeobacteria, Cyanobacteria, and remaining Gram-negative Bacteria
Volume 4 (1989)	Actinomycetes

Table 3.5 The Second edition of Bergey's Manual of Systematic bacteriology, Eds. George M. Garrity *et al.*, published by Springer, New York in 5 volumes:

Volume 1 (2001)	The Archaea and the deely branching and phototrophic Bacteria
Volume 2 (2005)	The proteobacteria
Volume 3 (2009)	The Firmicutes
Volume 3 (2009)	The Bacteroidetes, Planctomycetes, Chlamydiae, Spirochaetes, Fibrobacteres, Fusobacteria, Acidobacteria, Verrucomicrobia, Dictyoglomi and Gemmatimonadetes
Volume 5 (2010)	The Actinobacteria

In the following section, the classification of bacteria upto the generic level as published in the 9th edition of Bergey's *Manual of Determinative Bacteriology* is given. As described earlier, the manual contains 35 groups which are comparable with the 35 sections described in the four volumes of first edition of *Bergey's Manual of Systematic Bacteriology* (**Table 3.4**).

Classification of Bacteria based on Bergey's Manual of Determinative Bacteriology (9th edition)

Group 1 Spirochetes
Spirochaetales
Spirochaetaceae
Borrelia
Brachyspira
Cristispira
Spirochaeta
Treponema
Leptospiraceae
Leptonema
Leptospira
Serpulina
Other: Genera of insect gut Spirochetes
Hingdut spirochetes of Termites and *Cryptocercus punctulatus* (wood-eating cockroach)

Group 2 Aerobic/Microaerophilic, Motile, Helical/Vibrioid Gram-Negative Bacteria

Aleromonas
Aquaspirillum
Herbaspirillum
Marinomonas

Azospirillum
Bdellovibrio
Campylobacter
Cellvibrio
Halovibrio
Helicobacter
Micavibrio
Oceanospirillum
Spirillum
Sporospirillum
Vampirovibrio
Wolinella

Group 3 Non-motile (or rarely), Gram-negative Curved Bacteria

Spirosomaceae

Ancyclobacter
Brachyarcus
Cyclobacterium
Flectobacillus
Meniscus
Pelosigma
Runella
Spirosoma

Group 4 Gram-negative Aerobic Rods and Cocci

Subgroup 4A

Pseudomonadaceae
Frateuria
Pseudomonas
Xanthomonas
Zooglea

Azotobacteriaceae
Azomonas
Azotobacter

Rhizobiaceae
Agrobacterium
Bradyrhizobium
Phyllobacterium
Rhizobium

Methylococcaceae
Methylococcus
Methylomonas

Acetobacteraceae
Acetobacter
Gluconobacter

Halobacteriaceae
Halobacterium
Halococcs

Legionellaceae
Legionella
Legionella

Neisseriaceae
Acinetobacter
Kingella
Moraxella
Neisseria

Other Genera

Acidiphilium
Acidomonas
Acidothermus
Afipiaq
Agromonas
Alcaligenes
Alteromonas
Aminobacter
Aquaspirillum
Beijerinckia
Bordetella
Brucella
Chromohalobacter
Erythrobacter
Flavimonas
Flavobacterium
Francisella
Halomonas
Hydrogenophaga
Janthinobacterium
Lampropedia
Marinomonas
Mesophilobacter
Methylobacillus
Methylobacterium
Methylophaga
Phenylobacterium
Psychrobacter
Rhizobacter
Roseobacter
Rugamonas
Serpens
Sinorhizobium
Sphingobacterium
Thermomicrobium
Thermus
Variovorax
Volcaniella
Weeksella

Chrysemonas *Methylophilus* *Xanthobacter*
Cupriavidas *Methylovorus* *Xylophilus*
Deleya *Morococcus* *Zoogloea*
Derxia *Oceanospirillum*
Ensifer *Oligella*
Paracoccus

Subgroup 4B
Bacteroides *Taylorella* *Wolinella*

Group 5 Facultatively Anaerobic Gram-negative Rods

Subgroup 1 Enterobacteriaceae
Arsenophonus *Budvicia* *Buttiauxella*
Cedecea *Citrobacter,* *Edwardsiella*
Enterobacter *Erwinia* *Escherichia*
Ewingella *Hafnia* *Klebsiella*
Kluyvera *Leclercia* *Leminorella*
Moellerella *Morganella* *Obsembacterium*
Pantoea *Pragia* *Proteus*
Providencia *Rahnella* *Salmonella*
Serratia *Shigella* *Tatumella*
Xenorhabdus, *Yersinia* *Yokenella*

Subgroup 2 Vibrionaceae
Aeromonas *Enhydrobacter* *Photobacterium*
Plesiomonas *Vibrio*

Subgroup 3 Pasturellaceae
Actinobacillus
Haemophilus
Pasteurella

Subgroup 3 Pasturellaceae
Actinobacillus
Haemophilus
Pasteurella

Subgroup 4 Other Genera
Calymmatobacterium *Cardiobacterium*
Cedecea *Eikenella*
Gardnerella *Kluyvera*
Obesumbacterium *Rahnella*
Streptobacillus *Tatumella*
Xenorhabdus *Chromobacterium*
Zymomonas

Group 6 Gram-negative Anaerobic, Straight, Curved, and Helical Rods
Bacteroidaceae
Acetivibrio *Halobacteroides* *Ruminobacter*
Acetoanaerobium *Ilyobacter* *Sebaldella*
Acetofilamentum *Lachnospira* *Selenomonas*

Acetogenium
Acetomicrobium
Acetothermus
Acedaminobacter
Anacerobiospirillum
Anaerorhabdus
Anaerovibrio
Bacteroides
Butyrivibrio
Centipeda
Fervidobacterium
Fibrobacter
Fusobacterium
Haloanaerobium
Leptotrichia
Malonomonas
Megamonas
Mitsuokella
Oxalobacter
Pectinatus
Pelobacter
Porphyromonas
Prevotella
Propionigenium
Propionispira
Rikenella
Roseburia
Sporomusa
Succinimonas
Succinivibrio
Syntrophobacter
Syntrophosmonas
Thermobacteroides
Thermospipho
Thermotoga
Tissierella
Wolinella
Zymophilus

Group 7 Dissimilatory Sulphate- or Sulphur-Reducing Bacteria

Subgroup 1
Desulfotomaculum (also endospore-forming)

Subgroup 2
Desulfobulbus
Desulfuomonas
Desulfovibrio
Thermodesulfobacterium
Desulfomicrobium

Subgroup 3
Desulfobacter
Desulfobacterium
Desulfococcus
Desulfomonile
Desulfonema
Desulfosarcina

Subgroup 3
Desulfurella
Desulfuromonas

Group 8 Anaerobic Gram-negative Cocci
Veillonellaceae
Acidominococcus
Megasphaera
Syntrophococcus
Veillonella

Group 9 Rickettsias and Chlamydias
I. Rickettsiales
Rickettsiaceae
Cowdria
Coxiella
Ehrlichia
Neorickettsia
Rickettsia

Rickettsiella
Rochalimaea
Wolbachia
Bartonellaceae
Bartonella
Grahamella
Anaplasmataceae
Aegyptianella
Anaplasma
Eperythrozoon
Haemobartonella

II. Chlamydiales
Chlamydiaceae
Chlamydia

Group 10 Anoxygenic Phototrophic Bacteria
Subgroup 1 Purple Sulphur Bacteria: Internal Sulphur Granules
Chromatiaceae
Amoebobacter
Chromatium
Lamprobacter
Lamprocystis
Thiocapsa
Thiocystis
Thiodictyon
Thiopedia
Thiospirillum

Subgroup 2 Purple Sulphur Bacteria: External Sulphur Granules
Ectothiorhodospiraceae
Ectothiorhodopsin
Ectothiorhodospira

Subgroup 3 Purple Nonsulphur Bacteria
Rhodobacter
Rhodyclus
Rhodomicrobium
Rhodopila
Rhodopseudomonas
Rhodospirillum

Subgroup 4 Green Nonsulphur Bacteria
Heliobacillus
Heliobacterium

Subgroup 5 Green Sulphur Bacteria
Anacalochloris
Chlorobium
Chloroherpeton
Pelodictyon
Prosthecochloris

Subgroup 6 Multicellular Filamentous Green Bacteria
Chloroflexus
Chloronema

Heliothrix
Oscillochloris

Subgroup 7 Anaerobic Chemotrophic Bacteria
Erythrobacter

Group 11 Oxygenic Phototrophic Bacteria
I. The Cyanobacteria

Subgroup 1 Chroococcales
Chamaespiphon
Cyanothece
Gloeobacter
Gloeocapsa
Gloethece
Microcystis
Myxobaktron
Synechococcus
Synechocystis

Subgroup 2 Pleurocapsales
Chroococcidiopsis
Dermocarpa
Dermocarpella
Myxosarcina
Pleurocapsa
Xenococcus

Subgroup 3 Oscillatoriales
Arthrospira
Crinalium
Lyngbya
Microcoleus
Oscillatoria
Pseudanabaena
Spirulina
Starria
Trichodesmium

Subgroup 4 Nostocales
Section A Nostocaceae
Anabaena
Aphanizomenon
Cylindrospermum
Nodularia
Nostoc
Section B Scytonemataceae
Scytonema
Section C Rivulariaceae
Calothrix

Subgroup 5 Stigonematales
Chlorogloeopsis
Fischerella
Geitleria
Stigonema
II. The Prochlorophytes (Prochlorales)
Prochloraceae
Prochloron
Prochlorothrix

Group 12 Aerobic Chemolithotrophic Bacteria and Associated Organisms

Subgroup 1 Colourless sulphur-oxidizing bacteria

Acidiphilium	*Thiodendron*
Macromonas	*Thiomicrospira*
Thermothrix	*Thiosphaera*
Thiobacillus	*Thiospira*
Thiobacterium	*Thiovulum*

Subgroup 2 Iron- and Manganese-oxidizing and/or -Depositing Bacteria

Siderocapsaceae

Aquaspirillum	*Ochrobium*
Bilophococcus	*Siderocapsa*
Gallionella	*Siderococcus*
Leptospirillum	*Sulfobacillus*
Metallogenium	
Naumaniella	

Subgroup 3 Nitrifying Bactria

Nitrobacteraceae

Section A Nitrate-oxidizing bacteria

Nitrobacter
Nitrococcus
Nitrospina
Nitrospira

Section B Ammonia-oxidizing bacteria

Nitrosococcus
Nitrosolobus
Nitrosomonas
Nitrosospira
Nitrosovibrio

Obligate chemolithotrophic Hydrogen bacteria

Hydrogenobacter

Group 13 Budding *and/or Appendaged Bacteria

Subgroup 1 Prosthecate Bacteria

*Ancalomicrobium**	*Hyphomonas*
Asticcacaulus	*Labrys**
Caulobacter	*Pedomicrobium**
Dichotomicrobium	*Prosthecobacter*
Filomicrobium	*Prosthecomicrobium**
Hirschia	*Stella**
*Hyphomicrobium**	*Verrucomicrobium*

Subgroup 2 Planctomycetales

Gemmata
Pirelulla
Isosphaera
Plactomyces

Subgroup 3 Other (Nonprosthecate) Budding and or appendaged Bacteria

Angulomicrobium

Blastobacter
Ensifer
Gemmiger
Isosphaera
Planctomyces

Nonbudding Bacteria
Asticcacaulis
Caulobacter
Gallionella
Nevskia
Prosthecobacter

Morphologically Unusual Budding Bacteria
Caulococcus
Kuznezovia
Metallogenium
Seliberia
Thiodendron

Group 14 Sheathed Bacteria
Clonothrix
Crenothrix
Haliscominobacter
Leptothrix
Lieskeela
Phragmidiothrix
Sphaerotilus

Group 15 Nonphotosynthetic, Nonfruiting Gliding Bacteria
Subgroup 1 Single-celled, Rod-shaped Gliding Bacteria
Cytophagales
Cytophagaceae
Capnocytophaga
Chitinophaga
Cytophaga
Flexibacter
Flexithrix
Microscilla
Sporocytophaga
Thermonema
Lysobacterales
Lysobacteriaceae
Lysobacter

Subgroup 2 Flattened, Filamentous Gliding Bacteria
Simonsiellaceae
Alysiella
Simonsiella

Subgroup 3 Sulphur-Oxidizing Gliding Bacteria
Beggiatoales

Beggiatoaceae
Achromatium
Beggiatoa
Thioploca
Thirospirillopsis
Thiothrix

Subgroup 4 The Pelonemas
Pelonemataceae
Achroonema
Desmanthus
Pelonema
Peloploca

Other Genera
Agitococcus
Isophaera
Saprospira
Vtreoscilla
Desulfonema
Leucothrix
Toxothrix
Herpetosiphon

Group 16 The Fruiting, Gliding, Bacteria: Myxobacteria
Archangiaceae
Archangium
Cystobacteraceae
Cystobacter
Melittangium
Stigmatella
Myxococcales
Myxococcaceae
Myxococcus
Polyangiaceae

Group 17 Gram-positive cocci
Aerobic, Catalase-positive Genera
Deinobacter
Deinococcus
Marinococcus
Micrococcus
Planococcus
Saccharococcus
Staphylococcus
Stomatococcus
Aerotolerant, Catalase-negative Genera
Aerococcus
Enterococcus
Gemella
Lactococcus
Leuconostoc
Melissococcus
Pediococcus
Streptococcus

Trichococcus
Vagococcus
Anaerobic, Catalase-Negative Genera
Coprococcus
Peptococcus
Peptostreptococcus
Ruminococcus
Sarcina

Group 18 Endospore-Forming Gram-positive Rods and Cocci
Amphibacillus
Bacillus
Clostridium
Desulfotomaculum
Oscillospira
Sporolactobacillus
Sporosarcina
Sulfidobacillus
Syntrophospora

Group 19 Regular, Nonsporing Gram-positive Rods
Brochothrix
Carnobacterium
Caryophanon
Erysipelothrix
Kurthia
Lactobacillus
Listeria
Renibacterium

Group 20 Irregular, Nonsporing Gram-positive Rods
Acetobacterium
Acetogenium
Actinomyces
Aeromicrobium
Agromyces
Arachnia
Arcanobacterium
Arthrobacter
Aureobacterium
Bifidobacterium
Brachybacterium
Brevibacterium
Butyrivibrio
Caseobacter
Cellulomonas
Clavibacter
Coriobacterium
Corynebacterium
Curtobacterium
Dermabacter
Eubacterium
Exigouibacterium
Falcivibrio
Gardnerella
Jonesia
Lachnospira
Microbacterium
Mobiluncus
Pimelobacter
Propionibacterium
Rarobacter
Rthia
Rubrobacter
Sphaerobacter
Terrabacter
Thermoanaerobacter

Group 21 Mycobacteria
Mycobacteriaceae
Mycobacterium

Groups 22–29 Actinomycetes

Groups 22 Nocardioform Actinomycetes

Subgroup 1 Mycolic acid containing bacteria

Gordona
Rhodococcus
Nocardia
Tsukamurella

Subgroup 2 Pseudonocardia and related genera

Actnobispora	*Intrasporangium*
Actinokineospora	*Kibdelosporangium*
Actinopolyspora	*Pseudoamycolata*
Amycolata	*Pseudonocardia*
Amycolatopsis	*Rhodococcus*
Faenia (Micropolyspora)	*Saccharomonospora*
	Saccharopolyspora

Subgroup 3 Nocardioides and Terrabacter

Nocardioides
Terrabacter

Subgroup 4 Promicromonospora and related genera

Jonesia
Promicromonospora
Oerkovia

Group 23 Actinomycetes with Multilocular Sporangia

Dermatophilus	*Frankia*	*Geodermatophilus*

Group 24 Actinoplanetes

Actinoplnes	*Ampullariella*	*Catelbatospora*
Dactylosporangium	*Micromonospora*	*Pilimelia*

Group 25 Streptomyces and Related Genera

Intrasporangium	*Kineosporia*	*Sporichthya*
Streptomyces	*Streptoverticillium*	

Group 26 Maduromycetes

Subgroup 1 Streptosporangium and related genera

Microbispora	*Microtetraspora*	*Planobispora*
Planomonospora	*Spirillospora*	*Streptosporangium*

Subgroup 2

Actinomadura

Group 27 Thermomonospora and Related Genera

Actinosynnema
Nocardiopsis
Streptoalloteichus
Thermomonospora

Group 28 Thermoactinomycetes

Thermoactinomyces

Group 29 Other Actinomycete Genera
Glycomyces
Kitasatosporia
Saccarothrix

Group 30 Mycoplasmas
Mycoplasmataceae
Mycoplasma
Ureaplasma
Acholeplasmataceae
Acholeplasma
Spiroplasmataceae
Spiroplasma

Other Genera
Anaeroplasma
Asteroleplasma
Thermoplasma

Group 31 Methanogens
Cell Walls With Pseudomurein
Methanobacteriales
Methanobacteriaceae
Methanobacterium
Methanobrevibacter
Methanothermaceae
Methanothermus
Cell Walls Without Pseudomurein
Methanococcales
Methanococcaceae
Methanococcaceae
Methanococcaceae
Methanolacinia
Methanococcoides
Methanolobus
Methanothrix

Methanomicrobiales
Methanomicrobiaceae
Methanomicrobium
Methanogenium
Methanospirillum
Methanoplanaceae
Methanoplanus

Other Genera
Methanocorpusculum
Methanoculleus
Methanohalobium
Methanohalophilus
Methanosarcina
Methanospheaera

Group 32 Archaeal Sulphate Reducers
Archaeoglobales
Archaeoglobaceae
Archaeoglobus

Group 33 Extremely Halophilic, Aerobic Archaeobacteria
Halobacteriales
Halobacteriaceae
Haloarcula
Halobacterium
Halococcus
Haloferax
Natronobacterium
Natronococcus

Group 34 Cell Wall-Less Archaeobacteria
Thermoplasma

Group 35 Extremely Thermophilic and Hyperthermophilic S°- Metabolizers

Subgroup 1 Cocci, Optimum Growth Below pH 4.0
Acidianus
Desulfurolobus
Metallosphaera
Sulfolobus

Subgroup 2 Rods That Use H_2 as Energy Source
Pyrobaculum
Thermoproteales
Thermoproteaceae
Thermoproteus
Thermofilum

Subgroup 3 Cocci or Disc-Shaped That Oxidize H_2S
Hyperthermus
Staphylothermus
Thermodiscus
Desulfurococcaceae
Desulfurococcus
Pyrodictium
Thermococcales
Thermococcaceae
Thermococcus
Pyrococcus

PROKARYOTE SYSTEMATICS: CURRENT PERSPECTIVE

The prokaryote taxonomy has undergone rapid growth, with over 2200 new species and 390 new genera having been described. More than 850 published new combinations have resulted in numerous rearrangements in groups. Advent in molecular sequencing of highly conserved regions of the prokaryotic genome, most notably genes coding for the RNA of the small ribosomal subunit, have lead to a natural classification that reflects the evolutionary history of Bacteria and Archaea, and to the development of new, universally applicable methods of identifying these organisms. Thus, the second edition of Bergey's Manual of Systematic Bacteriology has been completely restructured on the basis of phylogenetic relatedness to reflect the current state of prokaryotic taxonomy. It utilizes orthologous genes (genes that belong to different species but retain their original function) for construction of phylogenetic trees e.g. 16S ribosomal RNA gene. **Figure 3.15** shows an overview of major lineages in the three domain classification of Carl Woese (1977). **Figures 3.16 (a)** and **(b)** present detailed phylogenetic tree of major lineages or phyla of Archaea and Bacteria.

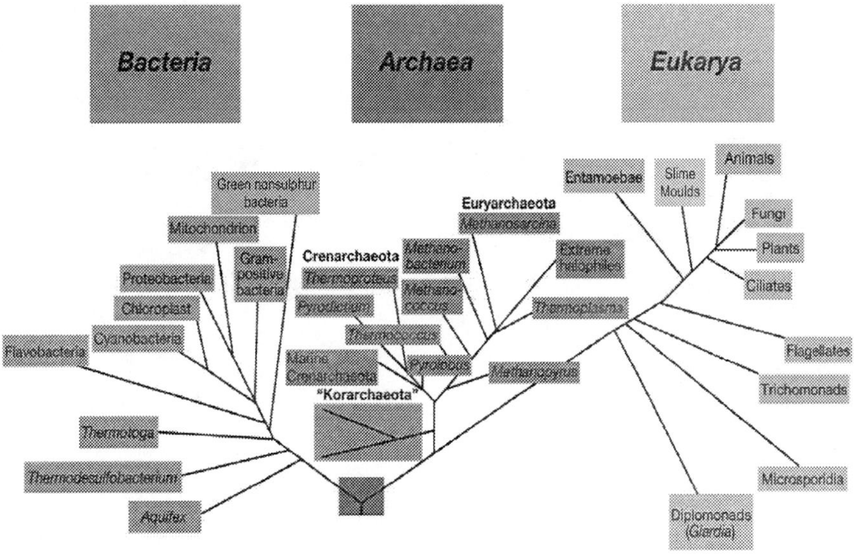

Figure 3.15 Universal phylogenetic tree as determined from comparative ribosomal RNA sequencing (Woese and Fox, 1977).

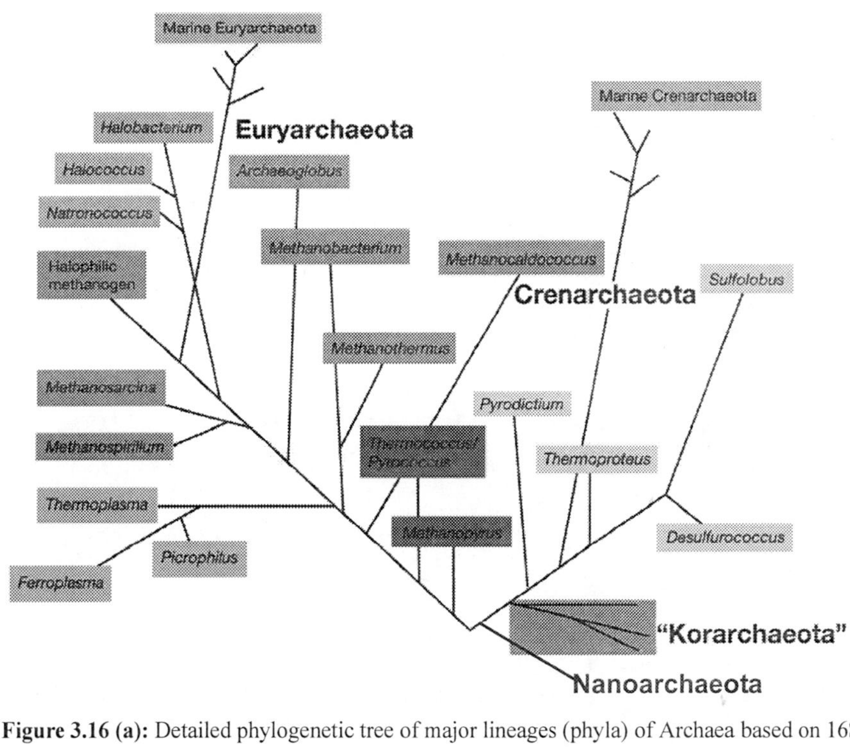

Figure 3.16 (a): Detailed phylogenetic tree of major lineages (phyla) of Archaea based on 16S ribosomal RNA sequence comparisons.

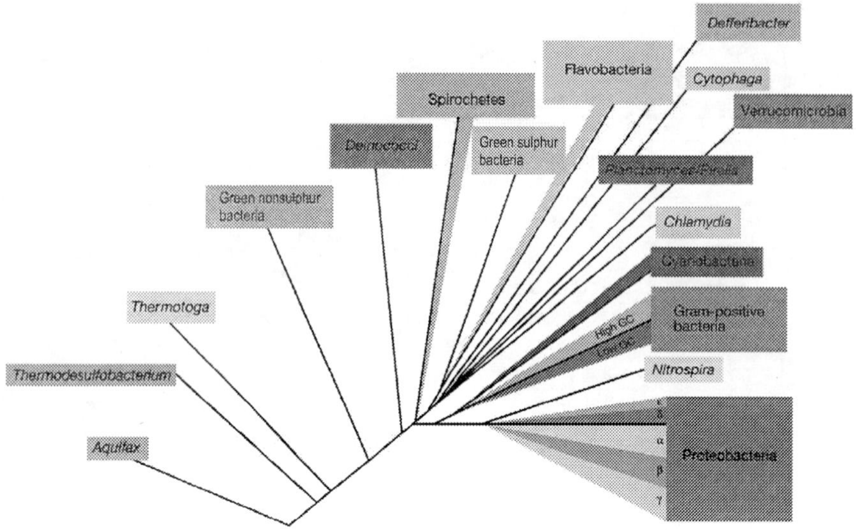

Figure 3.16 (b): Detailed phylogenetic tree of major lineages (phyla) of bacteria based on 16S ribosomal RNA sequence comparisons.

Table 3.6 An overview of the groups of bacteria as per the *Bergey's Manual of Systematic Bacteriology* (2nd edition). Numbers in parentheses indicate number of known species.

(1) Archaea

Euryarchaeota: This group is comprised of methanogens, halophiles, thermophiles.

(a) Extremely halophilic (*Halobacterium, Haloferax, Natronobacterium*)
(b) Methanogens (*Methanobacterium, Methanosarcina*)
(c) Thermoplasmatales (*Thermoplasma, Picrophilus*)
(d) Hyperthermophiles (*Thermococcus, Pyrococcus, Methanopyrus, Archaeoglobus, Ferroglobus*)

Crenarchaeota: This group is comprised of thermophilic archaea. They use sulphur compounds as electron donors or as acceptors.

(a) Terrestrial and volcanic hyperthermophiles (*Sulfolobus, Thermoproteus*)
(b) Submarine volcanic hyperthermophiles (*Pyrolobus, Ignicoccus, Staphylothermus*)

Nanoarchaeota : Tiny parasitic hyperthermophiles, *Nanoarchaeum* having smallest prokaryotic genome and 0.4 μm in diameter lives on *Ignicoccus*.

Korarchaeota : This group consists of uncultivable hyperthermophiles of hot springs.

(2) Proteobacteria (2086)

- Purple phototrophic bacteria (*Chromatium, Rhodospirillium*)
- The nitrifying bacteria (*Nitrosomonas, Nitrobacter*)
- Sulphur and iron oxidizing bacteria (*Thiobacillus, Beggiatoa*)
- Hydrogen oxidizing bacteria (*Ralstonia, Paracoccus*)
- Methanotrophs and methylotrophs (*Methylomonas, Methylobacter*)
- *Pseudomonas* and the Pseudomonads (*Pseudomonas, Burkholderia*)
- Acetic acid bacteria (*Acetobacter, Gluconobacter*)
- Free living aerobic nitrogen fixing bacteria (*Azotobacter, Azomonas*)
- *Neisseria* and *Chromobacterium*

- Enteric bacteria (*Escherichia, Salmonella*)
- *Vibrio* and *Photobacterium*
- *Rickettsia* (*Rickettsia, Wolbachea*)
- Spirilla (*Spirillum, Bdellovibrio, Campylobacter*)
- Sheathed proteobacteria (*Sphaerotilus, Leptothrix*)
- Budding and prosthecate/stalked bacteria (*Hyphomicrobium, Caulobacter*)
- Gliding Myxobacteria (*Myxococcus, Stigmatella*)
- Sulphate and sulphur reducing proteobacteria (*Desulfovibrio, Desulfobacter*)

(3) Firmicutes (1421)

- Nonsporulating, low GC, Gram-positive bacteria: Lactic acid bacteria and relatives
- Endospore forming, low GC, Gram-positive bacteria: *Bacillus* (673), *Clostridia* (536) and relatives (212).
- Cell wall less, low GC, Gram-positive bacteria: the *Mycoplasmas.*

(4) Bacteroidetes, Spirochaetes, Tenericutes (Mollicutes), Chlamydiae, and Planctomycetes, etc.

- *Chloroflexus (12)*
- Chlorobium (13)
- Cyanobacteria and prochlorophytes (82)
- *Aquifex* (12)
- *Thermatoga* (23)
- *Thermodesulphobacterium (4)*
- *Deinococcus / Thermus* (23)
- *Bacteriodes (288)*
- *Verrucomicrobium (6)* and *Prothecabacter*
- Planctomyces
- Chlamydia
- Spirochetes (96)
- *Fibrobacter*
- *Cytophaga*

(5) Actinobacteria (1626)

- High GC, Gram-positive bacteria: *Coryneform* and propionic acid bacteria
- High GC, Gram-positive bacteria: *Mycobacteria*
- Filamentous, high GC, Gram-positive bacteria: *Streptomyces* and other *Actinomycetes*

SUGGESTED READING

Balows, A., Truper, H.G., M., Harder, W. and Schleifer, K.H. (Eds.) 1991. *The Prokaryotes*, Springer Verlag, N.Y.

Brock, T.D. and Madigan, M.T. 1991. *Biology of Microorganisms*, 6th ed., Prentice Hall, Englewood Cliffs, N.J.

Buchanan, R.E. and Gibbons, N.E (Eds) 1974. *Bergey's Manual of Determinative Bacteriology* (VIII edition) Williams and Wilkinson Company, Baltimore, USA.

Woese, C. and Fox, G. 1977. *Phylogenetic structure of the prokaryotic domain: the primary kingdoms. Proc Natl Acad Sci USA* 74 (11): 5088–90

Kushner, D.J. 1978. *Microbial Life in Extreme Environment*, Academic Press.

Frazier, W. C. 1967. *Food Microbiology*, McGraw-Hill Book Company, London and New York.

Frobisher, M. 1968. *Fundamental Principles of Microbiology*, W. B. Saunders Company, London.

Holt J.G., Krieg, N.R., Sneath, P.N.A., Staley, J.J. and Williams, S.T. 1994. *Bergey's Manual of Determinative Bacteriology* (IX edition) Williams and Wilkinson, Baltimore, USA.

Pelczar M., Chan E.C.S. and Krieg, N.R. 1993. *Microbiology*, Tata McGraw-Hill Publishing Co. Ltd., New Delhi.

Razin, S. 1978. *The Mycoplasmas*, Microbiological Reviews, 42:414- 470.

Salle, A. J. 1973. *Fundamental Principles of Bacteriology*, McGraw-Hill, Inc., New York.

Schlegel, H.G. 1993. *General Microbiology*, 7th ed., Cambridge Univ. Press.

Stainier, R.V., Ingraham, J.L., Wheelis, M.L. and Painter, P.R. 1986. *The Microbial World*, Prentice-Hall of India (Pvt.) Ltd., New Delhi.

Wolk, C.P. 1980. *Cyanobacteria* (Blue-Green Algae) In: "The Biochemistry of Plants", Vol. I, 659-686. Academic Press N.Y.

REVIEW QUESTIONS

1. Give an account of important colonial and morphological features of bacteria.
2. Discuss principles involved in following biochemical tests:
   a) Indole production
   b) Voges-Proskauer (VP)
   c) Citrate utilization
   d) Triple sugar iron agar (TSIA)
3. Differentiate between cell walls of Gram-positive and Gram-negative bacteria.
4. Write detailed notes on the following:
   a) Taxonomy of prokaryotes
   b) Bergey's manual of determinative bacteriology
   c) Bergey's manual of systematic bacteriology
   d) Methods used in taxonomy of prokaryotes
   e) Phenetic and phylogenetic classification of prokaryotes

UNIT IV

General Characteristics of Fungi

Fungi (singular; fungus) are eukaryotic, achlorophyllous, thalloid, heterotrophic microorganisms which require organic compounds for their nutrition. The science or study of fungi is called **Mycology** (*Myco* = fungi; *logy* = to study) and is equally interesting to different scientists including dermatologists, biochemical engineers, geneticist, biochemists, who find them to be an appropriate tool for the study and application of several biological processes. The thallus or body of a fungus is generally branched and filamentous but may consist of a single cell as in the yeasts while some fungi are dimorphic that is, they occur in two distinct physical forms. The fungal cell is enclosed by a true cell wall made up of chitin and cellulose. As saprophytes, fungi decompose complex organic materials and act as decomposers in the ecological chain. Fungi are also major biodeteriorating agents of timber, textile and other materials. Some fungi are known to play crucial role in production of food, e.g. (ripening of cheese, leavening of bread) while some of them are edible (mushroom, mycoprotein). Many of them also contaminate and degrade food. Many members of this group are of immense industrial importance particularly in case of production of antibiotics (penicillin), and enzymes (α-amylase or taka-diastase). Some fungi have parasitic mode of nutrition and are found either living in or on the host (plant or animal). Plant pathogens, particularly those infecting agricultural cash crops have been reported to cause famines in the past (early blight of potato by *Phytopthora infestans*) and still remain one of the major concern. Moulds are also found as parasites on animals (zoophilic) and humans (anthropophilic). Apart from the economic importance, the beautiful and diverse world of fungi imposes a great challenge for us to enumerate, characterize, identify, catalogue and conserve them *in-situ* and *ex-situ* for future reference and application.

The thallus of a mold consists basically of two parts: the mycelium (plural, mycelia) and the spores. The mycelium is a complex of several filaments called hyphae (singular, hypha). New hyphae generally arise from a spore which on germination puts out a germ tube or tubes. Most fungi are characterized by long and slender filaments – hyphae which aggregate into mycelial mass or thallus. Each hypha is about is about 5 to 10 μm wide, as compared with a bacterial cell which is usually 1 μm in diameter. Hyphae are composed of an outer tube like wall surrounding a cavity, the *lumen*, which is filled or lined by protoplasm. Between the protoplasm and the wall is the ***plasmalemma***, a double-layered membrane. The fungal cell wall is composed of hemicelluloses or chitin; true cellulose occurs in the walls of lower fungi (see Unit X Carbohydrates). Wall matrix materials in which the microfibrils are embedded consist of proteins, lipid and other substances.

As we stride through the kingdom Mycota, the primitive forms including aquatic and bread molds show presence of multinucleate cells without any vertical walls while among higher forms, e.g. in most plant pathogens, there is partial compartmentalization of the cells that still allows passage of nucleus and other cell content. Such cross walls are called as septa which divide the filament into uni or multinucleate cells. Some fungi occur as unicellular yeasts that lack hyphae and characteristically reproduce vegetatively by budding.

Yeasts vary significantly in size, ranging from 1 to 5 μm or more in length. They are frequently egg or oval shaped, but some may be elongated and spherical. Each species has a characteristic shape, but even in pure culture there is considerable variation in size and shape of individual cells, depending on age and environment. Yeasts do not have flagella or other organelles of locomotion.

NUTRITION

Fungi are chemoorganoheterotrophs and use organic compounds as a source of carbon, electrons and energy. Fungi are non-chlorophyllous and thrive either as saprophytes on dead or decomposing organic matter or are parasitic that derive their nutrition from living organisms including plants and animals. Some fungi are also found as symbionts, e.g. in case of Lichens. The nutrients are taken up by fungi by diffusion. Simple nutrients such as monomeric sugars are directly absorbed through cell wall whereas polymeric complex substrates, e.g. starch, proteins, etc. are first degraded by action of hydrolytic enzymes released from fungal hyphae (amylase, protease) and then nutrition is derived from resulting simpler hydrolysates. Some species can use inorganic compounds of nitrogen, such as ammonium salts. But all fungi can use organic nitrogen; this is why culture media for fungi generally contain peptone, a hydrolyzed protein product. If the fungus has access to soluble food e.g. simple sugars (glucose, maltose, sucrose, etc.) and amino acids, the food is directly transported into the mycelium through their cell wall. On the other hand, if the available food is insoluble, i.e. large, complex, organic macromolecules such as starch, lignin or cellulose then it is broken down or digested by the specific extracellular enzymes elaborated by the fungus. The resulting simpler compounds are then taken up through the cell wall. Fungi usually are aerobic, however, some yeasts are facultatively anaerobic or microaerophilic and can obtain energy by fermentation.

CELL WALL

Most fungi except slime moulds have a well-defined cell wall made up of chitin and cellulose. The fungal cell wall provides mechanical strength to the cell and acts as a barrier to undesirable molecules. The barrier is more important in case of pathogenic yeasts and moulds where it impedes ingress of antifungal compounds produced by host defence.

GROWTH

Growth of a hypha is distal, near the tip or apex. The major area of elongation takes place in the region just behind the tip. The young hypha may become divided into the cells by cross-walls which are formed by centripetal invagination of the existing cell wall. These cross-walls constrict the plasmalemma and grow inward to form generally an incomplete septum (plural, septa) that has a central pore which allows for protoplasmic streaming. Even nuclei may migrate from cell to cell in the hypha. Growth in moulds is typically polarized as it occurs by elongation of hyphal tips. The growth is marked by movement and accumulation of vesicles carrying cell wall precursors and enzymes (synthetases) to the site of exocytosis at the apical dome of the hyphal tip (**Figure 4.1**). There is formation of chitin microfibrils at the apex of the hypha regulated by membrane-bound chitin synthetase. This is accompanied with movement of growth related organelles and cytoskeletal elements. In case of Baker's yeast (*Saccharomyces cerevisiae*), cell wall synthetases (e.g. chitin synthetase and β-1-3-glucan synthetase), are uniformly dispersed in the plasmalemma. The model of yeast morphogenesis suggests that the synthetase is active at sites where the wall is growing and inactive where it is inert. Growth and morphogenesis in molds involves movement of vesicles at the hyphal tip, activation of wall synthesizing enzymes and addition of newly synthesized cell wall components to the existing cell wall thus allowing elongation of the hypha.

Hyphae occur in mainly three forms-

1. Nonseptate or coenocytes. Such hyphae have no septa.
2. Septate with unicellular cells.
3. Septate with multinucleate cells. Each cell has more than one nucleus in each compartment.

Mycelia can be either vegetative or reproductive. Some hyphae of the vegetative mycelium penetrate into the medium so as to obtain nutrients: soluble nutrients are absorbed through the walls. Reproductive mycelia are responsible for spore production and usually expand from the medium into the air (aerial parts). The mycelium of a mould may be loosely woven network or it may be an organized, compact structure, as in mushrooms.

REPRODUCTION

Most of the moulds that produce septate vegetative hyphae reproduce exclusively by asexual means (fungi imperfectii), giving rise to airborne propagules called

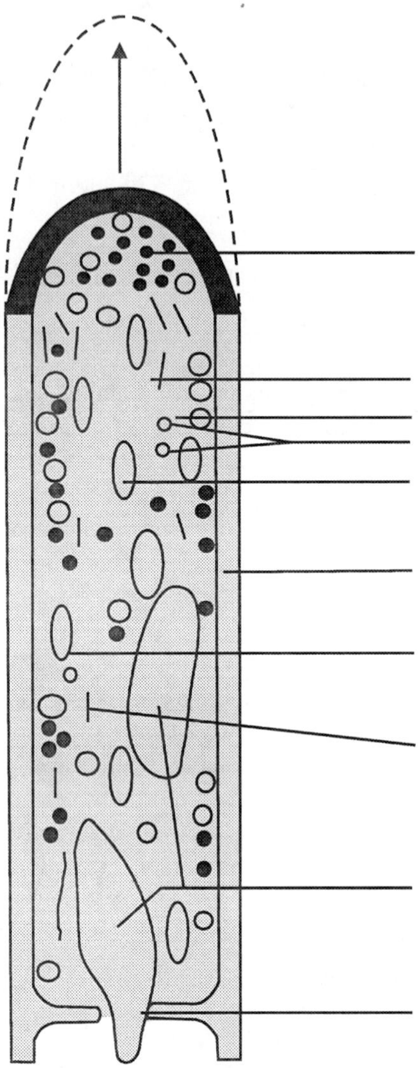

Figure 4.1: Apical growth of hypha in fungi.

conidia. On the other hand, elaborate mechanisms of sexual reproduction are also demonstrated by members of the Eumycota. Four distinct kinds of meiospores (products of karyogamy-meiosis-cytokinesis) are recognized: oospores (Oomycetes), zygospores (Zygomycetes), ascospores (Ascomycetes), and basidiospores (Basidiomycetes). The most common methods of asexual reproduction, in addition to simple budding in yeasts, are blastic development of conidia from specialized hyphae (conidiogenous cells), fragmentation of hyphae into conidia, and conversion of hyphal elements into conidia or chlamydospores (thick-walled resting spores).

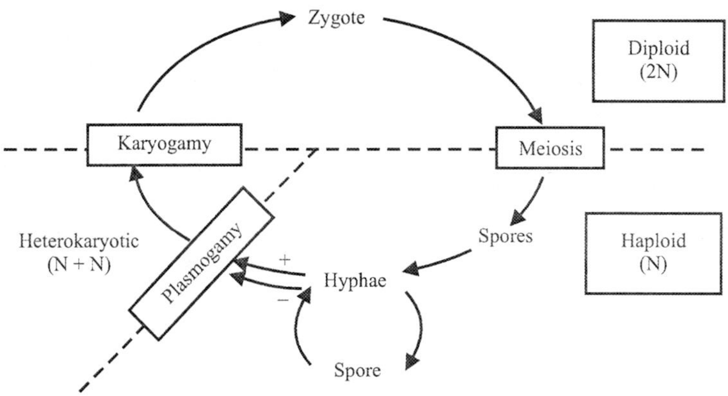

Figure 4.2: General life cycle of fungi.

Sexual reproduction in fungi involves fusion of two haploid (n) nuclei resulting in a dikaryon (2n). This dikaryon undergoes meiosis to produce spores (n). In case of homokaryosis, nuclei of similar thallus fuse together. Alternatively, in some forms, karyogamy and sexual reproduction is possible between only among different mating types (heterothallism). In such a case heterokaryon is preceded by conjugation of different mating types of thalli. Karyogamy is followed by plasmogamy, i.e. mixing of two hyphal protoplasm. The development of a dikaryotic mycelium results from simultaneous division of the two closely associated nuclei and separation of the sister nuclei into two daughter cells.

The above two types of reproduction methods are described in most fungi. In some fungi, as some *Deuteromycetes*, a third type- **parasexual method** is present. Here the true sexual cycle is absent and recombination occurs during mitosis crossing over. Plasmogamy, karyogamy and meiosis do not occur in a sequence, and there is no true meiosis (**Figure 4.3**).

Asexual Reproduction in Fungi

In these types of reproduction, asexual spores or mitospores are produced by mitosis and not the zygotes or meiospores. There are various methods of asexual reproduction in fungi.

1. **Fragmentation of mycelium:** A small fragment of the hypha detached from the parent mycelium has all the potential of developing itself into new colony. Fragmentation occurs frequently in nature due to shear in mycelial fungi of Ascomycota and Basidiomycota.
2. **Budding:** It is the production of a small bud from a parent cell or spore which eventually develops into a new individual. Simultaneous to the formation of bud, the nucleus of the parent cell divides into two, and one daughter nucleus migrates into the bud. The bud grows in size while attached to the parent cell and detaches as an individual.

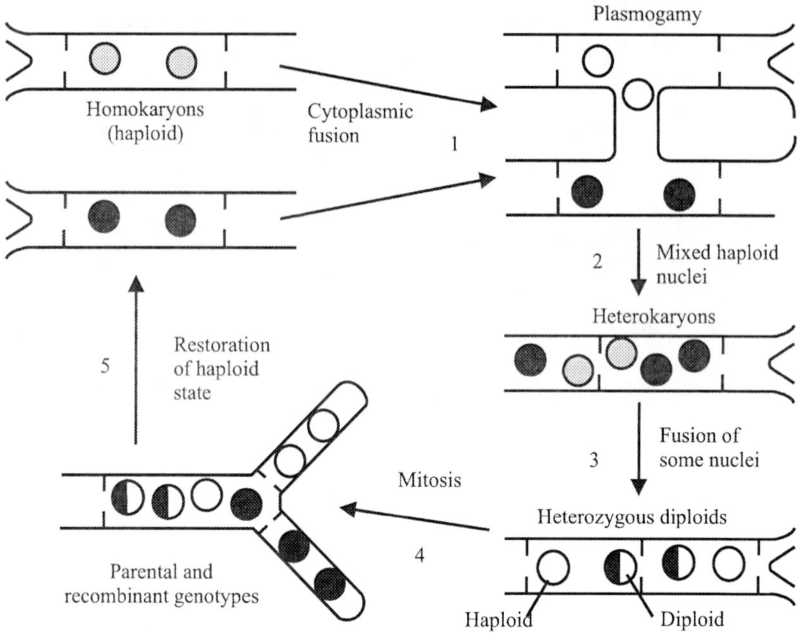

Figure 4.3: The parasexual cycle in fungi. Various stages of the parasexual cycle include (1) Hyphal conjugation (plasmogamy) (2) Heterokaryosis (3) Nuclear fusion (karyogamy) (4) Mitotic recombination and nondisjunction (5) Haploidization and nuclear segregation leading to homokaryosis.

3. **Fission:** Fission is simple division of a cell into two daughter cells by cleavage. The nucleus of parent cell divides mitotically along with the formation of a cell wall. This method is limited to some yeasts, e.g. *Schizosaccharomyces.*
4. **Production of spores by asexual means:** A prominent mode of fungal reproduction and dispersal is sporulation by asexual means. Spores, in strict mycological sense, are thick walled reproductive structures of fungi produced by division of hyphae in the sporangium (**Figure 4.4**). Spores vary in terms of being unicellular or multicellular, colourless or pigmented, thin-walled or thick-walled, motile or dispersed by wind, water or animal vectors. Spores among different classes vary greatly in form (circular, clubshaped, cylindrical and oval), colour, and size 1-150 μm in diameter.

Diverse and typical types of spores which also many a times make a characteristic and identifying feature of fungi are observed in many members of the kingdom mycota.

1. **Sporangiospores:** These single-celled spores are formed within sacs called sporangia (singular, sporangium) at the end of hyphae (sporangiophores).

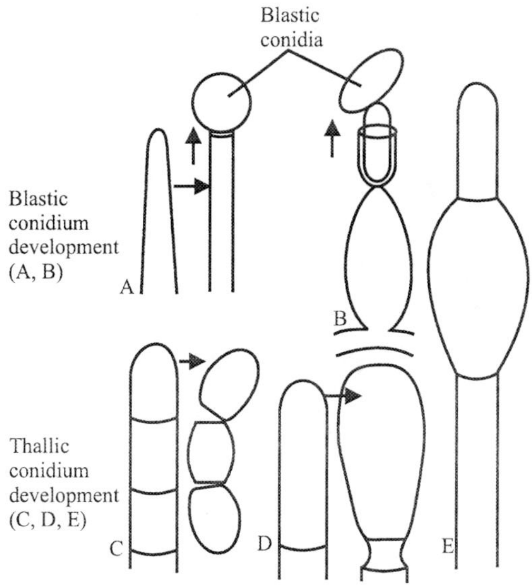

Figure 4.4: Methods of asexual reproduction and formation of conidia in fungi. (A) Terminal blastic conidium. (B) Repetitive blastic conidium formation from specialized conidiogenous cell (phialide). (C) Conidium formation by hyphal fragmentation. (D) Conidium formation by conversion of apical segment of hypha into single, asexual propagule. (E) Conversion of hypha into an intercalary chlamydospore.

Aplanospores are typically nonmotile sporangiospores. Zoospores are motile sporangiospores, their motility being due to the presence of flagella.

2. **Conidiospores:** It is called conidia (singular, conidium). Small, single-celled conidia are called microconidia. Conidia are formed at the tip or side of a hypha. They may occur singly or in chain (but not within sporangium). Large, multicelled conidia are called macroconidia.
3. **Oidia:** Oidia (singular, oidium) or arthrospores are single-celled spores and are formed by disjointing of hyphal cells. These spores are capable of immediate germination.
4. **Chlamydospores:** These thick-walled, single-celled spores are highly resistant to adverse conditions. They are formed from cells of the vegetative hypha.
5. **Blastospores:** These are spores produced by budding.

Sexual Reproduction in Fungi

In contrast to asexual reproduction where there is only one phase, sexual reproduction involves two distinct phases w.r.t. to chromosome number (i) Haploid phase with **n** number of chromosomes and (ii) Diploid phase with **2n** number of chromosomes. Sexual reproduction is marked by appearance of specialized cells (or sex organs) on the haploid thallus during conversion from haploid to diploid phase. These sex organs containing specialized cells or gametes are called **gametangia.** Morphologically similar gametangia are called as **iso-gametangia** whereas morphologically different sex organs are called as **hetero-gametangia.** Such distinct male (antheridium) and female (ascogonium) gametangia are characteristic features of ascomycota. Basidiomycota do not have very well differentiated gametangia while no such sexual structures are observed in members of dueteromycota.

Basically, sexual reproduction in fungi involves the mixing of cytoplasm (plasmogamy), followed by merger of two compatible nuclei (karyogamy) and eventually reductional division (Meiosis). The whole act of sexual reproduction consists of typically three phases which in tandem are plasmogamy, karyogamy and meiosis.

Among sexually reproducing fungi the mode of protoplasm fusion has been observed to be markedly different. **Planogametic copulation** is the fusion of two gametes, one or both of which may be motile. In most cases, at the time of fertilization, either one (male) or both of gametes are released out of their gametangia. The gametes then fuse in water. Planogametic copulation is observed in aquatic fungal forms having flagellated gametes. If the gametes are morphologically similar in all respects as in case of primitive moulds (Chytridiales - *Synchytrium, Catenaria*) the copulation is referred to as **isogamous** while fusion of heterogamous morphologically similar, but different in size with smaller male and larger female is called as Anisogamy (e.g. in *Allomyces*.). The third type of gametic fusion observed among moulds is heterogamous where the female gamete is larger and non-motile while the male gamete is smaller and motile. The male gamete enters the oogonium and fertilizes the egg (e.g. in Oomycetes).

In case of non-motile gametes there is **gametangial contact** and the gametes of both, male and female gametangia become reduced to undifferentiated protoplasts, each consisting chiefly of one nucleus. Unlike planogametic copulation, the gametic nuclei are not released outside and get transferred directly from one gametangium to another during gametangial contact. Two gametangia come into contact and one or more gamete nuclei migrate from the male (antheridium) to female (oogonium/ ascogonium). In some fungi, male nuclei enter the female gametangium through a pore at the point of contact between gametangial walls, but in other species through an especially developed fertilization tube. Most Oomycetes and Ascomycota exhibit this kind of sexual reproduction.

Gametangial copulation is the fusion of the entire contents of the two gametangia. This occurs either through **hologamy** where the entire male thallus attaches itself to the female gametangia and empties all its contents through a pore developed in the gametangial wall at the point of contact. (e.g. chytrids- *Polyphogus* and *Rhizophidium*). In another type of gametangial copulation called **Direct fusion**, the two gametangial cells fuse together into one after dissolution of the wall at the point of contact to form a common cell in which the two protoplasts mix. Example- *Rhizopus, Mucor, Sporodinia* (Zygomycota) and *Eremascus, Dipodascus* (Ascomycota).

PHYSIOLOGY

Fungi have typical water relationships. These can survive very well in the medium where there is lower availability of water (a_w). Thus, these can colonize on substrates containing high concentrations of sugar or salt such as pickles, jellies, etc. while bacterial cells can not. Yeasts and moulds can tolerate more acidic conditions than most other microorganisms. Generally, fungi are aerobic, however, some yeasts are facultative anaerobes and can grow under anaerobic or micro-aerophilic conditions as well.

Fungi grow over a wide range of temperature, with the optimum for most saprophytes ranging from 22 to 30°C; pathogenic species have a higher temperature optimum, generally 30 to 37°C suiting to the host body temperature, whereas some fungi can grow at or near 0°C (psychrophilic) and thus can cause spoilage of meat and vegetable in cold storage. On the other hand, some fungi are known to prefer high temperatures for growth. Cooney and Emerson (1964) used the term **thermophilic fungi** for moulds having maximum growth temperature at or above 50°C and minimum at or above 20°C while term thermotolerant was used to specify fungi with maximum temperature limit of growth at or above 50°C but could grow below 20°C. *Thermomyces lanuginosus* is only known eukaryote which can grow and survive in the temperature as high as 62°C (**Figure 4.5**). A summary of the physiological characteristics of fungi in comparison with those of bacteria is given **Table 4.1**.

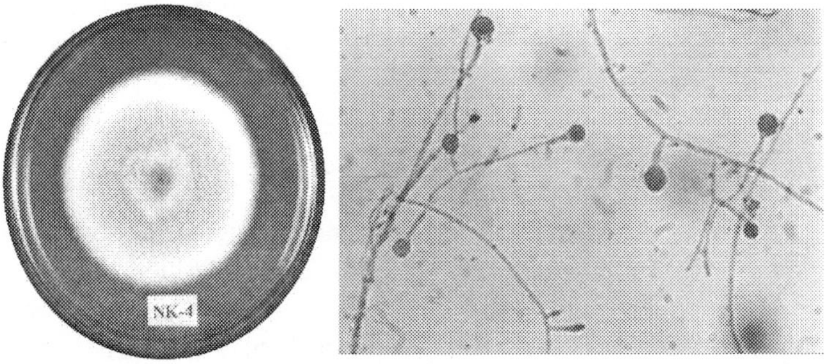

Figure 4.5: Colony and morphology of a thermophilic mould, *Thermomyces lanuginosus.*

CULTIVATION OF FUNGI

Most fungi grow aerobically on the routine culture media, e.g. nutrient agar, potato dextrose agar, Czapek's dox agar, etc. at temperatures between 20 and 30°C. The

Table 4.1 Comparison among physiological characteristics of Fungi and Bacteria

S.No.	Characteristic	Bacteria	Fungi
1.	DNA organization	Prokaryotic	Eukaryotic
2.	Cellular organization	Unicellular	Branched filamentous, multicellular
3.	Optimum pH	6.5–7.5	3.8–5.6
4.	Growth temperature	20–37°C mesophiles 55- < 100°C thermophiles	22°C–37°C mesophilic >20 °C – 50°C thermotolerant > 0°C - 4°C psychrophiles
5.	Oxygen requirement	Aerobic or anaerobic	Strictly aerobic (mold), microaerophilic and facultative (yeasts)
6.	Chlorophyll	Present in photosynthetic bacteria	None
7.	Water availability requirement (a_w)	> 0.95 a_w	Near about 0.85 a_w
8.	Carbon requirement	Inorganic and / or organic	Organic
9.	Cell-wall structure and Components	Peptidoglycan (N-acetyl muramic acid, N-acetyl glucosamine, peptides)	Chitin, cellulose, or hemicellulose
10.	Antibiotic susceptibility	Sensitive to penicillin, tetracyclines, Choloramphenicol	Sensitive to griseofulvin, amphotericin B

growth is slow as compared to bacteria. In order to avoid bacterial overgrowth or contamination it is better to use media that has high concentration of sugar or salt and low pH (4.0–5.0). High sugar concentrations and low pH are tolerated by molds but are inhibitory to many bacteria. One of the commonly used medium for the growth of fungi was devised by Sabouraud and contains dextrose and peptone as its principle ingredients (SDA- Sabouraud's Dextrose Agar). Basically, fungal colonies can be regenerated from any of its part (spores, mycelia). This is achieved by inoculating a small part of the parent colony on the fresh and sterile medium using a presterilized inoculation needle. The needle used is stiff and has a sharp tip for picking up the mycelium. Fungi are slow growers than bacteria and it takes 3-5 days for establishing a well grown colony in a Petriplate. Some broad spectrum antibacterial antibiotics may also be added to the medium to avoid bacterial growth.

CLASSIFICATION OF FUNGI

Fungi are most often associated with plants, commonly as decomposers, and pathogens, and as their benefactors, e.g. mycorrhiza. Once fungi were believed to be monophyletic and to be originated from an algal ancestor that lost its ability to photosynthesize. However, it is confirmed using molecular databases that the fungi are made up of a polyphyletic group of organisms that may be distantly related. Therefore, it should be understood that the organisms referred as fungi are not grouped together because they are closely related, but rather because they share a combination of morphological (and sometimes physiological) characteristics. The classification of fungi is based on the characteristics of the sexual spores and fruiting bodies present during the sexual stages of their life cycles. However, sexual (or perfect) stage and fruiting bodies in many fungi are witnessed only under specific environmental conditions and may not be seen ordinarily. Till today, the perfect stage is not seen in many a fungal forms. These imperfect forms have been described on the basis of characteristic and prominent features other than sexual stage. Morphological description of such forms includes asexual spores and thalli. However, taxonomic status of such imperfect forms is conditional and is subject to revaluation. All such imperfect higher fungi are grouped together in the form class *Deuteromycetes.* Most of the sexual stages discovered subsequently among members of this class have been of the ascomycete type. Whenever sexual stages are observed in these organisms, they are redesignated and placed in with the ascomycetes or basiodiomycetes as the case may be. Fungal classification still remains to be intricate and sometimes different viewpoints arise due to our insufficient comprehension of the morphology, development, and physiology of fungi. Thus, it is anticipated that fungal classification is subject to change with increase in our knowledge. Taxonomy of fungi follows the recommendations of the Committee on International Rules of Botanical nomenclature. Accordingly, the different taxa have following endings:

Division:	- mycota
Subdivision:	- mycotina
Classes:	- mycetes
Orders:	- ales
Families:	- aceae

Genera and species are Latin or Latinized but do not have any customary ending. These may bear the name of their discoverer or character. A number of classification systems have been devised by various mycologists from time to time. In the following part, classification schemes as proposed by Alexopoulos (1962), Bessey (1950) and Anisworth & Bisby (1971) are presented (**Table 4.2**).

A brief description of the major divisions of kingdom Mycota is given in the following paragraphs.

Myxomycota (Plasmodial slime moulds) (*Physarum, Plasmodiophora, Dictyostelium*)

These fungi have a multinucleate network of protoplasm (*plasm -odium*) and their mode of nutrition is phagocytosis by which they engulf bacteria and organic matter. Whole of the plasmodium is converted into sporangia and haploid spores are

Table 4.2 Some important systems of fungal classification

Ainsworth and Bisby (1971)	Bessey (1950)	Alexopoulos (1962)
Kingdom Fungi		**Kingdom** Plantae **Division** Mycota
Myxomycota		
• **Acrasiomycetes**	Mycetozoa	Myxomycotina
• **Hydromyxomycetes**		• Myxomycetes
• **Myxomycetes**		Eumycotina
• **Plasmodiophoromycetes**		• Plasmodiophormycetes
• **Myxomycetes**		
Eumycota		
Mastigiomycotina	Phycomyceteae	
• **Chytridiomycetes**		• **Chytridiomycetes**
• **Hyphochytridiomycetes**		• **Hyphochytridiomycetes**
• **Oomycetes**		• **Oomycetes**
Zygomycotina	Carpomycetae	
• **Zygomycetes**		• **Zygomycetes**
• **Trichomycetes**		• **Trichomycetes**
Ascomycotina	Ascomyceteae	Ascomycetes
• **Hemiascomycetes**		• **Hemiascomycetidae**
• **Plectomycetes**	**Pyrenomycetes**	• **Euascomycetidae**
• **Pyrenomycetes**		o **Plectomycetes**
• **Discomycetes**		o **Pyrenomycetes**
• **Laboulbenomycetes**		o **Discomycetes**
• **Loculoascomycetes**		o **Laboulbeniomycetes**
		• **Loculoascomycetidae**
Basidiomycotina	Basidiomyceteae	Basidiomycetes
• **Teliomycetes**	Teliosporeae	Heterobasidiomycetidae
• **Hymenomycetes**	Heterobasidiae	Homobasidiomycetidae
o **Phragmobasidio-mycetidae**	Hymenomyceteae	**Gasteromycetes**
	Gasteromyceteae	
o **Holobasidiomycetidae**		
• **Gastromycetes**		
Deuteromycotina	The Imperfect Fungi	Dueteromycetes
• **Blastomycetes**	• **Moniliales**	
• **Hyphomycetes**	• **Sphaeropsidales**	
• **Coelomycetes**	• **Melanconiales**	

dispersed by wind. The spores germinate to produce mobile swarmers or myxamoebae which fuse in pairs to make a diploid plasmodium. e.g. *Physarum polycephalum*. Three other group of organisms with naked protoplasmic nature have also been associated with fungi: cellular slime moulds (acrasiomycota, dictyosteliomycota) and plasmodiophoromycota. These groups have similarities with protozoans but resemble fungi in production of walled spores which are dispersed by wind. However, their spores are different from fungi with respect to their high galactosamine content. The life cycle of *Plasmodiophora brassicae* and *Dictyostelium discoideum* are given in **Figures 4.6 and 4.7.**

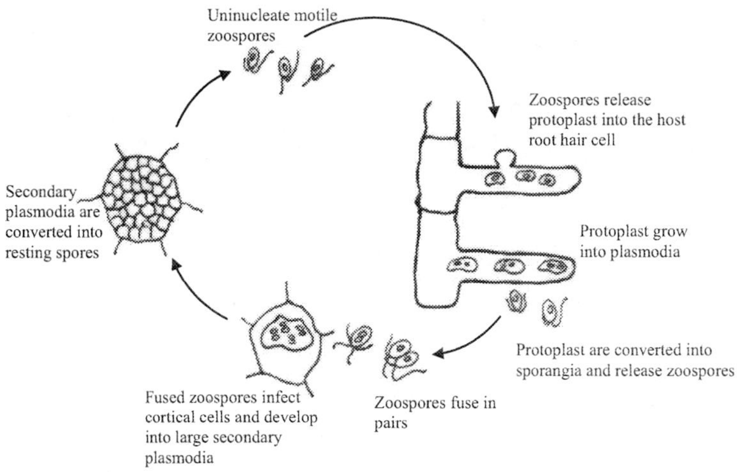

Figure 4.6: Stages in the infection cycle of *Plasmodiophora brassicae* (Myxomycota)

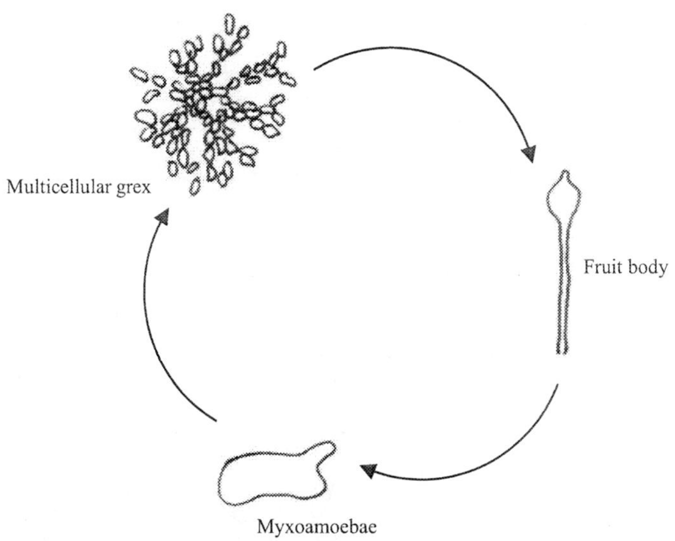

Figure 4.7: Life cycle of *Dictyostelium discoideum* (Myxomycota)

Table 4.3 Description of the major divisions and classes of the kingdom fungi (Alexopoulos, 1979)

Division / Class	Characteristics features	Representatives
Division *Gymnomycota* (slime moulds)	Fungi which ingest particulate nutrients and lack cell walls in the vegetative stage	
Class *Acrasiomycetes* (cellular moulds)	Free-living amoebae which aggregate to form a stalked sporocarp bearing spores in a mucilaginous thallus.	*Dictyostelium discoidium Polysphondilium violaceum*
Class *Myxomycetes* (acellular slime moulds)	A multicellular, wall-less plasmodium, which transforms into highly organized sporangia bearing sporangiospores.	*Physarum polycephalum Didiumum irdis*
Division: *Mastigomycota* (flagellated lower fungi)	Aquatic fungi producing motile, flagellated cells	
Class: *Chytridiomycetes*	Motile cells bearing a single, posteriorly positioned whiplash type flagellum	*Allomyces macrogynus*
Class: *Hyphochytridiomycetes*	Motile cells have a single, anteriorly positioned flagellum	*Rhizidiomyces arbuscula* *Hyphochrytrium catrenoides.*
Class: *Plasmodiophoromycetes*	Obligate parasites on higher plants; vegetative stage is a plasmodium; motile cell with two unequal anterior whiplash flagilla	*Plasmodiophora brassica*
Class: *Oomycetes*	Motile cell with two laterally inserted flagella, one tinsel anteriorly directed, the other whiplash posteriorly directed.	*Saprolegnia ferax*
Division: *Amastigomycota* (terrestrial fungi)	Flagella absent	
Class: *Zygomycetes*	Sexual reproduction by gametangial fusion; zygote transformed into a thick-walled resting spore, the zygospore; vegetative reproduction by means of mitospores produced within a sporangium.	Bread mould *Rhizopus stolonifer Phycomyces blakesleanus Mucor rouxii*
Class: *Ascomycetes*	Sexual spores produced endogenously in a sac like ascus typically produced in a well-differentiated ascocarp; vegetative reproduction by conidiospores	*Saccharomyces cerevisiae and other yeasts; Neurospora crassa,* Morels and truffles

Class: *Basidiomycetes*	Sexual spores produced exogenously on club like cells called basidia; basidia formed on well-differentiated basidiocarps (fruiting bodies)	Mushrooms, toadstools and puffballs Rusts, smuts and jelly fungi, Bracket fungi.
Form Class: *Deuteromycetes* (Fungi imperfectii)	Sexual reproduction not seen; vegetative reproduction by means of conidiospores arising from well-defined conidiogenous cells	Causal agents of (ringworm and athletes foot) and many plant diseases, such as wilts, mildews, etc. *Trichophyton rubrum, Candida albicans, Alternaria alternata*

Oomycota (cellulose walled fungi) (*Saprolegnia, Phytophthora, Pythium*)

This phylum consists of four orders viz. Saprolegniales (fresh water saprotrophs, e.g. *Achlya, Saprolegnia*); Leptomitales (fermentative aquatic fungi); Lagenidales (plant root parasites); and Perenosporales (plant pathogens). Peronosporales is the most important group as it contains important plant pathogens including *Pythium* spp., *Phytophthora infestans* (causing late blight of potato) and *Plasmopara viticola* (causing downy mildew of grapewines). Members have ceonocytic diploid mycelium without septa with characteristic cell wall made up of cellulose and glucose polymers. Asexual reproduction is marked by production of motile diploid zoospores by cytoplasmic cleavage in sporangium having anterior tinsel and one posterior whiplash flagellum. During sexual reproduction these produce different male (antheridium) and female (oogonium). Nuclei pass from the antheridium into oogonium where they fuse in pairs to produce thick walled oospores. The oospores germinate to produce diploid hyphae. The life cycle of *Phytophthora infestans* is given in **Figure 4.8.**

Chytridiomycota (*Allomyces, Rhizophlyctis, Olpidium, Coelomomyces*)

The phylum Chytridiomycota, the chytrids, represents a group of primitive aquatic fungi. The thallus is unicellular or dichotomously branched with rhizoids for absorption. Cell wall is made up of chitin and glucans. These are characterized by having zoospores that are motile by means of a posterior whiplash flagellum. Sexual reproduction involves fusion of motile male and female gametes to produce zygote. The group contains unicellular Chytridales and branched chains of cells (Blastocladiales: *Allomyces*). Most members are saprotrophs in freshwater and accumulate on baits such as seeds, pollens and insect skeletons. Some are strongly cellulolytic (*Rhizophlyctis rosea*) or chitinolytic (*Chytriomyces haylinus*) in soil. A few members are severe plant pathogens, e.g. *Synchytrium endobioticum* that

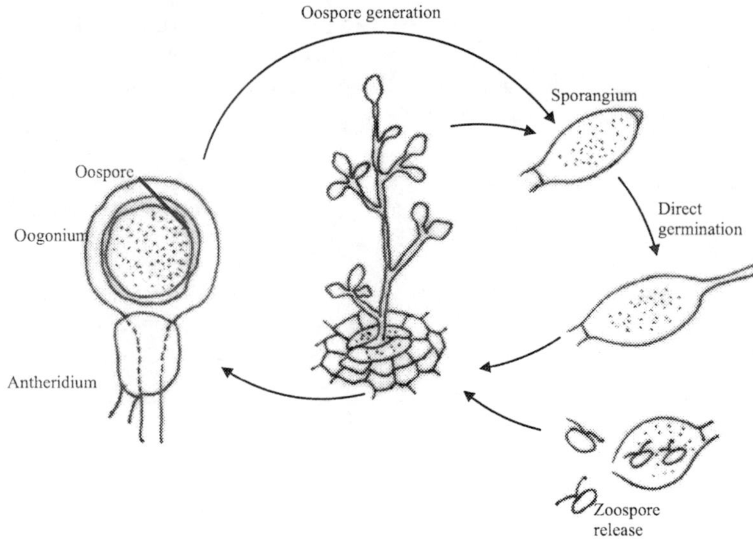

Figure 4.8: Life cycle of *Phytophthora infestans* (oomycetes), the causative agent of potato blight. Opposite mating strains inside the host tissues produce antheridia and oogonia; Fertilization leads to production of resting oospore. The oospore produces lemon shaped sporangia. The sporangia are dispersed by wind and reach on potato leaves where they germinate either by producing a germ tube or by releasing biflagellate zoospores. These infect the host leaf and the fungal sporangiophores emerge through the stomata.

causes potato wart disease. Some members can be used as biological control agent, e.g. *Catenaria anguillulae* parasitizing on nematodes and fungal spores while *Coelomomyces* spp. parasitizes on mosquito larvae. The life cycle of a chytridiomycetous fungus, *Allomyces* is given in **Figure 4.9**.

Zygomycota (conjugating fungi)
(*Mucor, Rhizopus, Pilobolus, Glomales, Mortierella*)

Species belonging to this phylum reproduce sexually by forming a zygospore, a thick-walled, diploid cell which contains thousands of nuclei. Phylum has about 600 species which are mostly terrestrial and feed on organic matter, although there are a few parasitic species. Members are coenocytic and have continuous mycelium containing hundreds of haploid nuclei without any distinct division. However, septa are seen in between their reproductive structures and the rest of their mycelium. The conjugating fungi have a life cycle that includes a sexual phase and an asexual phase. In the asexual phase, thousands of spores develop inside a sporangium, a small spherical structure. The sporangium grows on the tip of a sporangiophore, a specialized aerial hypha.

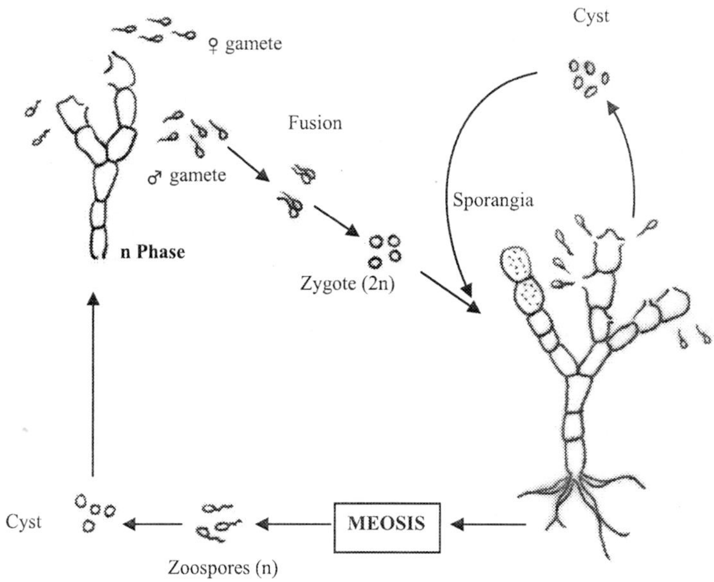

Figure 4.9: Life cycle of *Allomyces* (Chytridiomycota)

In the sexual phase of their life cycle, these fungi produce specialized hyphae, called gametangia, which are of two different strains (or sex), plus (+) and minus (-). The plus and minus strains are morphologically similar, but differ physiologically and biochemically. Plus and minus gametangia conjugate with one another and form a structure with hundreds or thousands of nuclei from each strain. After conjugation, a thick-walled structure, the zygospore, develops from the conjugated gametangia which is thick walled and dark in colour. Within this zygospore, as many as thousands of nuclei from the plus and minus strains pair and fuse together to form thousands of diploid nuclei.

The zygospore germinates to produce germsporangia which are born on germsporangiophores, structures morphologically similar to the asexual sporangium and sporangiophore. The germsporangium contains thousands of haploid germspores which are produced by **meiosis** of the diploid nuclei of the zygospore. Upon liberation each germspore germinates to a give rise to a new haploid mycelium. One common and interesting example is coprophilic (grows on animal faeces) zygomycetous mould *Phycomyces blakesleeanus*. Interestingly, the sporangiophores of *Phycomyces* respond to a variety of sensory stimuli including **light** (phototropism), gravity (gravitropism), wind (anemotropism), and nearby objects (avoidance response). The life cycle of *Mucor* is given in **Figure 4.10.**

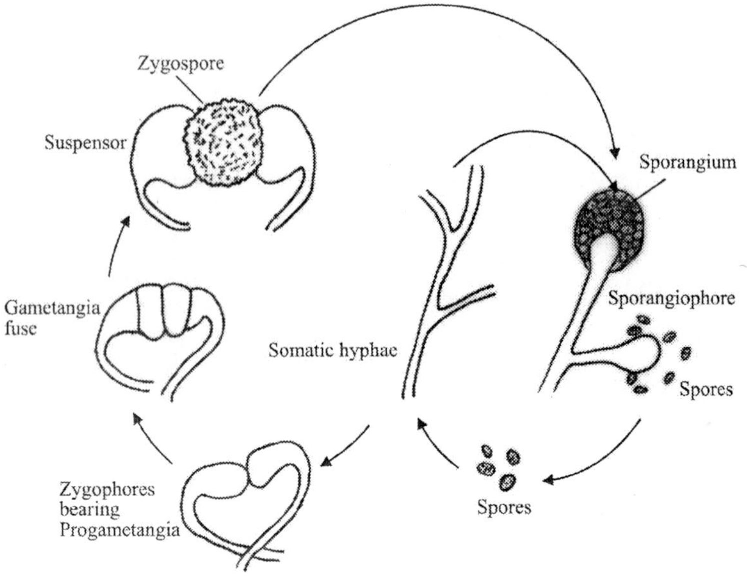

Figure 4.10: Life cycle of *Mucor* (Zygomycetes). Sexual reproduction occurs when hyphae of opposite mating groups produce aerial branches or zygophores which grow towards each other and fuse through swollen progametangia at their tips. The progametangia fuse to produce thick walled resting Zygospore. Zygospore undergoes meiosis and germinates to produce a hyphae or sporangiophore.

Ascomycota (sac fungi)

(*Saccharomyces, Pneumocystis, Pichia, Chaetomium, Erysiphe, Tuber*)

The main character of the members of this phylum is the sexual structure, i.e. an ascus (pl. asci), or a sac containing spores. The hyphae of the sac fungi are divided by septa with pores, that is, they have perforated walls between adjacent cells. Septa separate nuclei but pores allow cytoplasm to mix. Ascomycetous fungi also reproduce asexually by prolific production of spores, called conidia, which are born on specialized erect hyphae, called conidiophores.

The sac fungi exhibit a distinct sexual phase of their life cycles. In the first step of this process, compatible hyphae fuse together and the nuclei from the different hyphae come together in one cell to form a dikaryon (a cell with two distinct haploid nuclei). This dikaryon undergoes several cell divisions resulting in many cells with two different haploid nuclei per cell. Two haploid nuclei fuse in one of these cells called as the ascus mother cell which develops into an ascus. Meiosis occurs in the diploid ascus cells and, depending on the species, four or eight haploid ascospores are formed inside the ascus. In some species, such as the fleshy and edible morels, a large number of asci are massed together to form an ascocarp (fruiting body). This large phylum of fungi includes many species which are of immense economic importance to humans. Yeasts are unicellular ascomycetous fungi which reproduce by budding. Different yeasts in the genus *Saccharomyces* which can ferment sugars are employed by bakers, brewers, and vintners (wine makers), in production bread, beer or wine.

Truffles are edible fruiting bodies of subterranean ascomycetes which grow in association with **tree** roots. Some members of this group are plant pathogens e.g. *Claviceps purpurea*, the ergot fungus, infects agricultural grains including wheat, and when ingested can cause intense hallucinations or fatality due to the presence of LSD (D-Lysergic acid diethylamide). Another very famous ascomycete which became tool of investigating genetic traits is the red bread mould or *Neurospora crassa*. The mould produces eight spores in highly ordered manner during sexual reproduction that has allowed geneticists to construct a map of the genes on its chromosomes and the movement of genetic traits during sexual reproduction. The life cycle of yeast, *Sachharomyces* and *Neurospora crassa* is given in **Figures 4.11 and 4.12,** respectively.

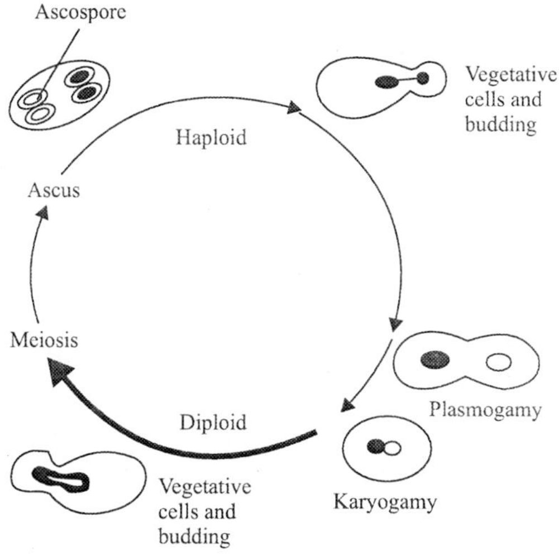

Figure 4.11: Life cycle of *Saccharomyces* (Yeast)

Basidiomycota (club fungi)

(*Agaricus, Amanita, Pleurotus, Ganoderma, Lycoperdon, Puccinia*)

Members of this phylum reproduce sexually by forming spores affixed to a typically club-shaped structure called basidia. Similar to the ascomycetes, these fungi also have septate mycelia and a dikaryotic phase in their life cycle. The septum of the club fungi is characteristically different from those of sac fungi and is called a **dolipore septum** bearing a pore in its centre. The club fungi reproduce asexually by producing asexual spores or by shear or fragmentation of mycelium.

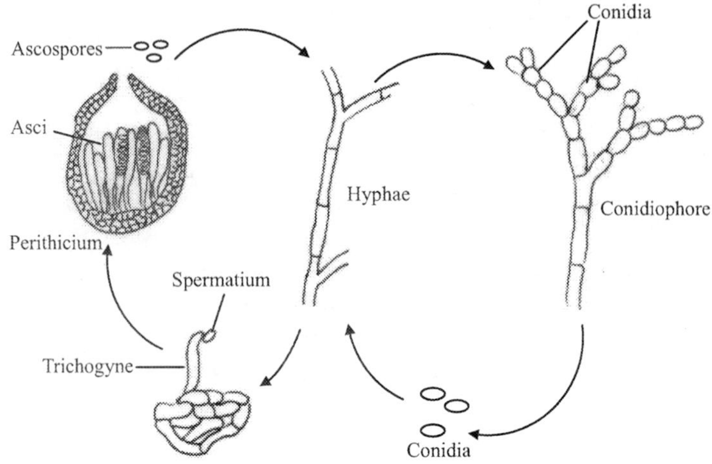

Figure 4.12: Life cycle of *Neurospora crassa* (Ascomycota). The somatic hyphae bear conidiophore which bear chain of conidia. These conidia germinate to produce new somatic hyphae. The somatic hyphae may also reproduce by producing spermatium (♂ spore) and trichogyne (♀ receptive hyphae). These special structures fuse to produce asci covered with a perithicium. Nucleus fuse in each asci and undero meiosis and finally mitosis to produce ascospores. The asci mature and release the ascospores which germinate to produce somatic hyphae.

The sexual reproduction phase of the club fungi involves three developmental stages including germination of a haploid spore into a germ tube, which develops into mycelium. The mycelium initially contains a single haploid nucleus. Then, its haploid nucleus divides and septa are formed between the nuclei. A secondary mycelium is formed upon conjugation of two sexually compatible hyphae. The secondary mycelium is dikaryotic, and has two haploid nuclei, one from each parent. As the dikaryotic mycelium grows, the cells divide and more septa are formed between the new cells. Each of the new cells in the secondary mycelium has one haploid nucleus from each parent. In this stage **clamp connections**, which are loop-like hyphae that connect the cytoplasm of neighbouring cells and allow nuclei to move during **cell division** can be seen. During cell division, one nucleus divides directly into the newly formed cell; the other nucleus divides inside the clamp connection and the two daughter nuclei migrate through the clamp connection in opposite directions to the two daughter cells. The tertiary mycelium is simply an organized mass of secondary mycelium. It is a morphologically complex tissue and forms structures such as the typically mushroom-shaped basidiocarps commonly seen in nature.

Sexual reproduction of the club fungi begins with fusion of two primary hyphae to form a club-shaped structure, known as a basidium. The two haploid nuclei inside the basidium fuse together to form a diploid zygote which undergoes meiosis to form two haploid nuclei. Later, these two haploid nuclei undergo **mitosis** to form a total of four haploid nuclei. These four nuclei then migrate into projections onto the tip of the basidium. These projections then develop into four separate haploid spores, each with a single nucleus. The life cycle of an edible mushroom, *Agaricus* is explained in **Figure 4.13.**

This phylum includes hundreds of species which are known as mushrooms, toadstools, earthstars, stinkhorns, puffballs, jelly fungi, coral fungi, etc. Some species, such as the rusts and smuts, are pathogens which attack agricultural grains. Other species, such as the fly agaric (*Agaricus muscaria*) and some species in the genus *Psilocybe,* produce chemical **hallucinogens** and have been used by numerous cultures in their religious ceremonies. *Agaricus bisporus*, is the common edible button mushroom which is commercially produced and marketed worldwide. Another important aspect of the club fungi is the great diversity of metabolites and

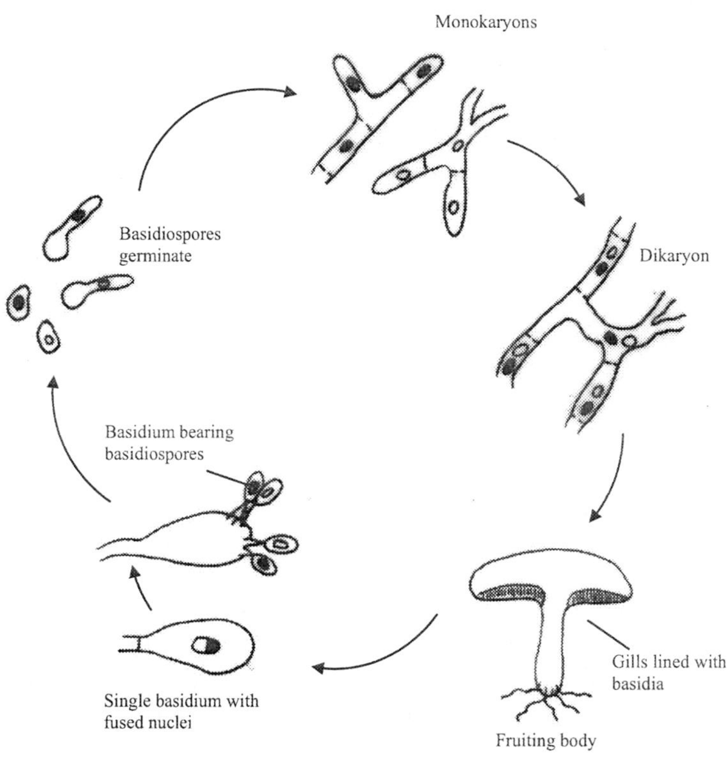

Figure 4.13: Life cycle of a mushroom, *Agaricus*. Hyphae of two monokaryons (n) fuse to produce dikaryon having one nucleus of each parental type in each hyphal compartment. Dikaryon produces fruiting body (mushroom) in which the gills are lined with basidia. Nuclei fuse inside the basidium and the diploid nucleus (2n) undergoes redutional division or meiosis to produce four haploid nuclei which enter developing basidiospores (n). Basiodiospores develop into monokaryon hyphae.

bioactive molecules such as alkaloids and psychogenic chemicals produced by many of its members. While mushrooms are the most common representatives of basidiomycetes, many other club fungi grow underground as symbionts with plant roots (*myco-* fungus; *rrhizae-* roots). Mycorrhizae result from a symbiosis between a plant root and a fungus where the fungus enhances the absorption surface and supplies nitrogen-containing compounds to the plant, and the plant supplies carbohydrates and other organic substrates to the fungus. In particular, mycorrhizae are very important for the growth of orchids and gymnosperms. These fungi occur as ecto- , endo- and vesicular arbuscular – mycorrhizae (VAM) in association with plant roots.

Deuteromycota (imperfect fungi)
(*Aspergillus, Alternaria, Aureobasidium, Cladosporium, Humicola, Penicillium, Candida, Trichoderma, Fusarium, Beauveria, Phoma, Collectrichum*)

The Deuteromycota is a heterogeneous group of unrelated species having just one feature in common – the missing perfect or sexual stage (**Figure 4.14**). For this reason, these fungi are often referred to as fungi imperfectii. It is proposed that these fungi may have lost their sexual phase through the course of evolution or possibly we could not observe their sexual phase due to absence of required environmental and physiological conditions. Other features of these fungi, such as production of asexual spores (conidia) on erect hyphae (conidiophores) and ceonocytic thalloid body, are similar to sac fungi. The best known genera of this phylum are *Penicillium* and *Aspergillus.* Former is well known for the production of first ever discovered antibiotic while the latter is known for production of amylase (diastase), citric acid and many industrial products. An example of a pathogenic deuteromycete is *Candida albicans*, a dimorphic fungus responsible for skin and scalp infections in humans.

Mycophycophyta (lichens)

Lichen is a symbiotic association between a fungus and an alga and also between a fungus and a photosynthetic cyanobacterium. Lichens constitute a very diverse and polyphyletic group of organisms and are classified together simply because they all occur as fungus-alga symbionts. In most lichens, the fungal species belongs to the Ascomycota and the green alga belongs to Chlorophyta. The photosynthetic species supplies carbohydrates to the fungus and the fungus supplies nitrogen and other nutrients to the alga. The overall morphology of lichen differs from its constituent species. Lichens produce specialized reproductive and dispersal structures called as soredia in which the algal component is enclosed by fungal mycelium. Alternatively, the fungal component of the lichen can produce and disperse spores which germinate and recombine with the algal components in nature.

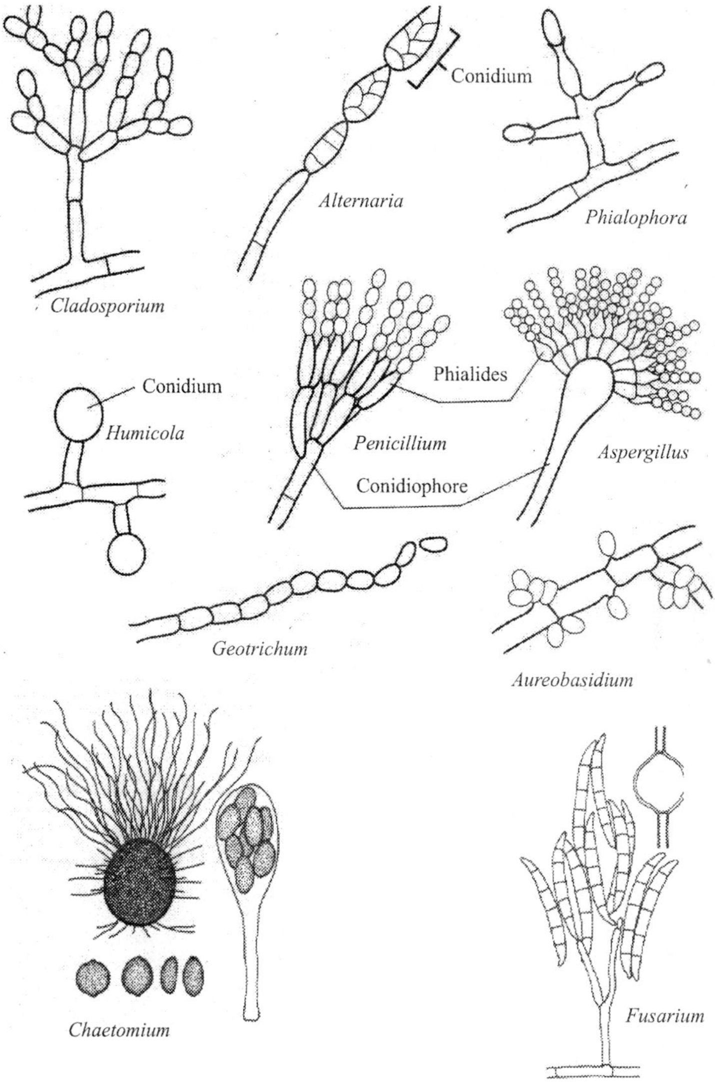

Figure 4.14: Some members of the group fungi imperfectii (Deuteromycetes)

Lichens are believed to be primary colonizers in ecological succession. They can survive and colonize otherwise inhabitable niche due to their symbiotic photosynthetic and saprophytic ability. Lichens are highly sensitive to air pollutants such as **sulphur dioxide** and are used as **indicator species** for **air pollution**; this symbiotic association of lichens initiated the idea of **Symbiogenesis** which suggests that novel life forms evolve from the symbiotic association of two or more independent species. Modern biologists believe that symbiogenesis of various bacterial cells led to the origin of eukaryotic cells. The chloroplast of the modern plant cell must have originated as an association between a bacteria and photosynthetic bacteria where the latter transformed into chloroplast over decades. Similarly, different organelles (or *small organs of the cell*) having specialized functions must have evolved over a long time scale in the similar manner.

IMPORTANCE OF FUNGI

Moulds and yeasts together make one of the most important groups among microorganisms. Primarily, their biggest contribution is in geochemical recycling of elements in nature. Fungi contribute a great deal to decomposition of dead organic matter. Fungi together with bacteria account for most of recycling of dead plant and animal matter; else, there would be heaps of cadavers, woods logs and decaying organic matter around us.

The symbiotic association between plant roots and fungi is known as Mycorrhizae (myco- fungus; rrhizae- roots). Mycorrhizae in the form of ectomycorrhizae, endomycorrhizae and vesicular arbuscular mycorrhizae (VAM) contribute to the development of crops and several plant groups (conifers). Apart from this, the association between plants and fungi also prevents soil erosion.

Fungi contribute a lot to making of various foods and beverages. Mushrooms are one of the sought after foods and are considered as nutraceuticals. Fungi immensely contribute to production of cheese, bread, cakes, beers, wines, soya fermented foods (miso, sofu, tempeh) and many oriental fermented foods. Thus, various food industries including bakery, brewery and dairy make use of a variety of biochemical abilities of moulds and yeasts in production of desired flavour or texture of the food material. However, caution must be taken with regards to poisonous or hallucinogenic mushrooms and toxin producing fungi.

The first antibiotic, penicillin, made a difference to the world by saving life of millions. This discovery also opened a possibility of finding out other antibiotics from other microorganisms and made a significant change in healthcare. Fungi produce several therapeutic enzymes (diastase, protease, arginase and lipase) which are given to convalescent patients with weakened digestive system. Some of the fungi are now well known for their rejuvenating and health boosting activity. *Ganoderma lucidum* powder is a well known and much sought after health rejuvenating tonic. *Cordyseps sinensis* that grows on an insect is known to confer health rejuvenating benefits. Several such compounds and chemicals are produced by various fungi which make base or precursor for several drugs.

Fungi are also being used as very promising biological control agents. Biocontrols are an important intervention to sort out problems of chemical pesticides. Species of *Trichoderma* such as *Trichoderma viride* and *Trichoderma harzianum* are being used as biocontrol agents. Propagules (spores) of these fungi are sprayed on crop pests such as spittle bags, wasps and leaf hoppers. The spores parasitize on the insects and eventually kill them. *Beauveria bassiana* is one such fungus that parasitizes mosquito larvae.

Fungi also have a negative aspect where they harm agricultural crops, insects, animals and human beings and deteriorate valuable paintings, furniture, food, stored cereals, etc. Most of our common cash crops which yield our staple food are liable to fungal attacks; fungal parasites attack crops standing in the field where they rapidly multiply and bring about severe losses in the crop productivity in terms of both quality and quantity. Fungal diseases of crops such as late blight of potato have been responsible for famines. During storage also, fungi may colonize upon cereal grains and bring about severe losses in quality of vegetables, fruits, jellies, jams and cereals. Most agricultural commodities are susceptible to fungal attack. Once infected the product looses its quality and becomes unfit for consumption.

Several fungi are known to infect and parasitize on animals (zoophilic) and humans also (Anthropophilic). Such diseases include ring worm, tenia pedis (atheletes foot), aspergillosis, candidasis, thrush, dermatitis and other subcutaneous infections. Most of these infections are difficult to cure because there are only few antifungal antibiotics or chemotherapeutics. In particular, convalescent patients and immunocompromised persons are more susceptible to such opportunistic infections. Some of the major pathogens include dermatophilic moulds such as *Trichopyton sp*., (*T. rubrum, T. mentagrophytes*), *Epidermophyton, Chrysosporium,* etc. As a whole it can be concluded that fungi as a group have a wide array of applications in human welfare.

SUGGESTED READING

Ainsworth, G.C. and Sussman, A.S. (Eds.) 1965-1973. *The Fungi: An Advanced Treatise* (4 volumes), Academic Press, N.Y.

Bilgrami, K.S., Jamaluddin and Rizwi, M.A. 1979. *Fungi of India Part I, List and References*. Today and Tomorrow, New Delhi.

Alexopoulos, C.J., Mims, C.W. and Blackwell, M. 1996. *Introductory Mycology*, John Wiley & Sons Inc., N.Y.

Carile, M. and Waltkinson, S. 1994. *The Fungi*. Academic Press.

Bessey, E.A. 1950. *Morphology and Taxonomy of Fungi*, Vikas Publishing House Pvt. Ltd. New Delhi.

Mehrotra , R.S. and Aneja, K.R. 1990. *An Introduction to Mycology*, New Age International Press, New Delhi.

Moore- Landecker, E. 1982. *Fundamentals of the Fungi*. Prentice-Hall Inc. Englewood Cliffs. N.J.

Webster, J. 1985. *Introduction to Fungi.* Cambridge University Press. Cambridge, U.K.

Gilman, J.C. 1967. *A Manual of Soil Fungi.* Oxford and IBH Publishing Company, New Delhi.

Onions, A.H.S., Allsopp, D. and Eggins, H.O.W. 1981. *Smith's Introduction of Industrial Mycology*, (VII ed.). Edward Arnold Pub. Ltd.

REVIEW QUESTIONS

1. Describe general characteristics and classification of fungi.
2. Describe various sexual and asexual methods of reproduction in fungi.
3. Write notes on the following giving suitable examples:
   a) Myxomycetes
   b) Oomycetes
   c) Zygomycetes
   d) Ascomycetes
   e) Basidiomycetes
4. What are fungi imperfectii? Discuss their salient features and importance.
5. Discuss the general importance and scope of mycology.

UNIT V

General Characteristics of Viruses

Viruses are obligate intracellular parasites composed of nucleic acid and protein. Their RNA or DNA genome is covered with a protein coat. Viruses are also looked upon as highly evolved mobile genetic elements that may have cellular origin. Propagation of viruses solely depends on hijacking of host cell biosynthetic machinery for the supply of their metabolic requirements. A complete virus particle is called a virion. Virion or infectious virus delivers its DNA or RNA into the host cell and takes over the host cell machinery so that the viral genome is expressed (transcribed and translated) inside the host cell. The host cell is then filled with multicopies of newly produced viral genomes and proteins coded on the viral genome. Finally, viral genome and proteins are assembled and nucleic acid is packaged in the protein capsid. This assemblage is referred as nucleocapsid. Nucleocapsids of enveloped viruses are surrounded by a lipid bilayer with an outermost layer of glycoproteins derived from the host cell membrane.

Viruses exhibit typical morphological and physical characters including a great variety in size and shape, chemical composition, structure of the genome and mode of replication. The nucleocapsids of many filamentous viruses show helical morphology that consists of a helical arrangement of capsid proteins (protomers) wrapped around a helical filament of nucleic acid. The nucleocapsids of spherical viruses bear icosahedral morphology where the arrangement of the capsomeres (subunits of icosahedron) fit together to make a sphere. The number of capsomeres is a characteristic feature of icosahedral viruses. Some viruses have outer envelopes made of lipid and glycoproteins (enveloped viruses) while some are naked.

The strandedness of genome of a virus is a characteristic feature. The viral genome may consist of DNA or RNA, which may be single stranded (ss) or double stranded (ds). Further, the genome may either be linear or circular. The entire genome may occupy either one nucleic acid molecule (monopartite genome) or several nucleic acid segments (multipartite genome). The genomic RNA strand of a ssRNA virus is referred as positive or sense strand if it functions as an mRNA for translation of viral protein particles. It is referred as negative or antisense if it

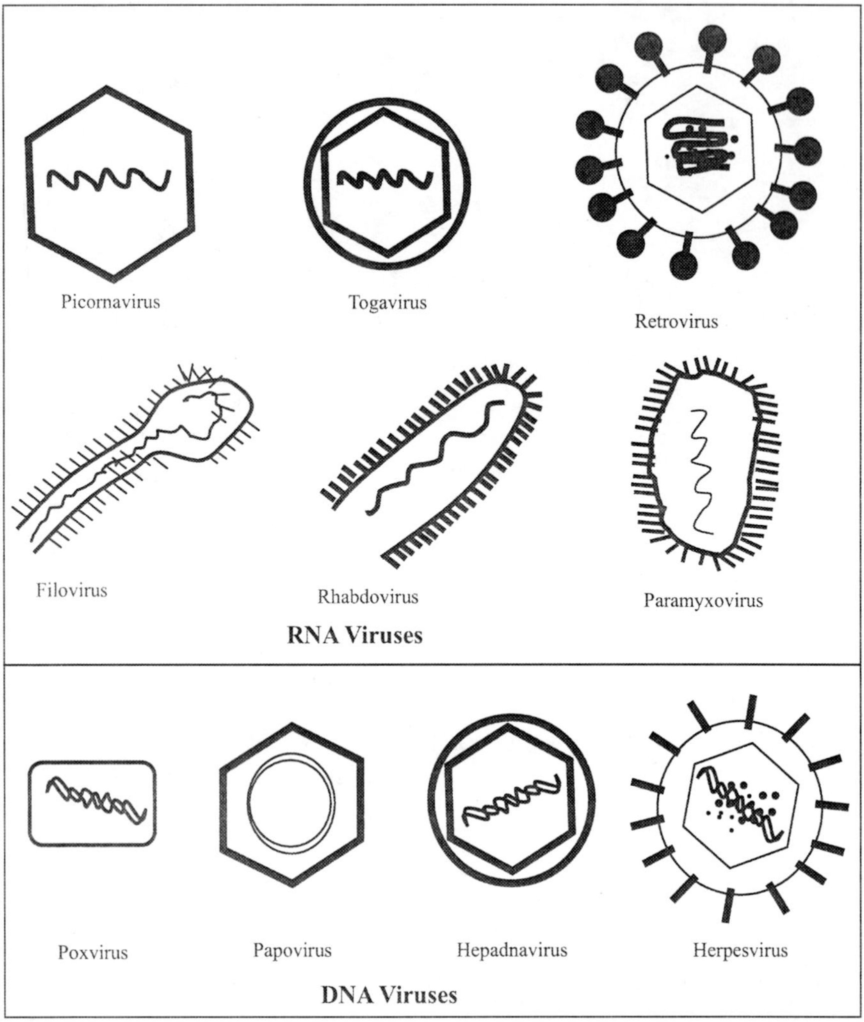

Figure 5.1: Morphological features of some RNA and DNA viruses.

is not translated directly but functions as template for synthesis of the actual mRNA complementary strand synthesized by its own RNA transcriptase.

Viral nomenclature and classification makes use of all the above properties of its physical structure, genome structure and mode of replication as the basis. Viruses are obligate intracellular parasites and outside the cell they cannot generate energy, can not replicate and hence remain inactive. Viruses can be crystallized *in vitro* which makes them unique among all life forms. Any virus has two fundamental components viz. nucleic acid (single- or double-stranded RNA or DNA) and the capsid, which protects the viral genome from nucleases. During infection, the capsid affixes the virion to specific receptors of the host cell. The genome of virion is small and codes for its structural proteins, i.e. capsid and few regulatory proteins of viral replication.

TAXONOMY OF VIRUSES

Initially, viruses were classified and grouped on the basis of their pathogenic properties and were named and placed according to the organ infested by them, e.g. Hepatitis B (organ tropism). Later on with the isolation reports of several new viruses in 1950s and early 1960s, it was felt that other viral properties such as virus structure and composition should be included while classifying viruses.

The International Committee on Nomenclature of Viruses (ICNV) was established in 1966 during the International Conference of Microbiologists in Moscow. The committee laid down norms on classifying viruses that included several characteristics and their weightage. The seventh report of the ICTV (International Committee on Taxonomy of Viruses) presented a universal taxonomy scheme consisting of three orders, 56 families and 233 genera with more than 4000 members. With the advancement in the techniques in molecular biology, electron microscopy and serology a number of viral characteristics such as virion morphology, nature of genome, symmetry, size, physical and chemical properties, structural proteins, lipids and glycoproteins have been included in the taxonomy of viruses. The important morphological features of a virion include size, shape,

Table 5.1 Classification of viruses based on the genetic contents with examples

Class	Nucleic acid	Examples	Envelopes	Genomes Size (Kb)
I	dsDNA	Herpes Virus	Yes	120–220
		Poxvirus	Yes	130–375
		Adenovirus	Yes	3.0–4.0
		Papillomavirus	No	5.3– 8.3
II	ssDNA	Adeno-associated virus	No	5.0
III	dsRNA	Reovirus*	No	18–31
IV	(+)ssRNA	Togavirus	Yes	9.7–11.8
		Poliovirus	No	7.4
		Foot and Mouth disease virus	No	7.5
		Hepatitis A virus	No	7.5
		Hepatitis C Virus	Yes	10.5
V	(–)ssRNA	Influenza virus*	Yes	12–15
VI	Reverse RNA	HIV	Yes	9.7
VII	Reverse DNA	Hepatitis B Virus	Yes	3.1

ss = single strand; ds = double strand; (+) RNA is the one which can function as mRNA for the synthesis of proteins; (–) RNA cannot function as mRNA; *Reovirus and influenza virus have segmented RNA genomes and the total length of genome is shown here.

presence of peplomers, presence of envelope, capsid symmetry and structure, etc. The important physicochemical properties include molecular weight of intact virion, sedimentation coefficient, stability towards pH, temperature, detergent, solvent, irradiation, etc. Genotypic features of viron include type of nucleic acid, size, strandedness, linear or circular, sense or antisense nature, nucleotide sequence, presence of repetitive sequences, G+C content, presence or absence of 5' capping and 3'poly A tail, linkage with proteins, etc.

Number and size of structural and non-structural proteins, their activities (reverse transcriptase, hemagglutinin, neuraminidase, and fusion), amino acid sequence, and antigenic (epitope) properties are also important in ascertaining viral taxons. Viral lipids and carbohydrates are also accounted in viral classification. Lastly, biological properties manifested by the virus *in vivo* such as host range, mode of transmission in nature, vector relations, geographic distribution, pathogenesis, organ tropism and histopathology are also important.

It is very difficult to comment upon phylogenetic relationships of viruses due to the lack of a fossil record and the dispute over status of living or non-living of these smallest creatures. A domain name of Acytota indicating their acellular nature has also been used. This domain puts viruses at par with the other domains of Eubacteria, Archaea, and Eukarya. Not all families are currently classified into orders, nor all genera classified into families.

As an example of viral classification, the chicken pox virus belongs to family Herpesviridae, subfamily Alphaherpesvirinae and genus Varicellovirus. However, it remains unranked in terms of order.

The general scheme of viral systematics is as follows:

Order (-virales)
Family (-viridae)
Subfamily (-virinae)
Genus (-virus)
Species (-virus)

On the basis of common characters viruses have been grouped into orders and families. Three orders (suffix *virales*) have been approved by ICTV:

(1) **Mononegavirales** (families: Bornaviridae, Paramyxoviridae, Rhabdoviridae and Filoviridae which includes single-stranded, antisense, non-segmented enveloped RNA viruses)
(2) **Nidovirales** (families: Coronaviridae and Arteriviridae which include single strand positive sense enveloped icosahedral RNA viruses)
(3) **Caudovirales** (consists of two bacteriophage families)

The groupings of viral genera sharing common features and are different from other genera are called as viral families (suffix **viridae**). Most families have distinct virion morphology, genome structure and replication methods. Some examples are Picornaviridae, Togaviridae, Poxviridae, Herpesviridae and

Paramyxoviridae. Some sub-families (suffix **virinae**) have also been constituted to form groups of individual members with complex relationships, e.g. Herpesviridae is further divided into three sub-families – Herpesvirinae (herpes simplex virus), Betaherpesvirinae (cytomegalovirus) and Gammaherpesvirinae (Epstein-Barr virus). Viral genera are groupings of viral species which share common characteristics and are distinct from members of other genera. The genera are denoted by suffix **virus** (genus: *Simplexvirus, Varicellovirus*). A viral species is defined as *a polythetic class of viruses that constitutes a replicating lineage and occupies a particular ecological niche.* The term polythetic refers to members of a group which are defined by more than one property. Further, viral species are classified into strains, e.g. HIV strains.

THE BALTIMORE CLASSIFICATION OF VIRUSES

The Baltimore system of virus classification provides a useful guide with regard to the various mechanisms of viral genome replication. The central theme here is that all viruses must generate positive strand mRNA from their genomes, in order to produce proteins and replicate themselves. The precise mechanisms whereby this is achieved differ for each virus family. These various types of virus genomes can

Table 5.2 Taxonomy, characteristics and examples of some virus families (DNA Viruses)

Family	Characteristics	Typical Members	Diseases Caused
Poxviridae	dsDNA, brick shaped particles, largest viral size	Vaccinia, Variola	Cowpox Smallpox
Herpesviridae	dsDNA, Icosahedron, capsid enclosed in an envelope, Latent infection in hosts	Herpes Simplex, Varicella Zoster Cytomegalovirus Epstein-Barr Virus	Cold, genital infections Chickenpox, Shingeles Febrile illness Glandular Fever and malignancies such as Burkitt's lymphoma
Adenoviridae	dsDNA, Icosahedron with fiber structures, non-enveloped	Adenovirus	Respiratory and eye infections, causes tumor in experimental animals
Papillomaviridae	circular dsDNA, 72 capsomeres in capsid, non-enveloped	Human papilloma viruses	Warts, associated with some cancers
Hepadnaviridae	One complete DNA minus strand with 5' terminal protein, DNA circularized by an incomplete plus strand, 42 nm enveloped particle	Hepatitis B-virus	Serum hepatitis associated with hepatic cancer

be broken down into seven fundamentally different groups, which obviously require different basic strategies for their replication. "The Baltimore Classification" of virus genomes was proposed by David Baltimore (1971). By convention, the top strand of coding DNA is written in the $5' \rightarrow 3'$ direction. The replication strategy of the virus depends on the nature of its genome. Viruses can be classified into seven groups:

(i) **dsDNA viruses**
(ii) **ssDNA viruses**
(iii) **dsRNA viruses**
(iv) **(+)-sense ssRNA viruses**
(v) **(–)-sense ssRNA viruses**
(vi) **RNA reverse transcribing viruses**
(vii) **DNA reverse transcribing viruses**

where, "ds" represents "double strand" and "ss" denotes "single strand".

(i) **Double-stranded DNA** (Example: adenoviruses; herpesviruses; poxviruses) some members of this group replicate in the nucleus, e.g. adenoviruses using cellular proteins. Poxviruses replicate in the cytoplasm and make their own enzymes for nucleic acid replication (**Table 5.2**).

(ii) **Single-stranded (+) sense DNA** (Example: parvoviruses) Replication occurs in the nucleus, involving the formation of a (–) sense strand, which serves as a template for (+) strand RNA and DNA synthesis.

(iii) **Double-stranded RNA** (Example: reoviruses; birnaviruses) These viruses have segmented (multipartite) genomes. Each genome segment is transcribed separately to produce monocistronic mRNAs.

(iv) **Single-stranded (+) sense RNA** (Example: picornaviruses; togaviruses, etc). These viruses have polycistronic mRNA, e.g. Picornaviruses; hepatitis A. Naked RNA alone is infectious and there is no virion associated polymerase. Translation results in the formation of a polyprotein product, which is subsequently cleaved to form the mature proteins. Some other viruses have to undergo complex transcription and two or more rounds of translation are necessary to produce the genomic RNA, e.g. Togaviruses (**Table 5.3**).

(v) **Single-stranded (–) sense RNA** (Example: orthomyxoviruses, rhabdoviruses). These viruses always have a virion associated RNA directed RNA polymerase. The genome may be (a) segmented, e.g. Orthomyxoviruses. First step in replication is transcription of the (–) sense RNA genome by the virion RNA-dependent RNA polymerase to produce monocistronic mRNAs, which also serve as the template for genome replication and (b) non-segmented, e.g. Rhabdoviruses. Replication occurs as above and monocistronic mRNAs are produced.

(vi) **Single-stranded (+) sense RNA with DNA intermediate in life cycle** (Example: retroviruses) The genome is (+) sense but does not serves as

Table 5.3 Taxonomy, characteristics and examples of some virus families (RNA Viruses)

Family	Characteristics	Typical Members	Diseases Caused
Paramyxoviridae	ssRNA, enveloped, covered with spikes	Parainfluenza virus, Measles Virus Respiratory Synctial Virus	RTI (Respiratory tract infection) Bronchiolitis Measles
Orthomyxoviridae	Eight segments of ss RNA, enveloped, covered with spikes, helical nucleocapsid	Influenza virus	Influenza
Reoviridae	10–12 segments of dsRNA, icosahedron and non-enveloped	Rotavirus	Infantile Diarrhea
Picornaviridae	ssRNA, 22–30 nm particle of cubic symmetry, non-enveloped	Poliovirus Coxsackie virus Rhinovirus Hepatitis –A virus	Poliomyelitis Myocarditis Common cold Infectious hepatitis
Togaviridae	ssRNA, enveloped particles, icosahedron nucleocapsid	Rubella virus	German measles
Flaviviridae	ssRNA, enveloped particles, icosahedron	Yellow fever virus Hepatitis C virus	Yellow fever Hepatitis C
Rhabdoviridae	ssRNA, enveloped particles, bullet shaped	Rabies virus	Rabies
Filoviridae	ssRNA, pleomorphic enveloped (slender filaments)	Marburg virus	Hemorrhagic fever
Bunyaviridae	ssRNA, enveloped spherical or pleomorphic, helical	Hantavirus	Hemorrhagic fever
Retroviridae	ss RNA, enveloped particles with icosahedral nucleopcapsid, have reverse transcriptase enzyme to make copy of genome using host cell machinery	Human T lymphotropic virus Human immunodeficiency virus	Adult T cell leukemia and lymphoma Acquired Immuno Deficiency Syndrome (AIDS)

mRNA, but acts as a template for reverse transcription for DNA intermediates.

(vii) **Double-stranded DNA with RNA intermediate** (Example: hepadnaviruses) this group of viruses also relies on reverse transcription, but unlike the retroviruses, this occurs inside the virus particle on maturation. On infection of a new cell, the first event to occur is repair of the gapped genome, followed by transcription.

MORPHOLOGY OF VIRUS

Viruses are extremely small (20–200 nm) obligate intracellular parasites which can only be seen by electron microscopes. Simplest viruses have protein coat surrounding a strand of nucleic acid. Complex viruses have an extra structure, an envelope made up from viral glycoproteins. These glycoproteins are coded by viral genome and are produced in the host cell during maturation of the virus. The complete fully assembled virus is called as virion. Virion may have glycoprotein envelope with peplomers (projections) surrounding the viral particle. The protein coat in touch with nucleic acid is called capsid made up of capsomere. The shape and type of capsomere depends upon the overall shape of viral capsid. The nucleic acid and surrounding protein complex is often referred as nucleocapsid.

Viruses are seen by negative staining under a negative contrast electron microscope. Viral particles are mixed with a heavy metal solution (e.g. sodium phosphotungstate) and dried on a support film. The negative stain provides an electron opaque background against which viruses are seen at the level of 50 Å resolution. View of sections of negatively stained infected cells allows us to see viral-structures made during viral maturation and assembly. The ultimate technique for deciphering viral structure is X-ray diffraction of crystallized viruses. Results of these techniques allow mathematical processing of virus structure for prediction of configuration. The capsid of virions have two kinds of symmetries: helical and icosahedral.

Helical Symmetry

In viruses with helical symmetry, identical protein subunits (protomers) are arranged into a helical array surrounding the nucleic acid, which follows a similar helical path. Such viruses appear as rigid elongated rods or flexible filaments;

These helical rods can be enveloped or naked and are characterized by length, width, pitch of the helix, and number of protomers per helical turn. The most extensively studied helical virus is tobacco mosaic virus (TMV). The structural features of this tobacco plant virus have been elucidated using x-ray diffraction technique (**Figure 5.2**).

Icosahedral Symmetry

An icosahedron is a polyhedron having 20 equilateral triangular faces, 30 edges and 12 vertices and exhibits 2, 3 and 5- fold symmetry. Lines through opposite vertices define axes of fivefold rotational symmetry: all structural features of the polyhedron repeat five times within each 360° of rotation about any of the fivefold axes. Lines through the centres of opposite triangular faces form axes of threefold rotational symmetry; twofold rotational symmetry axes are formed by lines through midpoints of opposite edges. An icosahedron (polyhedral or spherical) with fivefold, threefold, and twofold axes of rotational symmetry is defined as having 532 symmetry (**Figure 5.3**).

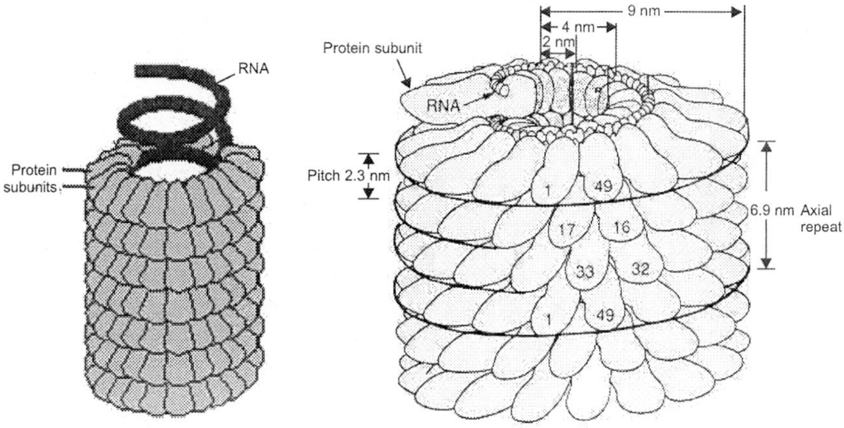

Fiigure 5.2: The helical structure of the tobacco mosaic virus. Individual 17,400-Da protein subunits (protomers) assemble in a helix with an axial repeat of 6.9 nm (49 subunits per three turns). Every turn contains 16-1/3 subunits, making a pitch of 2.3 nm. The RNA (2×10^6 Da) is seen between ajacent turns of capsid protein, forming a RNA helix of 8 nm diameter with three nucleotide bases in contact with each subunit. Some 2,130 protomers per virion cover and protect the RNA. The complete virus is 300 nm long and 18 nm in diameter with a hollow cylindrical core 4 nm in diameter.

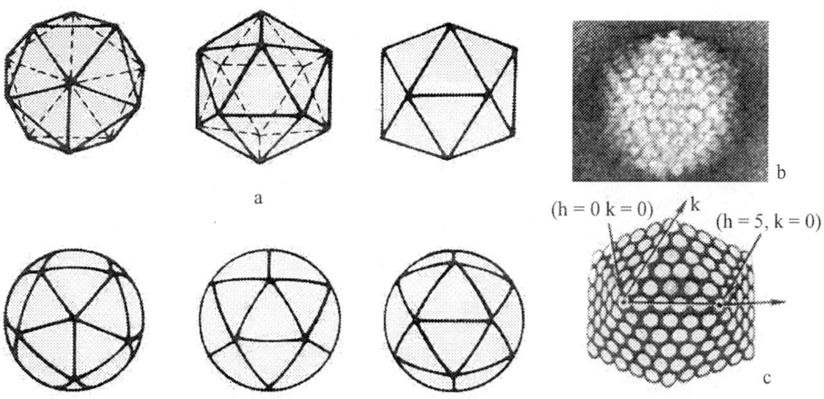

Figure 5.3: Icosahedral symmetry in viral structures. (a) Icosahedrons (polyhedral or spherical) with fivefold, threefold, and twofold axes of rotational symmetry (called as 5, 3, 2). Both polyhedral (upper) and spherical (lower) forms are represented by different virus families. (b) **Adenovirus after negative stain electron microscopy.** The structure reveals the typical isometric shell made up from 20 equilateral triangular faces. The 252 cashmeres, 12 pantones and the 240 hollow heron cashmeres are arranged in a T = 25 symmetry pattern (c) **Adenovirus model.** Capsomeres are depicted as circles surrounded by an electron dense stain. The inclined axes, h and k, are indicated. The second vertex has indices h = 5, k = 0.

Viruses were first found to have 532 symmetry by x-ray diffraction studies and subsequently by electron microscopy with negative-staining techniques. In most icosahedral viruses, the protomers, i.e. the structural polypeptide chains, are arranged in oligomeric clusters called capsomeres, which form the closed capsid shell. The arrangement of capsomeres into an icosahedral shell is used for the classification of such viruses by capsomere number and pattern.

The electron microscopic view of influenza viruses looks like small spheres (100 to 150 nm) with a distinctly visible membrane and numerous spikes. Spikes are particles of the receptor-binding protein called hemagglutinin (ha). The external part of this protein is responsible for binding to the respective cellular receptors (**Figure 5.4**). In addition to these, two other proteins are anchored in the membrane: the neuraminidase (NA) enzyme and protein M2 which makes an ion channel between the environment and the interior of the virus. The membrane consists of fat particles. The viral capsid consisting of the layer formed by protein M1 particles is found within the membrane. The capsid envelops the viral interior. The genetic material of the virus is eight single parts of RNA (multipartite genome). Each RNA segment is combined with an enzyme (polymerase) and nucleoprotein. The whole complex is called ribonucleoprotein (RNP).

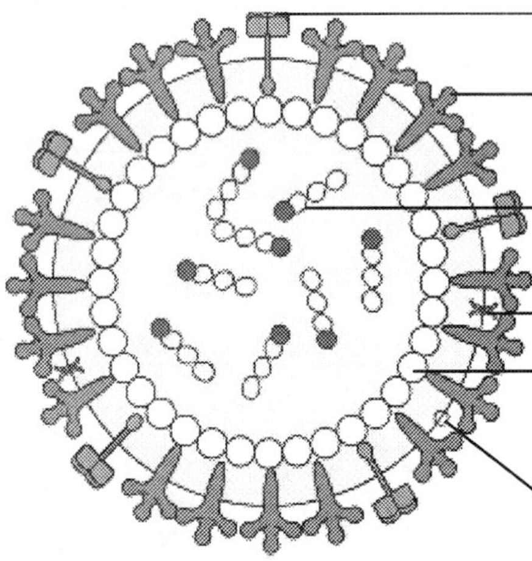

Figure 5.4: Morphological features of influenza virus.

Virus Core Structure

Except in helical nucleocapsids (e.g. TMV), little is known about the packaging or organization of the viral genome within the core. Small virions are simple nucleocapsids containing 1 to 2 protein species. The larger viruses contain a core having the nucleic acid genome complexed with basic protein(s) and protected by a single- or double layered capsid (consisting of more than one species of protein)

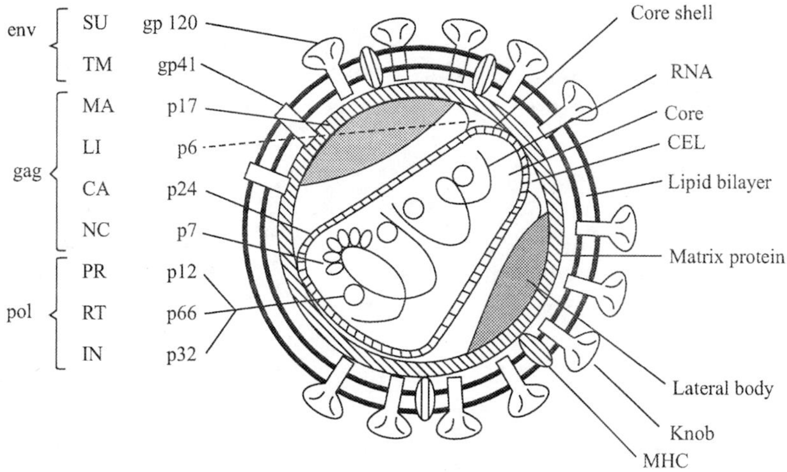

Figure 5.5: Diagram of HIV-1 showing its structural components. SU- outer surface glycoprotein, TM - transmembrane gp, MA - membrane associated or matrix protein, LI-core-envelope-link, CA- capsid, and NC - nucleocapsid protein. PR, RT and IN represent the virus-coded functional enzymes, protease, reverse transcriptase and integrase, respectively.

or by an envelope. The organization of RNA genome with in the virus core in influenza virus and HIV is shown in **Figures 5.4** and **5.5**, respectively.

Viral Proteins

A viral genome may code for five to hundred genes. The product of viral genes, the viral proteins may either be structural, making a part of the nucleocapsid or virion associated functional proteins (enzymes). Some other non-structural proteins coded by viral genome are found only in the host cell. Many such proteins are multifunctional in nature and may be classified as essential and non-essential. Essential proteins are indispensable and virus can not complete its replicative cycle without them while the deletion of viral genes coding for non-essential proteins do not have any detrimental affect on virus growth.

Virus Envelopes

In addition to a capsid, many viruses have a surrounding envelope. Most enveloped viruses bud out from a cellular plasma membrane e.g. influenza virus, or from nuclear membrane, e.g. herpes simplex virus. The envelope contains a lipid-protein bilayer with a number of virus-encoded glycoproteins. Viral envelopes can be amorphous or loosely defined, e.g. herpes virion or regular and tightly bound to capsid. The lipids of the envelope are largely derived from the host cell while glycoproteins are encoded in viral genomes.

Virus Culture

For studying virus in detail it is rather compulsory to culture them *in vitro*. Being an obligate intracellular parasite viruses can be cultured only on cell cultures. The cell culture techniques were developed only in 1950s and hereafter a spurt of interest in virus research gained a thrust. Today, thousands of cell lines which suit virus culturing are available.

Initially, cells are cultured in glass bottle, plastic flasks or large vessels of different sizes under highly aseptic conditions. For this, single cell suspensions are inoculated into pre-sterilized media in a bottle (Roux bottle) in horizontal or flat positon under appropriate growth medium and conditions (usually 37°C). Most cell culture media are chemically well defined and usually contain 2–5% serum for necessary stimulation of cells for division. Media are prepared in isotonic salt solution supplemented with amino acids, vitamins, glucose, etc. Eagle's minimal essential medium (MEM) is one such example of chemically defined media for cell culture. It also contains a pH indicator (phenol red) for monitoring of pH change during growth and antibiotics (penicillin, streptomycin) for preventing any bacterial contamination. Usually, the pH of the medium is maintained around 7.0–7.4. pH shifts due to oxygenation in open flasks and production of lactic acid due to high growth and activity of cells are managed using a bicarbonate-CO_2 buffer system. In case of open flasks the provision of CO_2 is made by purging the incubator with CO_2 gas. Reagents used in media preparation are sterilized by autoclaving or by membrane filtration if thermolabile. The cell lines thus generated may be epithelial (containing cuboid shaped cells) or fibroblastic (containing spindle shaped cells) or a mixture. Continuous cell lines capable of number of cell passages, e.g. Hela cells (epithelial), BHK cells (baby hamster kidney cells, fibroid), etc. are derived from tumours or cancerous tissues. Normal diploid cell lines are derived from fetal lung tissue. Cells adhere to the surface of the medium and begin to multiply forming a monolayer of cells. These cells can be treated with trypsin or versene solution for separating them into single cells which may be used for seeding fresh medium (cell passage). Each cell continues to divide further and the monolayer becomes confluent with high number of cells.

Viruses are grown in cultures to create stock of virus particles which may be stored at low temperatures (–70°C). Such stocks carry 0.1–0.01 infectious units per cell with very low multiplicity of infection. After a few days of infected cell culture viruses can be harvested from the extracellular medium surrounding the cultured cells or by rupturing cells (10–10000 virions per cell).

For culturing some viruses, fertilized chick embryo is preferred over cell culture method. The fertilized embryonated egg has layers of membranes and cavities which support the growth of virus inoculated in allantoic cavity. Its simple and relatively much cheaper than cell culture. Small aliquot of influenza virus is inoculated in the allantoic cavity of the egg and incubated at 37°C. The virus then attaches itself to the epithelial cells and starts replicating. Subsequently, virus is released into the allantoic fluid and can be harvested after 48 hrs.

Centrifugation of a liquid sample containing viruses through a sucrose gradient-density centrifugation results in a sharp band of virus at a specific location. This is collected for further studies. The nature of proteins of viruses is studied using SDS-Page (molecular weight), western blotting, and immunoprecipitation (with homologous antibodies). Viral particles can be localized with in the infected cells using differential staining, immunofluorescence and confocal microscopy. The incorporation of a radiolabelled amino acid or nucleotide in the virion during its growth is helpful in further studies.

REPRODUCTION

A virus must attach and enter itself into a living cell to follow a reproductive cycle. The virus hijacks the cellular machinery and orchestrates it towards its own metabolic necessities and halts the host cell protein and nucleic acid synthesis. This results in production of many progeny virions that leave the host cell by cell lysis or by budding out. The virus reaches out to infect other cells and repeats the growth cycle. Such a growth cycle of virus is referred to as **one step growth cycle (Figure 5.6).** The cycle has several distinct phases including attachment, penetration, nucleic acid synthesis, transcription, protein synthesis, maturation,

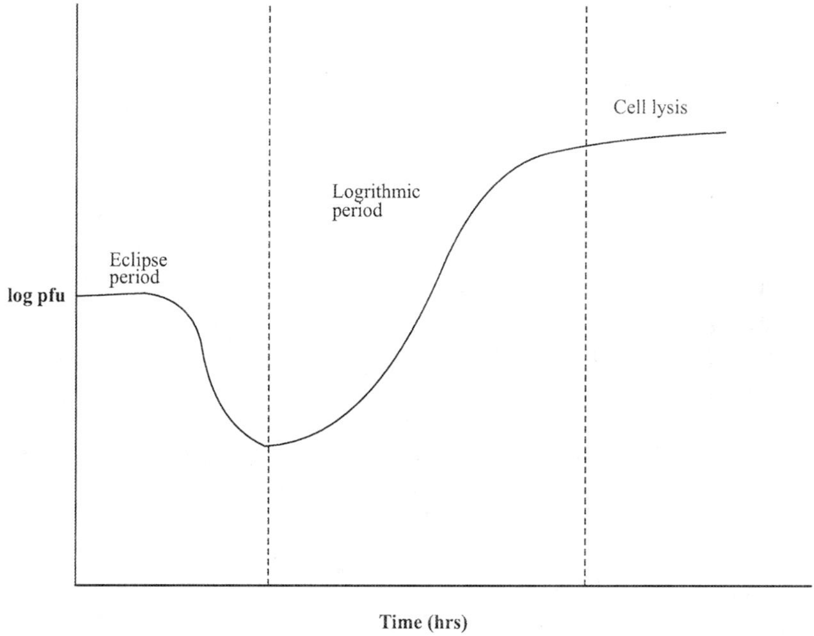

Figure 5.6: One step growth curve of viral reproduction.

assembly and release. The growth pattern of the virus can be followed by lysing the infected cells at various time intervals and titrating the infectious virus particles by counting plaque forming units (pfus) on bacterial lawn. Virus can be quantified by plaque assay technique where the number of plaques produced in an otherwise healthy cell line represents number of infectious virions in the given sample. Plaques are visibly localized discrete foci of viral infection formed due to cell lysis in a monolayer of cells. Each plaque is formed by infection due to a single virion and hence the number of virions in given sample can be counted precisely. The semi-logarithmic curve plotted between the pfus and time results in a characteristic curve having an eclipse period followed by exponential or logarithmic growth phase that continues until the death of host cell. The growth cycle lasts only 60 minutes in bacteriophages while in case of many animal viruses it takes more than 24 hrs to complete one cycle. The strategy of multiplication varies strikingly in different viruses. In case of bacteriophages, it can follow either a lytic or a lysogenic mode. Alternatively, a lysogenic prophage can shift to lytic mode (**Figure 5.7**).

Attachment, Penetration and Uncoating

Attachment of virus particle on cell membrane involves presence of a receptor that determines cell tropism and species tropism of the virus. For example, Herpes simplex virus (HSV) binds to a heparin sulphate proteoglycan plus mannose-6-phosphate receptor via envelope glycoproteins; HIV attaches to the CD4 receptor of T_4 lymphocytes through its major envelope glycoprotein (gp 120). The generalized life cycle of a retrovirus is explained in **Figure 5.8.** In these viruses, penetration is achieved through fusion of viral envelope with the cell membrane. In case of poliovirus, the naked viral particle is taken up by the cell by endocytosis or invagination.

Transcription and Translation

The transcription and translation in different viruses varies altogether. In case of **poliovirus** which has sense RNA strand that also acts as mRNA, transcription starts immediately after penetration and results in formation of polyproteins which are cleaved into different structural and non-structural viral proteins including protease, and RNA polymerase (**Figure 5.9**). Replication is completed in the cytoplasm under the direction of RNA polymerase and ssRNA genome of the virus is replicated through replicative intermediates. Newly formed RNA strands then act as template for synthesis of new genomic ssRNA. In case of **HIV,** the viral genome, i.e. ssRNA undergoes reverse transcription under the effect of viral reverse transcriptase leading to formation of circular DNA molecule. The life cylce of HIV is described in the **Figure 5.10.** This DNA is integrated as **provirus** with the host cell under the effect of viral integrase. The provirus replicates with the routine host cell replication and all the progeny cells contain HIV genome as provirus. The viral genome bears the control of transcription in long term repeat (LTR) regions

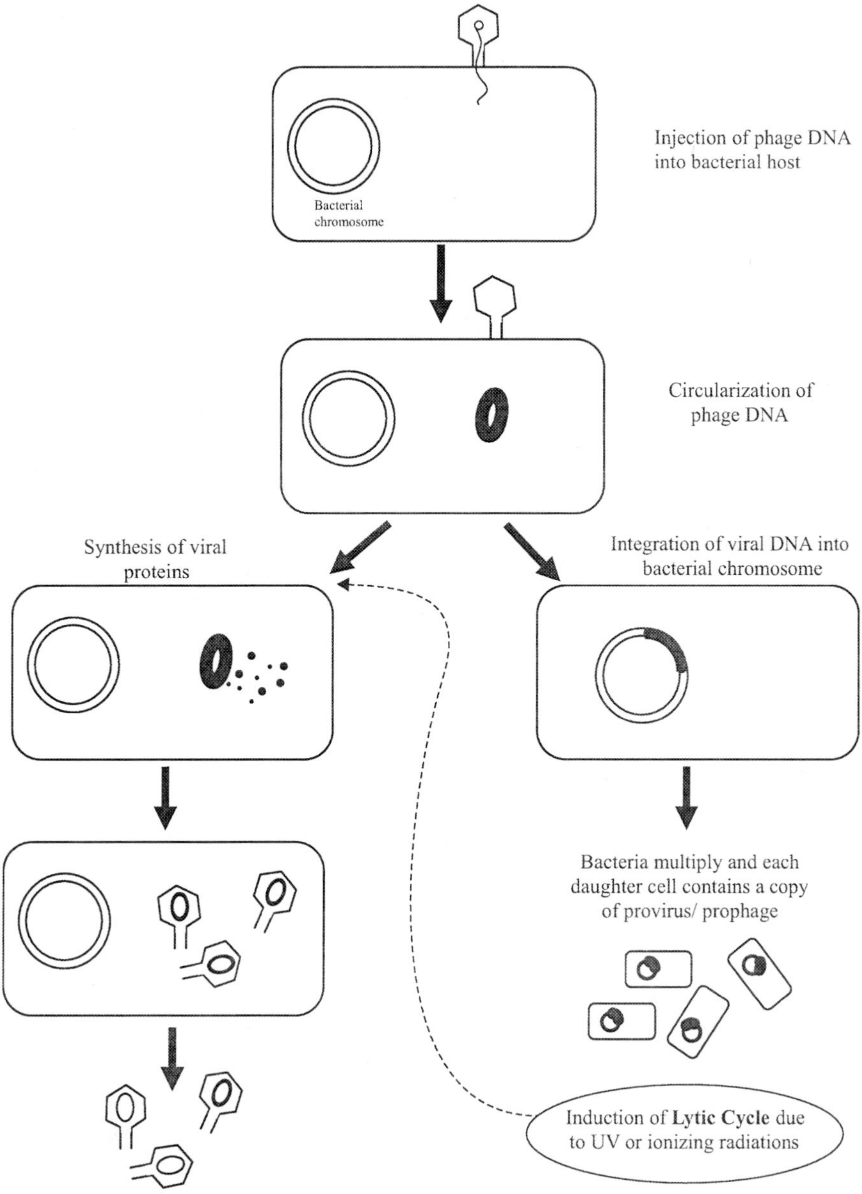

Figure 5.7: Replication of bacteriophage; lytic cycle and lysogeny.

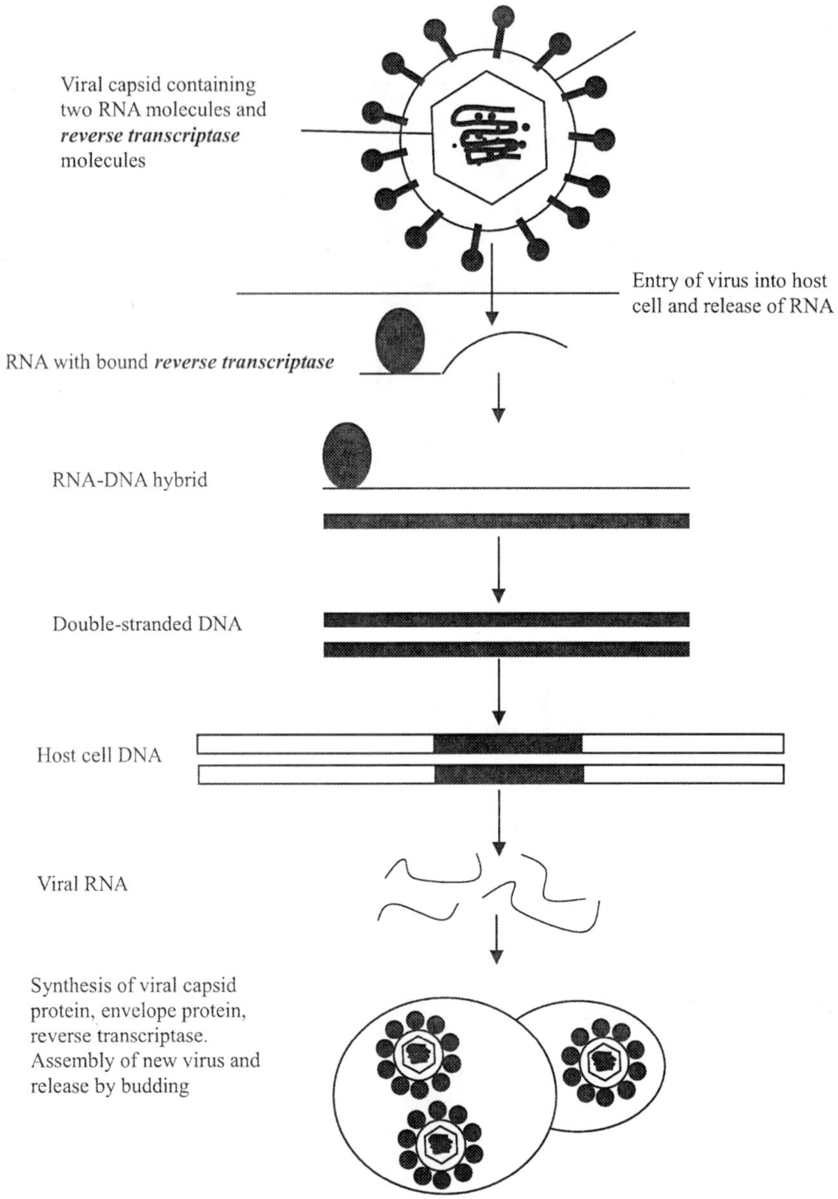

Figure 5.8: General life cycle of a retrovirus.

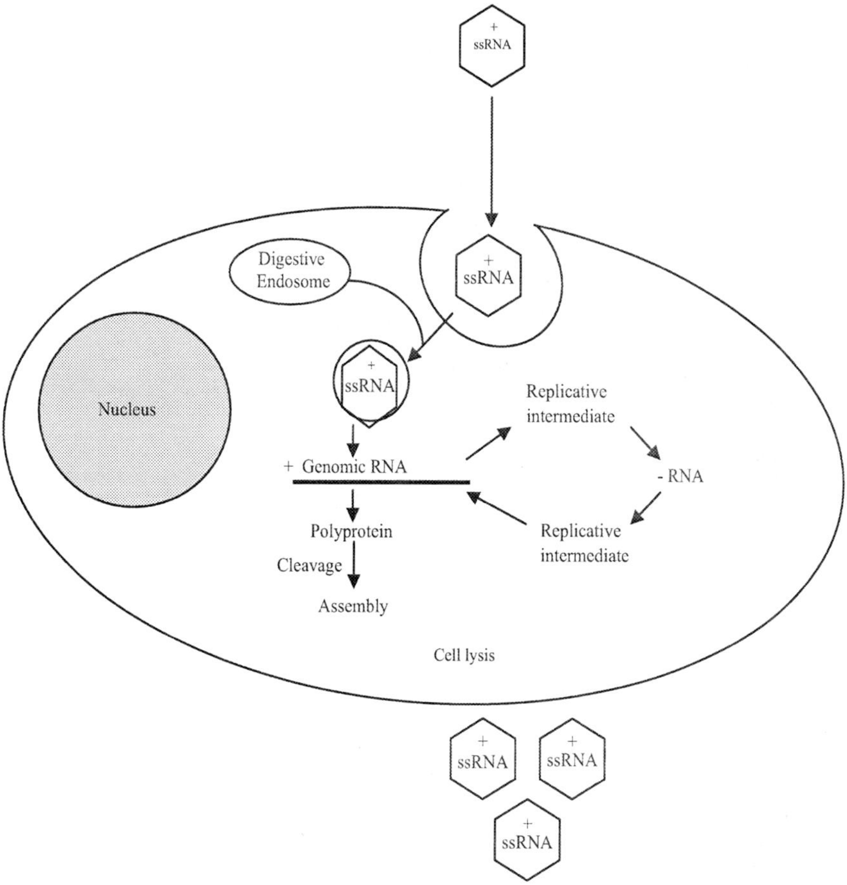

Figure 5.9: Replication of polio virus, an ssRNA virus.

of its genome and triggers production of viral mRNA molecules of various sizes. These are translated into proteins which are sometimes cleaved by viral proteases to yield desired HIV structural and functional proteins. Progeny HIV genomic RNA is formed by the full length transcription of the provirus. These RNA strands are encapsulated to produce new infectious virions.

Herpes simplex genome is a large DNA molecule encoding dozens of virus specific proteins including DNA polymerase and capsid, envelope and tegument proteins (**Figure 5.11**). Since these enzymes and structural proteins are needed in varying amounts throughout the replication cycle, the viral DNA exhibits a great deal of transcription control. The viral genes based on the time of their expression are classified as immediate early (α), early (β) or late (γ). The transcribed mRNA also encodes regulatory proteins. All the viral proteins are produced in the cytoplasmic ribosomes while the viral genome is replicated in the host cell nucleus. The proteins are glycosylated or phosporylated in cytoplasmic endoplasmic

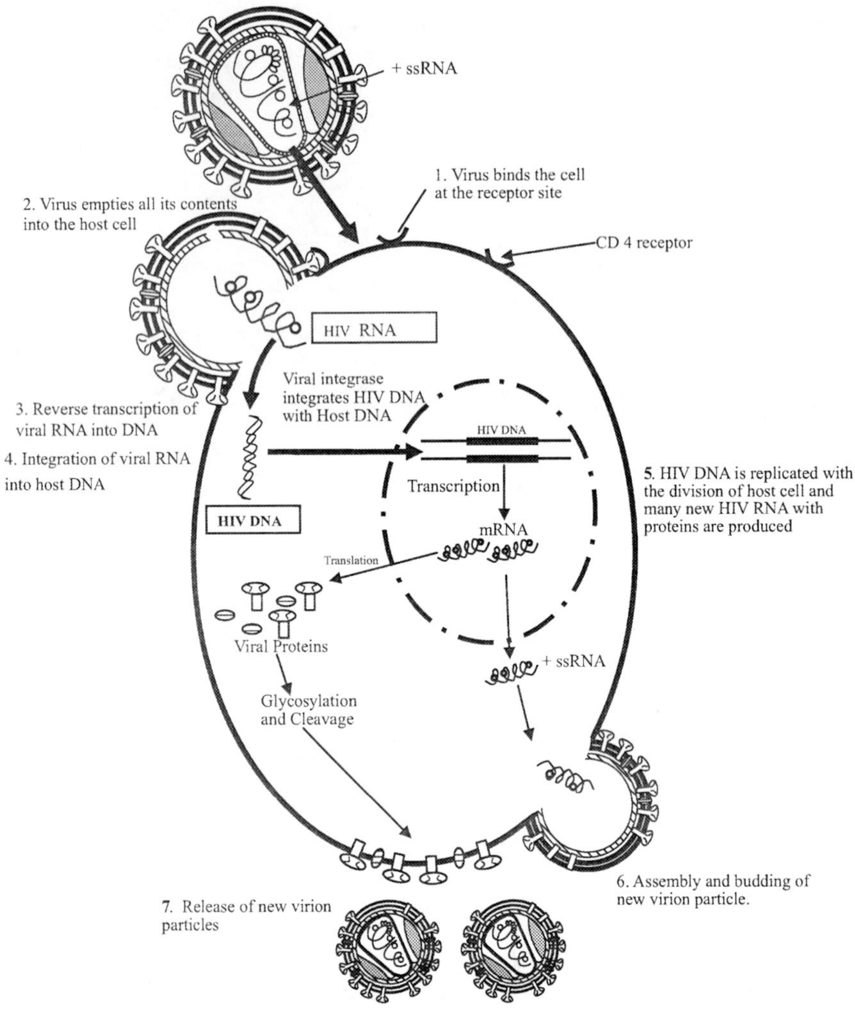

Figure 5.10: Replication of HIV (ssRNA) virus, a retrovirus that replicates through reverse transcription.

reticulum and are directed (chaperoned) to the nucleus for assembly with the viral genome.

Maturation, Assembly and Release

After the synthesis of genomic nucleic acid and proteins within the infected cell, these are migrated to specific locations for assembly of virions. In case of poliovirus, the icosahedral capsid is made of just four proteins (VP 1–4). The capsid of the poliovirus self-assembles without the need of any scaffolding protein to produce an empty capsid that takes up the genomic RNA later. The empty capsid shell

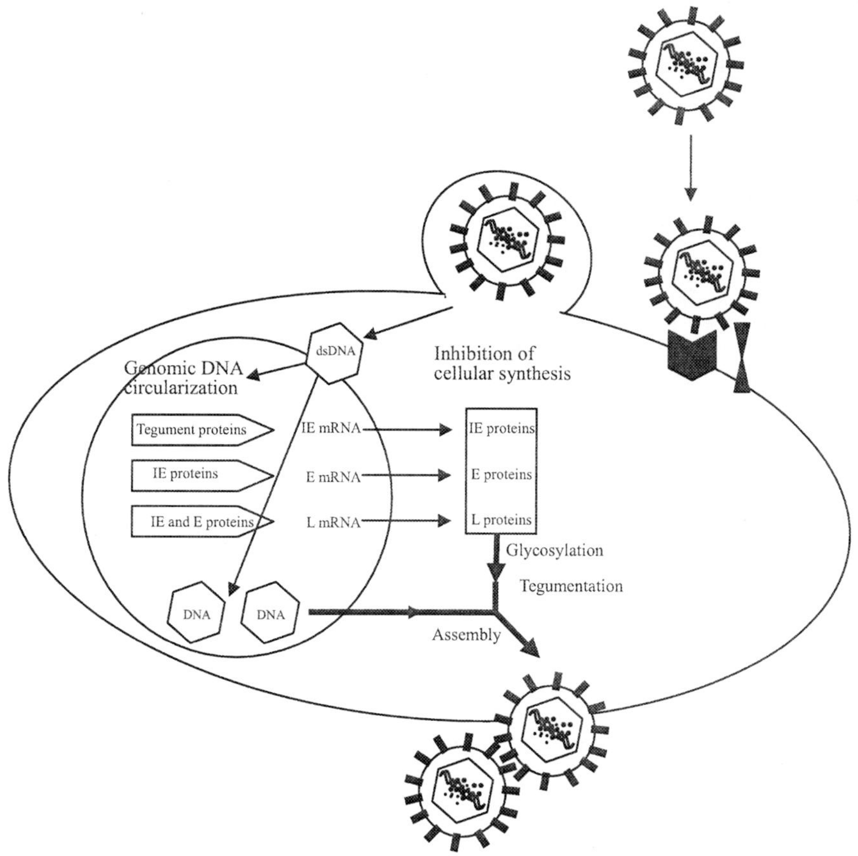

Figure 5.11: Replication of Herpes simplex virus, a dsDNA virus in the host cell.

contains VP0, VP2 and VP3 proteins. As soon as the RNA enters capsid, the VP0 splits into VP2 and VP4 resulting in mature vrion particle.

The icosahedral capsid of HIV is assembled in the cytoplasm through a well-defined controlled mechanism. The regulation is important to package several components, including two identical genomic RNA strands, reverse trancriptase, integrase, protease complex and capsid proteins to make a complete virus.

The capsids of herpes simplex are assembled in nucleus via a series of precursor forms. This packaging is assisted by **scaffolding proteins** which are essentially not a part of final form but facilitate protein-protein and protein-nucleic acid bonding during packaging. The envelope of the mature nucleocapsid is acquired from the nuclear membrane as it buds through it. Complete virions move through the endoplasmic reticulum to the cell membrane and are released. Progeny HSV invade adjacent cells by fusing with the infected cell. HSV infection therefore results in multinucleate fused host cells called as synctia.

RELATED SUB-VIRAL AGENTS: PRIONS AND VIROIDS

The central dogma DNA → RNA → Protein is the basic concept of molecular biology of life. Transcription and translation are almost every minute business of a living cell whereby the cell produces a multitude of proteins for a variety of its structural and functional purposes. These processes are highly regulated by the cell machinery in order to conserve the energy and maintain homeostasis. Along wtith enormous number of proteins, a protein called as PrP 33–35 is also synthesized in human cells. This protein is accumulated in the neural cell membrane where it serves to sequester Cu^{++} ions. In abnormal ("infected") cells, the PrP 27–30 is produced from the PrP 33–35 protein. The PrP 27–30 triggers a series of reactions that produce more PrP 27–30 proteins, i.e., PrP 27–30 induces its own synthesis.

Prions are infectious agents composed exclusively of one kind of protein, more precisely, sialoglycoprotein called PrP 27–30 (Mr: 27–30 kD) and is composed of 145 amino acids with glycosylation at or near amino acids 181 and 197. They do not contain any nucleic acid. The Carboxy or C-terminus of this protein is marked by presence of phosphatidylinositol glycolipid made up of ethanolamine, phosphate, myo-inositol and stearic acid. This protein can polymerize on its own to produce rods having all the properties of an amyloid. Amyloid refers to any optically homogenous, waxy, translucent glycoproteins deposited intercellularly and/or intracellularly in many human diseases including Creutzfeldt-Jakob disease, Fatal familial insomnia, Gerstmann-Straussler syndrome and Kuru Leprosy.

Replication

The prion protein is a product of a human gene, referred the PrP gene located on 20th chromosome. This gene contains two exons separated by a single intron. Exon I and Exon II are transcribed and the two RNAs are ligated into a single mRNA. This mRNA contains an open reading frame (ORF) or protein coding region which is translated into the PrP protein. The PrP protein is a precursor of the prion protein. It is termed PrP 33–35. The PrP 33–35 undergoes several post-translational events to become the prion protein PrP 27–30:

1. Glycosylation – at two sites.
2. Formation of a disulfide bond between two cysteine residues.
3. Removal of the N-terminal signal peptide.
4. Removal of the C-terminal hydrophobic segment.
5. Addition of a phosphatidylinositol glycolipid at the C-terminal.
6. Removal of the N-terminal first 57 amino acids.

In addition to the post translational modifications, the PrP 27–30 protein differs from the PrP 33–35 protein in a single amino acid residue. Residue 178 in the PrP 27–30 contains an asparagine residue whereas the PrP 33–35 protein has an aspartate residue at this position. This causes a conformational change in the PrP 27–30 protein from an α-helix to a β-sheet. It is by virtue of this conformational change that the PrP 27–30 protein acquires abilities of a Prion. The PrP 27–30 protein acqires ability to induce the same α-helix to β-sheet conformation in the normal PrP 33–35 protein. The change is permanent and hence it is said that the PrP 27–30 protein induces its own "replication." All such β-sheet-forming polypeptides get amassed in the cells to form amyloid fibrils. These amyloid fibrils initiate the denaturation of thalamus neurons through apoptosis or the programmed cell death. Now a number of such diseases are known to be induced by prions in human and animals. All these diseases are basically neurodegenerative diseases including Creutzfeldt-Jakob disease (CJD), Fatal Familial Insomnia, Gerstmann-Straussler syndrome and Kuru. Any prion disease does not include any symptoms of common infections such as fever, inflammation, itching, etc. suggesting that prions do not elicit any immune response. This is because prions are self-proteins.

All pathological features of prion induced diseases are limited to the central nervous system (CNS). Prion proteins aggregate selectively and enormously in CNS neurons. The etiological, histopathological and clinical symptoms of prion induced diseases suggest that they are closely related and variants of the similar disorder. PrP 27–30 accumulation leads to astrocyte gliosis (an increase in the number of astrocytes), depletion of dendritic spines in neurons, formation of numerous vacuoles in the cerebellar cortex (spongiform encephalopathy) and amyloidosis – deposition of amyloid in the cerebellar cortex, thalamus, brain stem and in the lumen of blood vessels within the brain. These amyloid plaques consist of discrete eosinophilic glassy-appearing masses, often having radiating amyloid fibrils at their periphery.

The clinical symptoms of a prion induced disease appear over a long time (several years). Patients find difficulty in muscle coordination indicating a functional disorder of the cerebellum. They develop dementia and problems of memory loss, diminished intellect and poor judgment along with insomnia (sleeplessness) with a reduction or loss of the slow-wave and rapid-eye-movement phases.

Transmission

Spread of the disease is via horizontal transmission, i.e. transmission from one person to another, either directly or by fomites or by ingestion of contaminated meat.

Diagnosis

Diagnosis of prion induced diseases is made by histopathological examination of brain tissues of patients in advanced stages of the disease. In 1999, it was found that prion protein accumulates in the tonsils and could be detected by an immunofluorescence test on tonsilar biopsies. A sensitive method of diagnosing prion protein is based on western blot. Another capillary electrophoresis based method uses laser induced fluorescence and is sensitive enough to detect 10–18 moles in the blood samples.

VIROIDS

Viroids are infectious agents made up of only a single segment of circular, single stranded RNA having some double-stranded regions (**Figure 5.12**). Mostly these are plant pathogens that consist of a short stretch of naked circular, single-stranded, highly complementary RNA. The smallest viroid is 220 nucleotide scRNA (small cytoplasmic RNA) found associated with the rice yellow mottle sobemovirus (RYMV). The genome of the smallest known viruses that can cause an infection is about 2000 nucleotide base (2 KB) in size. Viroids were discovered and given this name by Theodor Diener in 1971. The first viroid to be identified was the potato spindle tuber viroid (PSTVd). Since then Some 33 species have been identified. Viroids are mainly indicated in plant diseases but recently have been reported to cause human disease as well. These are catalytic RNAs (similar to ribozymes) have ability to make and break covalent bonds and can cleave RNA to produce

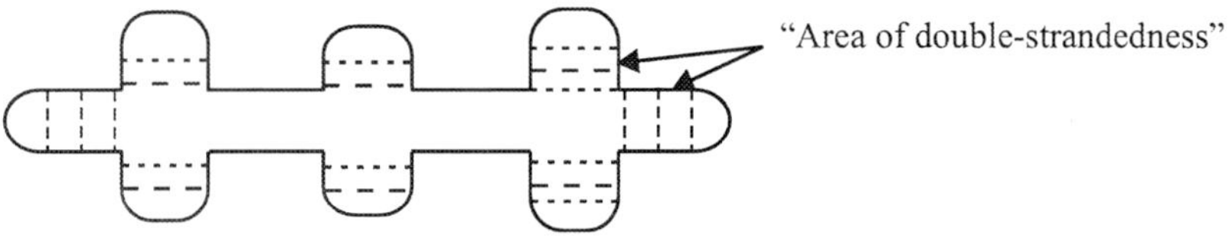

Figure 5.12: Structure of viroid.

fragments containing a 5'-hydroxyl and a 2', 3'-cyclic phosphate. This is a non-hydrolytic reaction in which the same numbers of phosphodiester bonds are maintained and the trans-esterification reaction is theoretically reversible. This reaction is considered to play an essential role in the replication of these RNAs *in vivo*. Such reactions are all intramolecular and hence partially-catalytic with single turnover.

Replication

Circular, pathogenic viroid RNAs are replicated by a rolling circle mechanism with the help of RNA polymerase II enzyme. This enzyme is normally associated with synthesis of mRNA but when infected with viroid, it uses viroid RNA as template instead of DNA, and carries out its replication. Some viroids are ribozymes, having catalytic properties which allow self-cleavage and ligation of unit-size genomes from larger replication intermediates. Site-specific cleavage of this strand produces a monomer that is circularized by a host RNA ligase and then copied by the RNA polymerase to produce a concatameric plus strand. Cleavage of this strand produces monomers which, on circularization, produce the progeny circular, plus RNA, the dominant form *in vivo*.

Human Diseases Caused by Viroids

Viroids are reported to cause hepatitis D in humans. The agent of this disease is a viroid enclosed in a hepatitis B virus capsid. For occurrence of hepatitis D infection the cell must be infected with both the hepatitis B virus and the hepatitis D viroid. There is extensive sequence complementarity between the hepatitis D viroid RNA and human liver cell 7S RNA. This small cytoplasmic RNA is a component of the signal recognition particle, the structure involved in the translocation of secretory and membrane-associated particles. The hepatitis D viroid causes death of the liver cell either by sequestering this 7S RNA or shearing it into pieces. The hepatitis D viroid enters into a human liver cell as an inclusion within capsid of Hepatitis D virus. The delta agent then enters the bloodstream and is further transmitted via blood or serum transfusions or contacts.

SUGGESTED READING

Birge, E.A. 1981. *Bacterial and Bacteriophage Genetics*, Springer-Verleg, W.Y.

Casjens, S. 1985. *Virus structure and assembly*, Jones and Barlett, Boston.

Diener, T.O., McKinley, M.P. and Prusiner, S.B. 1982. *Viroids and Prions*, Proceedings Nat. Acad. Sci., U.S.A. 79: 5220–5224.

Kaplan, A.S. 1982. *Organization and replication of viral DNA*. CRC Press Inc., Boca Raton, Fla.

White, D.O. and Fenner, F.J. 1986. *Medical Virology*, Acad. Press, Oelando, Fla.

Luria, S.E.. 1956 *General Virology*. NY, Wiley.

Fields, B.N., Knipe, D. M. and Howley, P. M. 1996 *Fundamental Virology*. Lippincott Williams & Wilkins.

REVIEW QUESTIONS

1. Discuss the variety of morphological characters witnessed among viruses.
2. Discuss classification of viruses in detail.
3. Write notes on the following giving suitable examples:
   a) Structure of tobacco mosaic virus (TMV)
   b) Icosahedral symmetry
   c) Nucleocapsid
   d) Virion
4. Discuss the methods of *in vitro* cultivation of viruses.
5. Describe life cycles of following:
   a) Human Immunodeficiency Virus (HIV)
   b) Adenovirus
6. Write notes on the following infectious agents:
   a) Viroids
   b) Prions

UNIT VI

Microbial Growth and Culture Media

The term growth when used in context of bacteria or other microorganisms refers to change in total population (cell number) rather than increase in size or mass of an individual cell. Bacteria multiply or reproduce by various methods. Transverse binary fission in which a single cell divides into two after developing a transverse septum is the most common method of bacterial reproduction. Apart from this bacteria also reproduce by budding (e.g. *Hyphomicrobium, Rhodopseudomonas acidophile*), fragmentation (e.g. *Nocardia*), and formation of conidiophore (e.g. *Streptomyces*). Reproduction of bacteria by binary fission leads to increase in the cell number in geometric progression.

$$1 \rightarrow 2 \rightarrow 4 \rightarrow 8 \rightarrow 16 \rightarrow 32 \rightarrow 64$$

$$1 \rightarrow 2^1 \rightarrow 2^2 \rightarrow 2^3 \rightarrow 2^4 \rightarrow 2^5 \rightarrow 2^6 \rightarrow 2^n$$

where 'n' is the number of generations. The total population (N) at the end of a given time period can be given as

$$N = N_0 \times 2^n$$

where N_0 is the number of bacteria inoculated initially.

Solving this equation for n by taking log of both sides, we get

$$\log N = \log (N_0 \times 2^n)$$

$$\log N = \log N_0 + n \log 2$$

$$n = (\log N - \log N_0)/\log 2$$

Substituting the value of log 2 = 0.301 in the equation will give

$$n = (\log N - \log N_0)/0.301$$

or $$n = 3.3 (\log N - \log N_0)$$

By using this relationship, we can calculate the number of generations that have taken place (n) provided, we know the initial population (N_0) and final population (N). The generation time 'g' for any bacteria can be calculated by dividing total time (t) by number of generations (n) as:

$$g = t / n = \frac{t}{3.3 (\log N - \log N_0)}$$

Generation time for bacteria also depends upon the nutrient and environmental conditions.

Growth rate (R) of any bacterium is defined as number of generation per hour. This can be calculated by taking the reciprocal of generation time g.

$$R = 1 / g = \frac{3.3 (\log N - \log N_0)}{t}$$

GROWTH PATTERN OF BACTERIA

Any bacteria when allowed to grow in a batch culture follows typical growth pattern. Initially, number of cells remains unchanged because the bacteria in the new medium adjusts itself to new environment, grows in size, produces various enzymes and coenzymes required for metabolizing new nutrients. This apparently quiescent phase is called **lag phase**. At the end of this phase all the cells start dividing at a constant rate leading to exponential increase in the number of cells. In this phase all the cells are highly active and have similar physiological characters. The plot of logarithm of the number of cells versus time interval yields a straight line indicating a proportional increase in the cell number over time period.

The **exponential phase** of the bacterial growth proceeds through a curve to another straight line parallel to x-axis. This phase is called as **stationary phase**. The phase represents cessation of net growth due to depletion of nutrients. The line remains straight or the cell number remains constant because the increase in cells is compensated by equal number of deaths.

After **stationary phase** bacteria start dying rapidly and the death rate exceeds growth rate. This is due to exhaustion of nutrients, formation of toxic metabolites and autolytic enzymes in the culture. During the **death phase** the number of viable cells decreases exponentially (**Figure 6.1**).

QUANTITATIVE ESTIMATION OF BACTERIAL GROWTH

The term growth in context of bacteria and other microorganisms refers to the increase in cell number or population size. Measurement of growth is essentially required in case of relating growth with various other important parameters. This can be achieved by measurement of (i) cell numbers or cell counts, (ii) cell mass or biomass or wet weight, and (iii) by relating cell activity with population size. Number of cells can be counted by finding out viable cell growth in given amount of liquid medium by dilution plating. They can also be counted directly using Petroff-Hausser chamber. Number of viable cells can be counted by plating a

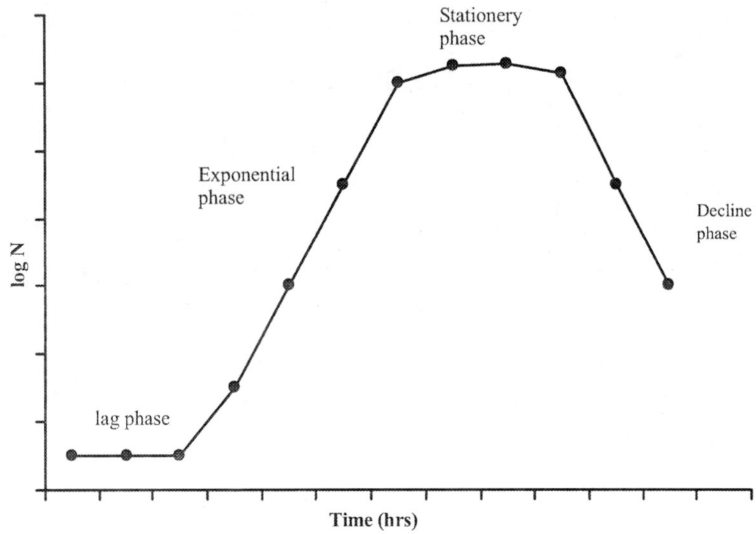

Figure 6.1: A typical bacterial growth curve of batch culture showing all the phases.

definite amount of appropriately diluted liquid on sterile medium. In dense bacterial cultures, growth can be estimated by measuring the cell mass. However, it is difficult to account for any extraneous material or extracellular polysaccharides in dry weight or wet weight of culture. Estimating production of any growth associated chemical or any chemical change which is associated with growth can also be used for quantitative estimation of bacterial growth. However, this requires a separate set of experiments for standardization of relationship between the two, the substance (metabolite) and growth.

REQUIREMENTS FOR GROWTH

The requirements for microbial growth can be classified into two main categories, viz. physical and chemical. Physical aspects consist of temperature, osmotic pressure, atmospheric pressure light, etc. Whereas chemical requirements consist of pH, water, sources of carbon, nitrogen, oxygen, mineral and organic growth factors.

PHYSICO-CHEMICAL REQUIREMENTS

Temperature

Microorganisms exhibit a great range of suitable growth temperature between 20-40°C. Many microorganisms grow well at moderate temperatures. However, certain microorganisms are capable of growing at extremes of temperature where most organisms will fail to survive. Each microbial species has its own minimum, optimum, and maximum temperature for growth. **The minimum growth** temperature is the lowest temperature at which the species can grow. **The optimum growth temperature** is the temperature at which the species shows maximum growth and **the maximum growth temperature** is the temperature maximum beyond which it will not be able to grow.

Based on temperature relationship microorganisms are classified mainly into three groups.

(a) Psychrophiles: These are cold-loving microbes that can not tolerate warm temperature (20 °C). They can grow at 0 °C, but have optimum growth temperature of about 15 °C or less.

(b) Mesophiles: These are moderate- temperature-loving microbes. They have optimum growth temperature range of 20-40 °C, and it is the most common range. It is believed that during evolution, microorganisms adapted to live in an optimum temperature close or similar to that of their host. Evidently, the optimum temperature for many pathogenic bacteria is about 37 °C. The mesophiles comprise nearly all of common spoilage and disease organisms. Apparently this is the reason why most bacterial incubators are maintained at 37 °C. Contrasting to this, most mesophilic fungi are grown at 28 °C.

(c) Thermophiles: These are heat-loving microbes. They are capable of growth at high temperatures having an optimum growth temperature between 50-60 °C. This temperature range is seen in tropics, deserts, thermal vents and hot springs, etc. Their temperature requirement may be strict and many thermophiles cannot grow at temperatures below 45 °C. Some extremophiles, members of archaea, have an optimum growth temperature of 80 °C. The temperature limits of growth of fungi have an altogether different reference and range. Cooney and Emerson (1964) used the term **thermophilic fungi** for moulds having maximum growth temperature at or above 50 ºC and a minimum at or above 20 ºC while term **thermotolerant** was used to specify fungi with maximum temperature limit at or above 50 ºC but could grow below 20ºC.

pH

pH value of a solution refers to its acidity or alkalinity. Most bacteria grow best in a narrow pH range near neutrality, between pH 6.5 and 7.5. Very few bacteria grow at an acidic pH near pH 4.0. This is why a number of foods, such as sauerkraut, pickles and cheeses, are preserved from spoilage by acids. Some bacteria are acidophilic and can tolerate acidity. Molds and Yeasts can tolerate a greater pH range than bacteria, but the optimum pH of moulds and yeasts is generally below that of bacteria, typically about 5.0-6.0. Alkaline pH can also inhibit microbial growth but is rarely used to preserve food because it alters the taste of food undesirably.

During *in vitro* culturing of microbes, they often produce acids from sugars that change the pH of the medium. For maintaining proper pH, chemical buffers are added in the medium. Peptone and amino acids in some media can also act as buffers. Phosphate salts also help in maintenance of pH during growth of most bacteria and provide phosphorus for the growth of microorganism.

Osmotic Pressure

Microorganisms derive all their nutrition from the surrounding liquid media with 80-90% water as chief constituent. Osmosis can be defined as movement of water through a semipermeable membrane from its own higher concentration to its lower concentration. High osmotic pressure results in removing essential water from the cell. When a microbial cell is in a solution that has a higher concentration of solutes than in the cell (hypertonic), the cellular water passes out through the plasma membrane to the high salt concentration. This osmotic loss of water causes plasmolysis or shrinkage of the plasma membrane.

Plasmolysis inhibits the growth and the plasma membrane is pulled off the cell wall. Therefore, increasing the concentration of salts in the solution, and the resulting increase in osmotic pressure, can be used to save food from the attack of microbes. Extreme salt loving bacteria or halophiles, have adapted to survive under high concentration of salt and have specialized mechanism of osmoregulation.

NUTRITIONAL REQUIREMENTS

Carbon

As for any other living being, carbon is one of the most essential requirements for microbial growth too. Reduced carbon compounds yield energy upon oxidation by respiratory mechanism. It is required for all the organic compounds that make framework of any living cell. Roughly, half the dry weight of the typical bacterial cell is carbon. Photosynthetic bacteria trap carbon from environment (CO_2) and reduce it by photosynthesis. However, most saprophytic bacteria depend upon other sources for their carbon requirement. Chemoheterotrophs derive their carbon requirement from organic material such as carbohydrate, protein, lipid, etc.

Nitrogen, Sulphur, and Phosphorus

Several other elements are required by microorganisms for synthesis of cellular material for growth. Protein synthesis requires extensive amounts of nitrogen as well as sulphur. The synthesis of DNA and RNA also require nitrogenous bases and phosphorus. Synthesis of ATP, the energy currency, also needs N-base and phosphorus. Nitrogen makes up about 14% of the dry weight of the bacterial cell, and sulphur and phosphorus together comprise about another 4%. Most of this nitrogen is used to form the amino group of the amino acids of proteins. Many microorganisms arrange for this requirement by decomposing other protein containing substrates and reincorporating the amino acids from hydrolysate. Some important bacteria such as Cyanobacteria, *Azotobacter, Rhizobium*, etc. can fix atmospheric nitrogen (N_2) directly from atmosphere, by phenomenon of **nitrogen fixation.**

Sulphur is used to synthesize sulphur-containing amino acids and vitamins such as thiamine and biotin. Important natural source of sulphur consist of the sulphate ion (SO_4^{-2}), hydrogen suphfide, and sulphur containing amino acids.

Phosphorus is essential for synthesis of nucleic acids and the phospholipids of the cell membrane. Phosphorus is found in the energy bonds of ATP. An important source of phosphorus is the phosphate ion (PO_4^{3-}). Potassium, magnesium, and calcium are also important elements required by microorganisms. These are required by many microbial enzymes as co-factors.

Trace Elements

Very small amounts of other mineral elements such as iron, copper, molybdenum, and zinc are required by microorganisms. These are referred as trace elements. Most are vital for the functions of certain enzymes, usually as cofactors. While these elements are sometimes added to a laboratory medium, they are naturally present in tap water and other natural media ingredients. In case of synthetic media such as Czapek's Dox medium trace elements (Cu, Zn) are added in a definite amount.

Oxygen

Based on the need of oxygen, microorganisms can be classified as aerobic, anaerobic and facultative anaerobes. Strict or obligate aerobes can not survive in the absence of oxygen while strict or obligate anaerobes die out in presence of oxygen. Thus, apart from media ingredients, it is customary to look into availability of oxygen. Microorganisms growing on Petri plates and shake flasks utilize environmental oxygen directly but in case of submerged cultures oxygen is provided by purging filtrated sterilized air into the medium through sparger or aeration device. In case of strict anaerobic microorganisms, it is mandatory to remove total oxygen. The simplest way to get rid of atmospheric oxygen is by burning a candle in an airtight bell jar so that all the oxygen is used up in combustion. Alternatively, anaerobic culturing jars and oil-layering methods can be used.

Organic Growth Factors

Essential organic compounds needed for proper growth of an organism are known as organic growth factors. One such group of organic growth factors for us is vitamins. Most vitamins function as coenzyme. Many bacteria can synthesize all their own vitamins and but many require it from an extraneous source. Amino acids, purine, and pyrimidines are some other organic factors needed by some bacteria for proper growth.

CULTURE MEDIA

A formulation of chemicals and nutrient rich materials prepared for the *in vitro* growth of microorganisms is called **culture medium**. Some microorganisms can grow well on *in vitro* nourishment of any general purpose culture medium; other with fastidious requirements need special media, some obligate parasites cannot grow on any non-living medium and hence are grown on cell cultures. The growth of any microorganism on a culture medium is referred as a **culture.** To support microbial growth, a medium must supply an energy source, as well as sources of carbon, nitrogen, phosphorous, sulphur, and specific essential growth factors.

Apart from nutrients, the medium should also provide sufficient moisture and proper pH. There should be adequate supply of oxygen, specially in case of submerged cultures and appropriate temperature for proper growth of microorganism. As of today, a wide variety of ready made media are available for the growth of microorganisms (**Table 6.1**). Media are frequently formulated, modified or revised for use in the isolation, culture and identification of microorganisms from food, water, soil and clinical samples.

Table 6.1 Culture media types and uses

S. No.	Type	Application
1.	Chemically defined media	Growth of chemoautotrophs, photoautotrophs and for microbiological assays
2.	Complex media	Growth of most chemoheterotrophic microorganisms
3.	Anaerobic media	Growth of obligate anaerobes
4.	Selective media	Suppression of unwanted microbes while encouraging desired microbes

It is necessary to culture microorganisms on a solid medium for localizing their single colonies and identifying their colonial characteristics. Agar-agar, a complex polysaccharide of the alga *Gelidium* is used as solidifying agent. Agar has some very important properties that make it indispensable in microbiology. It is inert and cannot be hydrolyzed by most microorganisms and melts at about the boiling point of water (95–100°C) and gels at about 40–45°C. Culture media can be classified into various types on the basis of the nature of components.

Chemically Defined Media

A chemically defined medium is one whose exact chemical composition is known and it is prepared by mixing known amount of standard and well-defined ingredients. For a chemoheterotroph, the chemically defined medium must contain organic growth factors that serve as a source of carbon and energy, a nitrogen source and metals, salts, etc. Chemically defined media are used for the growth of fastidious auxotrophic or autotrophic bacteria.

If required, organic growth factors must be provided in the chemically defined medium used to cultivate fastidious microorganisms (e.g. *Neisseria)*. Interestingly, the growth pattern of fastidious microorganisms (e.g. *Lactobacillus*) is used for bioassay of such essential growth factors (vitamins). Czapek's Dox agar having following composition is an example of chemically defined media in routine use.

Czapek's Dox Agar	
$NaNO_3$	2 g
K_2HPO_4	1 g
KCl	0.5 g
$MgSO_4 \cdot 7H_2O$	0.5 g
$FeSO_4 \cdot 2H_2O$	0.1 g
Sucrose	30 g
Agar	20 g
Distilled water	1000 ml

All the ingredients, excepting K_2HPO_4 are dissolved in half of water. K_2HPO_4 is dissolved separately and added to the rest and volume is made upto 1 litre.

Complex Media

Most heterotrophic bacteria and fungi are able to grow and colonize on **complex media** made up of complex nutrients such as extracts from yeast, meat, plants, or digest of proteins from these and other sources. Being natural in origin the exact chemical composition of such complex ingredients varies from batch to batch.

In complex media, various requirements of the microorganisms, e.g. the energy, carbon, nitrogen, and sulphur requirements are met by complex proteinaceous extracts or digests. Peptones are such soluble partial digests obtained by acids or enzymatic digestion of insoluble large proteins. These are readily utilized and assimilated by bacteria for robust growth. Vitamins, minerals and other organic growth factors are naturally present in extracts of plants and animal origin and are solubilized in liquid medium. These are taken up and utilized by microorganisms for growth.

Anaerobic Growth Media

To grow and maintain pure cultures of obligate anaerobic bacteria adequate provisions are made to put of oxygen in the media and surrounding environment. Anaerobic bacteria might be killed by exposure of oxygen, and therefore special media called reducing media must be used. These media contain ingredients, such as sodium thioglycolate, that chemically combine with dissolved oxygen and utilize the oxygen in the medium.

Figure 6.2: Anaerobic jar for cultivation of anaerobic microorganisms.

When anaerobe is to be cultured on solid media, special jars are used that hold several Petri plates and slants in an oxygen-free atmosphere. The oxygen is removed by using a packet of sodium bicarbonate and sodium borohydride moistened with water, in a sealed jar (**Figure 6.2**). Such ready to use packets are commercially available.

Selective and Differential Media

Apart from providing suitable nutrition, media can also be used for the purpose of selection of any particular microorganism or for differentiation between two microorganisms. This is specially useful in case of bacteria since till today bacterial identification relies on biochemical characterization based on their physiological properties. As per Koch's postulate's, diagnosis of a disease can finally be concluded by isolating and identifying the disease causing microorganism. Thus, it is most important to distinguish specific microorganism associated with disease before proceeding with treatment. Several selective and differential media have been devised and are used successfully for this purpose.

Selective media are devised to restrict the growth of unwanted bacteria and encourage the growth of the desired microorganisms under selection / inhibition pressure. For example, bismuth sulphite agar is used to isolate the typhoid bacterium *Salmonella typhi* from faeces of infected host. On this selective media, bismuth sulphite inhibits Gram-positive bacteria and most Gram-negative bacteria while only *S. typhi* is able to grow luxuriantly. Similarly, the Brilliant green agar media used to isolate the Gram-negative *Salmonella* employs brilliant green dye that inhibits Gram-positive bacteria. Wort agar with relatively low pH (3.6) allows exclusive growth of yeasts. Sabouraud's dextrose agar which has a pH of 5.6 is used to isolate fungi that outgrow bacteria at this pH.

Differential media are used to differentiate colonies of the specific microorganisms from other colonies growing on same plate either as mixed culture or as two separates streaks. This helps in ascertaining purity of cultures based on specific and identifiable reaction with differential media in plates. Blood agar is a dark, reddish-brown medium used to identify bacterial species that bring about haemolysis e.g. the bacterium that cause causes sore throat, shows a clear ring of haemolysis around its colony. Selective and differential media can also be compounded in a single media such as MacConkey agar. This medium contains bile salts and crystal violet dye, which inhibit the growth of Gram-positive bacteria. MacConkey agar also contains lactose, for supporting Gram-negative bacteria that grow on lactose. Thus, Gram-negative lactose fementers and nonfermenters can be distinguished using such media.

Enrichment Culture

Microorganisms have specific growth rates and are distributed non-uniformly in natural habitats. During isolation, microbes present in small numbers and having very slow growth rates or specific nutritional preferences lag behind other fast growing bacteria. The medium for enrichment culture provides specific nutrients and physico-chemical conditions that favour the growth of a particular desired organism. For example, for the isolation of microorganisms that can use xylan as a sole source of carbon, the soil sample may be inoculated in a liquid enrichment medium containing xylan as the only source of carbon. Any endeavour of isolating microorganisms that can degrade xenobiotics or any other recalcitrant molecule utilizes similar approach. Many such selection screens can be devised based on the specific requirement of desired microorganism.

SUGGESTED READING

Brige, E. 1992. *Modern Microbiology-Principles and Applications*, Wm. C. Brown Publishers.

Cooney, D.G. and Emerson, R. 1964. Thermophilic fungi: An account of their biology, activities and classification. W.H. Freeman and Co., San Francisco.

Edwards, C., 1981. *The Microbial Cell Cycle*, *Amer. Soc. Microbiol*. Washington, D.C.

Prescott, L.M., Harley, J.P. and Klein, D.A. 2003. *Microbiology*, Wm. C. Brown Publishers.

Schlegel, H.G. 1993. *General Microbiology*, 7th ed., Cambridge University Press.

Cappuccino, J. G. and Sherman, N. 2002. *Techniques for isolation of pure culture, In. Microbiology: a laboratory manual.* (VI Ed.) Pearson Education Inc. Singapore.

Collee, J.G., Duguid, J.P., Frasee, A.G. and Marmion, B.P. 1989. *Practical medical microbiology*, 13th ed. Longman, U.K.

Collins, C.H. and Lyne, R.M. 1976. *Microbiological Methods*. Butterworth, Mass, U.S.A.

REVIEW QUESTIONS

1. Discuss the nature and characteristics of microbial growth.

2. Discuss various methods used for quantification of the growth of microorganisms.
3. Draw a typical growth curve and discuss its various phases.
4. Describe major physico-chemical conditions affecting the growth of microorganisms.
5. Describe various kinds of media used for the cultivation of bacteria and fungi.
6. Write short notes on the following:
   a) Selective media　　b) Enrichment culture
   c) Differential media　　d) Synthetic media
7. What are the major components of a growth media? Compare chemical nature of the ingredients used for making complex and defined media.

UNIT–VII

Isolation and Maintenance of Microorganisms

The biodiversity of microorganisms is bewildering. As of today, we have seen just the tip of the iceberg and an enormous and huge microbial diversity is yet to be discovered. For last few decades, the advances in biotechnology revolve around recombinant DNA technology and more recently metagenomics has come to the centrestage. However, the statement of Buckland (1992) still remains meaningful that combined sales of four new microbial metabolites introduced in 1980s was greater than the sale of all recombinant products. Thus, it is worthwhile to isolate and enumerate novel microbial forms from various ecological niches.

Microorganisms are abundant and ubiquitous. They occur in air, soil, water, food, sewage, decomposing matter, and living organisms including plants and animals. Microbiology includes study of all the microscopic forms of life ranging from viruses, mycoplasmas and bacteria to algae, fungi, protozoan and helminths. They can be saprobic, parasitic or both; capable of colonizing dead organic matter or living with some form of life. All these forms of microbes exist as a mixed and dynamic population in nature. Identification and characterization of these tiny creatures is a very important task in microbiology. This can be achieved if we are able to study these microorganisms in total isolation. The technique of isolation basically involves segregation of microbial population by dilution and then localizing discrete colonies on solid media. Ultimately, these colonies are picked up and transferred on fresh and sterile media for growth and maintenance. All this procedure is done under religiously aseptic conditions and every effort is made to avoid contamination by any means. A culture containing a single neat species of only one kind of microorganisms or cells is called a **pure culture.** Several different techniques are used to isolate and study microorganisms in pure culture. Some of the important and prevalent methods of isolation are described here.

ISOLATION BY SERIAL DILUTION METHOD

A range of methods are known to isolate and enumerate microorganisms (bacteria, fungi, protozoa, algae, mycoplasmas and viruses) from soil, food, milk, water, etc. The serial dilution agar plating method or viable plate count method is one of mainly used procedures for isolation of fungi, bacteria and actinomycetes. This method is based upon the principle that when a liquid sample containing microorganisms is cultured, each of viable microorganisms will develop into a colony, hence the number of colonies appearing on the plates represent the number of living organisms present in the sample.

In serial dilution agar-plate method, a known amount (10 ml or 10 grams) of material is suspended or agitated in a known volume of sterile water blank (90 ml or so to make the total volume to 100 ml) to make a microbial suspension (**Figure 7.1**). Serial dilutions 10^{-2}, 10^{-3}, 10^{-7}, and so on are made by pipetting measured volumes (mainly 1 ml or 10 ml) into additional sterile water blanks of 9 ml or 90 ml. Finally, 1 ml aliquot of various dilutions is added to sterile Petri dishes (triplicate for each dilution) to which around 25 ml of the sterile molten (45 °C) agar media is added. The routine purpose media include Nutrient agar for bacteria, Sabouraud's agar medium or Czapek's Dox agar for fungi, Glycerol yeast agar for actinomycetes and Wort agar for yeasts. Generally, the dilutions 10^{-2} to 10^{-5} are selected for enumeration of fungi, 10^{-3} to 10^{-6} for actinomycetes and 10^{-4} to 10^{-7} for bacteria relative to their proportion in soil. Upon solidification, the plates are incubated, in an inverted position for 3–7 days at desired temperature (28–37 °C). The number of colonies appearing on dilution plates are counted, averaged and multiplied by the dilution factor to find the number of cells/spores or cfu (colony forming unit) per gram or millilitre of the sample. Normal saline (0.85% sodium chloride) or

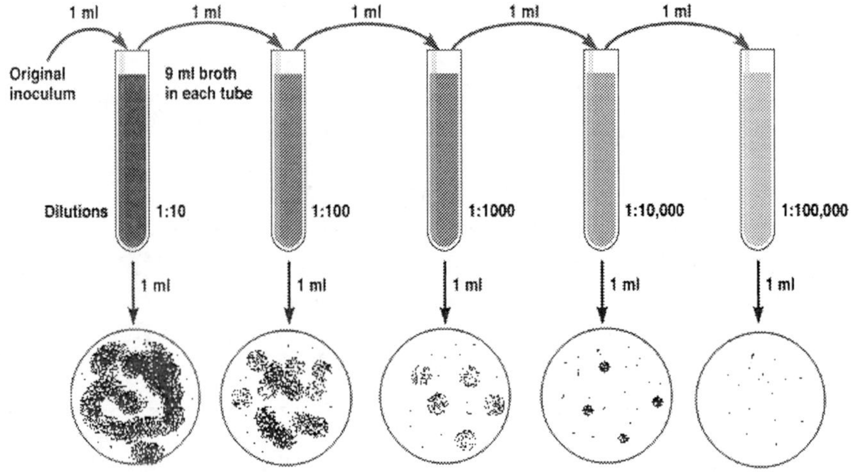

Figure 7.1: Serial dilution method for the isolation of microorganisms from soil.

buffers can also be used as diluents, in place of sterile water, for the preparation of suspension of soil sample. The total number of cells in original sample can be calculated by the given formula

$$\text{No. of cells/g} = \frac{\text{Number (average of 3 replicates) of colonies} \times \text{Dilution factor}}{\text{Dry wt. of soil}}$$

Isolation by Warcup Method

Direct soil plate or Warcup soil plate method was described by Warcup in 1950 and is superior to dilution plate method because of ease and need of less glassware. In this method about 0.005 to 0.15 g air dried soil is added in sterile Petri plates with help of a sterilized cooled loop or transfer needle, and 25 ml of melted, cooled (45°C) medium is poured. Optionally, one can supplement media with some antibiotic for selective isolation of fungi. Media is allowed to solidify and incubated at 28°C in inverted position and observed for appearance of colonies.

Plating Method

The use of media solidified with agar or some other appropriate jelling agent has been made successfully for many years in obtaining pure cultures of many microorganisms. The procedure is very helpful for bacteria, fungi, which produce a discrete colony on solidified media, and thus, allow us to pick up pure cultures

Streak Plate

A streak plate is prepared by dipping a loop or bent needle into a medium containing a suspension of microbial cells. The loop or needle is then used to make a series of parallel, non-overlapping streaks transversely on the surface of the medium. The number of cells is diluted with every successive streak to such a point that the final streaks will usually yield separate and distinct colonies (**Figure 7.2**).

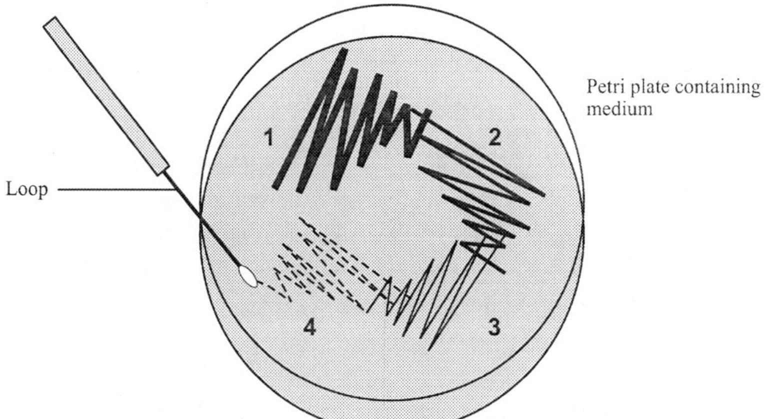

Figure 7.2: Streak plate method.

Pour Plate

This method consists of adding a portion of the medium containing the desired cells into molten agar medium at 45°C. If the cell concentration is too high, a portion of the seeded agar medium may be serially diluted with additional medium and finally poured into sterile Petri plates. This results in the formation of a solidified medium with colonies scattered throughout the medium allowing a fairly easy physical separation of microbial colonies.

Isolation of Algae

Algae occur in great abundance in water bodies including ocean, fresh water lakes, ponds, etc. Many are found growing in damp soil, rock, tree barks and other damp places. They range in size from microscopic unicellular forms smaller than bacteria to sea weeds that may reach many feet in length. The techniques used for the isolation of bacteria and fungi are applicable to microscopic algae too. Beneck's broth is commonly used for their isolation. In this method, sterilized Beneck's broth is filled in half-point bottle to a depth of 5 cm and 1gram soil is added. The bottle is incubated at 30–35 °C in an illuminated growth chamber for 15–20 days. After incubation algal growth is observed as green flocks and pellets.

Bait Technique

Soil has a variety of microorganisms with varying metabolic capabilities. Some microorganisms display a greater degree of inclination towards utilization of some specific substrates. Perhaps these microorganisms are equipped with battery of enzymes for degradation and assimilation of products from such substrates. Based on their preference towards substrates, microorganisms may be classified as cellulolytic, keratinolytic, amylolytic, proteolytic, etc. Keeping their dietary preference in mind, such microorganims can be isolated easily using bait technique. Such substrates are kept in proximity or contact of moist soil samples to allow specific organisms or group of organisms to attack the substrate (**Figure 7.3**). The technique is especially useful for the selective isolation of microorganisms generally present in soil in very small numbers.

A good example can be use of nails, feathers or hair for isolation of soil inhabiting keratinophilic moulds. The sterile Petri dishes are half filled with soil samples and then moistened with water. Pre-sterilized keratinic baits are pressed or buried in the soil samples. These plates are incubated at appropriate temperature and are examined for fungal growth after 4–5 days. The bait technique is also used for isolation of cellulolytic fungi and zoosporic fungi from water samples. A variety of baits such as paper, cotton, boiled seeds, leaves, stem pieces, insect wings and dead insects are used in this technique.

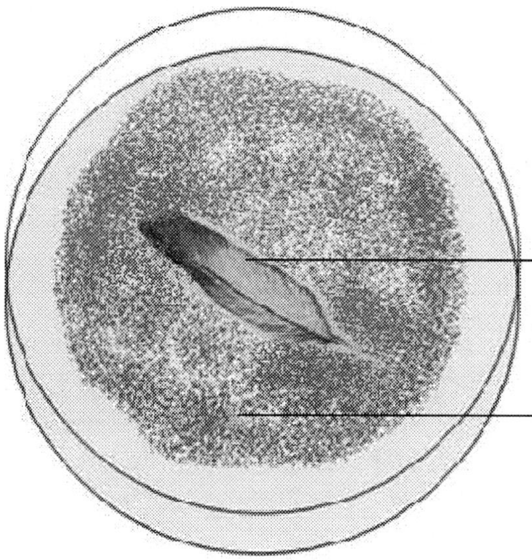

Figure 7.3: Bait technique for isolation of keratinophilic microorganism from soil.

PRESERVATION

Air, soil and water bodies are the reservoirs of microorganisms. Each of these reservoirs is dynamic and so the flora and fauna of these also keeps changing as per season, ecological succession and a variety of interactions among living forms. Thus, the microbial population of these niches also keeps on changing as per natural influences and interrelationships. Secondly, the microbial species, as any other organism, spontaneously mutate at a very slow rate and there may be very subtle but significant difference among members of the same species. Such differences may be exploited for the cause of scientific experimentation or for commercial purposes in industries. In nutshell, each microorganism is too unique to lose, as one may not be able to re-isolate or revive it again from the same sample. Looking at this, one can realize that it is very essential to preserve the microorganisms for their current and future use.

Maintenance and Preservation

Once a microorganism is isolated in pure form, it is subcultured on plate or slants at regular intervals to maintain its viability. The interval between subculturing may vary between four weeks or months depending on the storage condition and on the growth rate of microorganisms. The organism can be subcultured on media specially designed to slow down growth rates and expand the culture's life. It is a good practice to maintain two slants, one as the working culture to be used as a source for routine laboratory work and the other as stock culture from which new working cultures can be prepared whenever needed. Such a system limits the chances of contaminating the stock culture. Maintenance of stock cultures requires meticulous arrangement. To preserve microorganisms, it is necessary to reduce their metabolic rate to a minimum possible. As a result, the processes that lead to ageing and death are slowed down so that microorganism can be maintained in its inactive or inert state for longer periods. There are several methods available for maintenance of pure cultures; the choice of the method to be used depends upon the purpose, desired time of preservation, size of culture collection, nature of microorganisms and available infrastructure. Some of commonly used methods of preservation of microbial cultures are described in the following text.

Storage on agar slant

Perpetual transfer is a suitable method for maintenance of small culture collection with limited resource. It requires the regular preparation of agar slant tubes, transfer of strains to fresh slants, inspection of the tubes to assure growth, followed by storage at 4 °C for several months between transfers. This method requires little equipment including a refrigerator, incubator and routine instrumentation for preparing media and transferring cultures. It is, however, time consuming and labor intensive and thus suitable only for limited number of species. The demerits of perpetual transfer are the possibility of contamination of cultures by other microorganisms (bacteria, fungi) or mites, etc. Also, strains that are metabolically active are susceptible to mutation and loss of particular traits when subjected to regular transfer.

Storage under mineral oil or water

Moderate or intermediate length storage of cultures can be managed by growing the fungus and bacteria on an agar slant and then covering it with sterile water or mineral oil. Covering the cultures with sterile oil and water slows the rate of metabolism by significantly reducing available oxygen and at the same time by preventing the dehydration of the agar and microorganism. The strains are recovered by scrapping the agar slant surface with a wire loop and streaking the material onto suitable agar media. The demerit of this method is loss of viability, and enhanced chances of contamination with other fungi and bacteria due to presence of liquids.

Storage by lyophilization or freeze-drying

Lyophilization, or freeze-drying, is one of the most widely applied methods of storage of microbial cultures. In this method, bacteria and fungus are suspended in a defatted powdered milk (skimmed milk) solution or in sterile bovine serum, placed in small sterile ampoule. These are frozen rapidly in a dry-ice bath, and then water is dried under vacuum.

The small pellet formed is sealed hermetically inside ampoule, under vacuum, by heating the ampoule neck and is stored at low temperature. These lyophilized

Harvest culture

↓

Suspend culture in sterile bovine serum or
Skimmed milk power solution

↓

Add aliquots of 50-100 ml in prelabelled sterile ampoule

↓

Attach ampoule to manifold and freeze in
50% ethylene glycol-13% ethanol-dry ice
bath precooled to -40 to -50°C

↓

Apply vacuum

↓

Raise manifold when bath reaches 0°C (2-3 h)
to remove ampoule from ice bath

↓

Monitor drying with vacuum gauge

↓

Use bidirectional oxygen-natural gas torch to
Seal and cut off ampoules

Figure 7.4: Preservation of microbial cultures by lyophilization or freeze-drying.

cultures can be stored over years at room temperature. Cultures are revived by aseptically breaking open the ampoule, dispensing the pellet in sterile broth or distilled water, and then streaking the suspension onto suitable agar media. Lyophilized or freeze dried cultures have good viability after 40 or more years, if properly dried. There is no risk of mutation, or contamination in these metabolically inactive preparations. However, not all microorganisms can be stored in this manner. In general, the fungus must sporulate and form spores for successful preservation by lyophilization.

L-drying

A similar technique is L-drying in which the culture preparation in an ampoule is treated as described in the preceding section, but culture is not frozen prior to drying and the suspension is dried as such. The ampoule neck is heated, drawn out and a tight fitting cotton plug is inserted. Vacuum is applied and the culture is dried. The cotton plug and drawn-out neck reduces the evaporation rate, preventing the suspension from boiling under reduced pressure. This process requires a good quality vacuum pump, vacuum manifolds and an oxygen gas sealing torch for sealing ampoules under vacuum. The ampoules are stored at 4°C for increased longevity.

Storage in Silica Gel or Soil

An easier method of dehydrating spores and mycelium is through the use of anhydrous silica gel. In this method, fungal material is suspended in a defatted powdered milk solution, and put onto pre-sterilized anhydrous silica gel particles in screw-cap tubes. The caps of the tubes are sealed and stored at 4 °C. When an active culture of the strain is needed, a few crystals of the silica gel are taken out of the tube and inoculated on a suitable agar plate. The silica gel tube can be resealed and used many times without the loss of viability. Another similar method of preservation of microorganisms uses sterile soil in place of silica gel. A suspension of microbial culture is pipetted onto the sterile soil and sealed and stored. The cultures are recovered by putting soil particles on fresh agar plate and allowing the microorganism to grow.

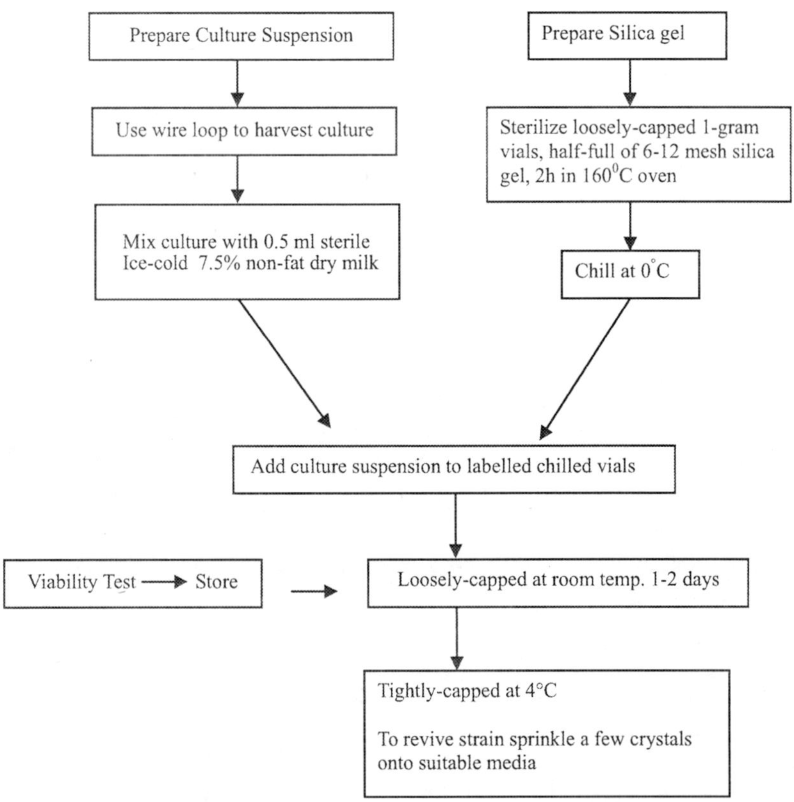

Figure 7.5: Preservation of microbial cultures by storage in silica gel.

Deep Freezing

Another appropriate method of preserving cultures is by freezing them either in a mechanical freezer at –10°C to –20°C or in a liquid nitrogen freezer at –176°C to –196°C. Some species can be maintained in mechanical freezer at –10°C to –20°C after growth on an agar slant. The strain is recovered by thawing the slant and transferring the culture to fresh agar. This method has the apparent advantage of reduced handling and is effective over period of several years. It is a good option for short term and cost-effective preservation of microbial cultures.

Prepare homogenous suspension in 10% glycerol

↓

Immediately dispense aliquot into pre-sterilized, labelled, screw-cap cryovials

↓

Allow cells to equilibrate in cryoprotectant for 15-30 minutes at room temperature

↓

Place vials in -55 to -80°C freezer

↓

Transfer vials to liquid nitrogen freezer

↓

To revive strain, remove cryovial from liquid Nitrogen freezer and rewarm with constant agitation in 37°C water bath until contents completely thawed

↓

Transfer part or all of cryovial contents to suitable agar or broth medium

↓

If a cryovial is returned to liquid nitrogen freezer, passively cool as above and record this event since viability may be affected

Figure 7.6: Cryopreservation for preservation of microorganisms at very low temperatures.

Cryopreservation

Cultures can be preserved at very low temperature in liquid nitrogen. Cultures are suspended in 10% glycerol or dimethyl sulphoxide, cooled gradually to –30 °C (1°C per minute), and then placed in the liquid freezer. Cultures are revived by rapid thawing in a 37°C water bath and plating of the culture on suitable media.

Tubes wrapped up in liquid nitrogen must be sealed completely and must be tested for leakage prior to immersion in liquid nitrogen. Tubes with any leaks will partially get filled with liquid nitrogen and can boil explosively when rapidly thawed. This method allows the use of convenient screw-cap plastic tubes rather than heat–sealed glass ampoule. The temperature in the vapor phase of the freezer is about –176°C as compared to –196 °C in the liquid phase. The costs of the freezers and supply of liquid nitrogen makes this method somewhat more expensive than other methods.

SUGGESTED READING

Aneja, K.R.. 1996. *Experiments in microbiology, plant pathology, tissue culture and mushroom cultivation.* (II Ed.), New Age International Pvt. Ltd., New Delhi.

Booth, C. 1971. *Methods in Microbiology*, Vol. 4 Academic Press London.

Cappuccino, J. G. and Sherman, N. 2002. *Techniques for isolation of pure culture In. Microbiology: a laboratory manual.* (VI Ed.) Pearson Education, Inc. Singapore.

Collee, J.G., Duguid, J.P., Frasee, A.G. and Marmion, B.P. 1989. *Practical medical microbiology*, 13th ed. Longman, U.K.

Collins, C.H. and Lyne, R.M. 1976. *Microbiological methods.* Butterworth, Mass, U.S.A.

MTCC 1988. *The Laboratory Manual : Actinomycetes-isolation, screening, identification and gene cloning in Streptomyces.* Institute of Microbial Technology, Chandigarh.

Vanbreuseghem, R. 1952. *Etiological technique for isolation of dermatophytes from soil.* Annals. De. La. Belge. De. Medicine. Tropical., 32: 173–178.

Waksman. S.A. 1927. *Principles of soil microbiology.* Baltimore: Williams and Wilkins Co.

Warcup, J. H. 1950. *The soil plate method for isolation of fungi from soil.* Nature London. 166: 117–118.

REVIEW QUESTIONS

1. What do you understand by isolation of microorganisms? Discuss the general methods of isolation of microorganisms from environmental samples.
2. What are the aims and objectives of different methods of isolation of microorganisms?
3. Write notes on the following:
   a) Serial dilution method b) Warcup plate method
   c) Pour plate method d) Bait technique
4. Discuss various methods of preservation and maintenance of microorganisms.
5. Why it is important to preserve and maintain pure cultures of microorganisms?

UNIT VIII

Sterilization and Control of Microorganisms

Sterilization is the complete removal of all life forms from a given area. From microbiological point of reference, it refers to any process that kills or eliminates transmissible agents including fungi, bacteria, viruses and microbial spores from any surface, area, equipment, solution, culture medium, etc. Sterilization, in strict microbiological sense, is the killing of all forms of microbial life, including endospores, which are the most resistant forms. Technically, sterilization refers to various processes that effectively kill or eliminate transmissible agents such as fungi, bacteria, viruses, prions and spore forms, etc. from a surface, equipment, food, medication, or biological culture medium.

Sterilization can be achieved by different means that may be referred as drastic and not so drastic or moderate. The choice of method depends upon the object of sterilization. Improper selection of sterilization method may lead to undesirable results and can alter the properties of object being treated undesirably rendering it useless. In microbiology, sterilization is done routinely to remove contamination (undesirable organisms) of culture media, Petri plates, fermenter vessel, antibiotics so as to allow the desired organism to grow under aseptic conditions. It is important to ensure the purity of microbial cultures for the very purpose of studying their properties, growth, etc. Similarly, asepsis is to be maintained throughout the run of fermenter to avoid loss due to growth of undesirable microbes during a bioprocess.

On the other hand, microbial control and eradication is also important in medical practice where use of contaminated surgical instruments, needles, bandages, aprons, etc. can be deleterious and may result in infection. Microbes bring about severe losses in food quality during food storage and transport. The consumption of contaminated food may lead to food poisoning. Milk is one of the complete foods that is easily utilized by several bacteria for their growth. All these considerations necessitate the need of a variety of sterilization methods suiting to particular objectives of sterilization. There are some milder methods (e.g. Pasteurization) that allow selective removal of harmful microorganisms without damaging the object. **Sterilizing agents or microbial control agents** inhibit or kill microorganisms by either of the two modes of action:

(i) Denaturation of protein (enzymes) and nucleic acid

Physical agents (mainly heat), and certain chemicals break the intra-molecular bonds of protein molecule (H-bonds, S-S bonds) that determine structure of a functional protein resulting in its denaturation. Heat also denatures the helical structure of DNA and RNA conformations while chemicals agents such as intercalating dye (ethidium bromide) disturb the routine transcription and replication of DNA. Such damage to nucleic acid leads to halt of cellular machinery and biosynthesis of necessary macromolecules and thus result in death of microorganisms.

(ii) Alteration of membrane permeability

Plasma membrane is made up of proteins and lipid bilayer. Integrity of plasma membrane is of utmost importance in cell organization and regulation of transport between cell and surroundings by virtue of its selective permeability. Some agents damage the lipid and protein content of cell membrane thus altering permeability of membrane resulting in total chaos in transport leading to death of the cell.

All the methods of sterilization can be put under two distinct types:

1. Physical Methods
2. Chemical Methods

PHYSICAL METHODS OF STERILIZATION

From time immemorial, humans were practicing some physical methods of microbial control for preservation of food. Desiccation and osmotic pressure were probably among earliest techniques. There are many types of physical methods used for sterilization.

Heat

Heat appears to kill microorganism by denaturing their enzymes. Heat resistance varies among different microbes and the difference can be construed through the idea of thermal death point. Thermal death point is the lowest temperature at which all the microorganisms in a liquid suspension will be killed in 10 minutes. The heat used in sterilization can be applied in the form of moist heat or dry heat.

Moist Heat

Moist heat destroys microorganisms primarily by the coagulation of proteins, which is caused by the breakage of the hydrogen bonds that hold the proteins in their three-dimensional structure. Protein coagulation or denaturation occur more quickly in the presence of water. One type of moist heat sterilization is boiling (100°C).

Boiling in water for 15 minutes will kill most vegetative forms of bacteria and viruses, but boiling is ineffective against prions, some viruses and many bacterial and fungal spores; therefore, boiling is not reliable for sterilization. However, since boiling kills most vegetative microbes and viruses, it is useful for reducing viable levels of contaminants. Boiling is a simple process, and is an option available to almost anyone anywhere, requiring only water, enough heat, and a container that can withstand the heat. Boiling is easy but may be hazardous to microbiological media and result in loss of water.

Higher temperatures can be achieved by steam under pressure in an **autoclave**. **Autoclaves** commonly use steam heated to 121°C (250°F), at 103 kPa (15 psi) above atmospheric pressure (**Figure 8.1**). Solid surfaces are effectively sterilized when autoclaved for at least 15 minutes. However, liquids and instruments packed in sheath of paper require a longer time to achieve effective sterilization. After, autoclaving liquids must be cooled slowly to avoid boiling over when the pressure is released. Under these conditions, steam at a pressure of about 15 psi (121°C) will kill all microorganisms including their heat resistant endospores in about 15 minutes. Autoclaving is routinely used to sterilize culture media, instruments, intravenous equipment, dressing, syringes, transfusing equipment and numerous other items that can bear high temperature and high pressure. Direct contact of steam with the object to be sterilized should be facilitated and wrapping materials impervious to steam such as aluminium foil should be avoided.

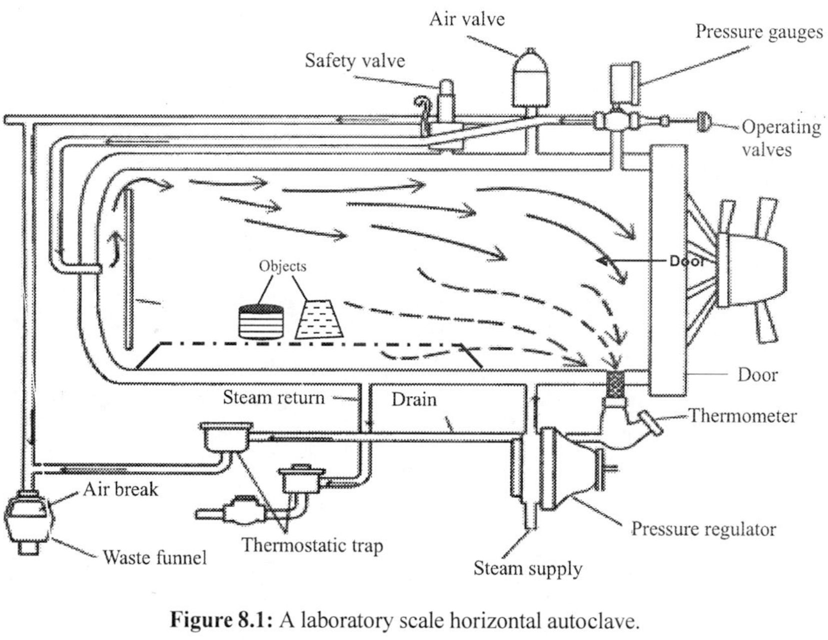

Figure 8.1: A laboratory scale horizontal autoclave.

Death Rate at High Temperatures

Microbial population subjected to high temperatures in oven (dry heat) or autoclave (moist heat) does not get killed at once but follow a definite pattern of first order kinetics. Thus to understand sterilization at high temperatures it is important to express microbial death quantitatively. One such quantitative expression is decimal reduction time or D-value. The D-value represents heat sensitivity of a microorganism and is defined as the time taken to decrease the microbial population by 10 fold (or one decimal) at a given temperature. D-value can be confirmed experimentally by incubating a known number of microbial cells at a lethal temperature and determining the viable population after periodic time intervals using viable plate count. The semi logarithmic plot between log values of microbial populations and time interval gives a linear relationship between the two and the negative reciprocal of the slope of the straight line gives the decimal reduction time. Once the D-value for a microbe in a specified medium is known it is possible to determine the total time required for its complete elimination for a given population of cells.

It is obvious that D-value of any microbial population holds only for a specific medium and specified temperature. Thus, a generalized concept of **thermal death time** (TDT) and the *Z* value was developed to calculate *D*- values at any lethal temperature. The TDT is defined as the time taken for a population of cells to reach 10^0 (considered as effectively zero) at a given temperature. This can be determined by multiplying the $\log_{10}$ of an initial cell population by the *D* value or by determining the X intercept from a *D*-value experiment.

The *Z* value measures the sensitivity of a microbe to a change in temperature and is defined as the amount of temperature increase necessary to decrease the TDT by ten-fold. To determine the *Z* value for a microbe, the TDT is plotted (on a log scale) against its temperature and a line is drawn through the points. The *Z* value is the negative reciprocal of the slope of the line. With a *Z* value in hand, it is possible to predict the microbes death rate in the specific medium at any lethal temperature.

Knowledge of D value, TDT and *Z* is important in food and pharmaceutical industries, where one wants the treatment to appropriately reduce the microbial population with minimum negative effect on the pharmaceutical and food preparations. For finding out D-value, a culture is incubated at a specific temperature and aliquots are removed at various time intervals. The population in each sample is determined by viable cell count. The data is plotted with the population on the y-axis and time on the x-axis. The D value is the negative inverse of the slope (**Figure 8.2**).

Pasteurization

Louis Pasteur used milk heating, which was sufficient to kill the microorganisms that caused the particular spoilage problem without seriously damaging the taste

Table 8.1 Relationship between the pressure and temperature of steam under pressure.

Pressure* (psi in excess of atm pressure)	Temperature (°C)
0 psi	100
5 psi	110
10 psi	116
15 psi	121
20 psi	126
30 psi	135

*(psi – pound per square inch or lb/in^2)

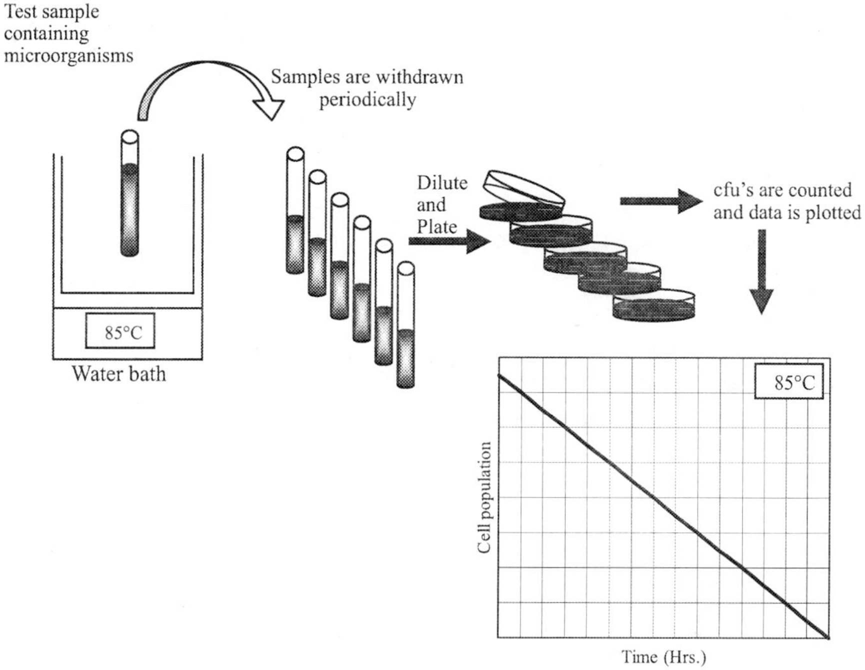

Figure 8.2: Effect of high temperature on microorganisms and calculation of thermal death time for a population of microorganisms. The D-value (decimal reduction time) is the negative inverse of the slope of the straight line.

of the product (**Pasteurization**). The same principle was later applied to production of pasteurized milk. Many heat resistant bacteria survive pasteurization, but these are unlikely to cause disease.

In the classic Pasteurization treatment of milk, the milk was exposed to 63°C for 30 minutes, but today most milk pasteurization use high temperature and this process is called **high–temperature short–time (HTST) pasteurization.** HTST pasteurization lowers total bacterial counts, so the milk keeps well under refrigeration and the shelf life is increased. Pasteurization is also used for preservation of wine and alcohols. Milk can also be sterilized by **ultra-high temperature (UHT) treatment** so that it can be stored without refrigeration.

Tyndallization

This process was devised by **John Tyndall.** It is a lengthy process designed to reduce the level of activity of sporulating bacteria that are left by a simple boiling water method. The process involves boiling for a period (typically 20 minutes) at atmospheric pressure, cooling, incubating for a day, boiling, cooling, incubating for a day, and repeating these steps for next two days and finally boiling it. The three incubation periods are to allow heat-resistant spores surviving the previous boiling period to germinate to form the heat-sensitive vegetative (growing) stage, which can be killed by the next boiling step. This is effective because many spores are stimulated to grow by the heat shock. The limitation of this procedure is that it only works for media that can support bacterial growth – it will not sterilize plain water and also it is ineffective against prions.

Dry Heat

Dry heat destroys microorganisms by oxidation effects. One of the simple methods of dry heat sterilization is direct flaming or incineration**. Flaming** is done to loops and straight-wires in microbiology laboratories by putting the loop in the flame of a Bunsen burner or alcohol lamp until it glows red to ensure that contaminants are killed. Flaming is used only for small metal or glass apparatus. It is possible that, during the initial flaming infectious material may spurt from the wire and contaminate nearby surfaces and objects. Therefore, special heaters have been developed that surround the inoculating loop with a heated surround, ensuring that such sprayed material does not contaminate the region. A similar principle is used in incineration that burns down any organism or organic matter to ash. It is used as a biosafety measure for disposal of biohazardous waste before it is discarded or released into environment. Routine glasswares in microbiology laboratories are sterilized in **Hot-air oven**. Generally, a temperature of about 170°C for about 2 hours ensures sterilization. Alternatively, glassware can be sterilized at 80°C for overnight.

Filtration

Filtration is the passage of a liquid or gas though screens like material with pores small enough to retain microorganisms. Filtration is used to sterilize heat-sensitive material such as some culture media, enzymes, vaccines and antibiotic solutions. A vacuum that is created in the receiving flask helps gravity pull the liquid through the filter. Some operating theaters and rooms occupied by burn patients receive filtered air to lower numbers of air borne pathogen. **High-efficiency Particulate Air (HEPA) filter** removes all microorganisms lager than 0.3 μm in diameter. HEPA filter is also used in laminar airflow for supply of germ-free air (**Figure 8.4**). Membrane filters made of cellulose ester or plastic polymers are used in

Table 8.2 Common methods of sterilization

Method	Temperature	Remarks
Incineration	>500°C	Burns organic material on non-flammable surfaces.
Boiling	100°C	Thirty minutes of boiling kills vegetative forms of bacteria but may not kill bacterial endospores. There are also toxins that are not inactivated at 100°C.
Intermittent boiling	100°C	Three 30-minute intervals of boiling followed by periods of cooling kills bacterial endospores.
Autoclave		
(steam under pressure)	121°C for 15 minutes at 15 psi	Kills all forms of life including bacterial endospores. The substance being sterilized must be maintained at the effective temperature for the entire time.
Dry heat (hot air oven)	160°C for 2 hours	Used for materials that must remain dry. Good for glassware, metal, but not suitable for most plastic or rubber items.
	170°C for 1 hour	Same as above. Increasing the temperature by 10°C shortens the sterilizing time by 50 %.
Pasteurization (batch method)	63–66°C for 30 minutes	Kills most vegetative bacterial cells, including pathogens such as *Streptococci*, *Staphylococci* and *Mycobacterium tuberculosis*.
Pasteurization (flash method)	72°C for 15 seconds	Effect on bacterial cells is similar to batch method. For milk, this method has fewer undesirable effects on quality or taste.

industry and laboratory. The pores of membrane filter include 0.22 μm and 0.45 μm, intended for filtering out bacteria, but some flexible bacteria like spirochetes or cell-wall lacking mycoplasma (PPLO) will sometimes pass through such filters. Filters are available with pores as small as 0.01 μm that retains virus and some protein molecules.

Low Temperature

Low temperature retards or slows down activity of microorganisms and hence growth is severely retarded. At temperature of ordinary refrigerators (0–7°C), the metabolic rates of most microorganism are reduced so, they cannot synthesize toxins. Ordinary refrigerators have bacteriostatic effect, however; psychrotrophs grow slowly in refrigerators, thus altering the taste of food. Pathogenic bacteria have optimal temperature range between 30–38°C and hence cannot multiply in refrigerator. Slow freezing can be lethal for bacteria, and the ice crystals formed inside the cell grow and disturb their cellular machinery and molecular structures.

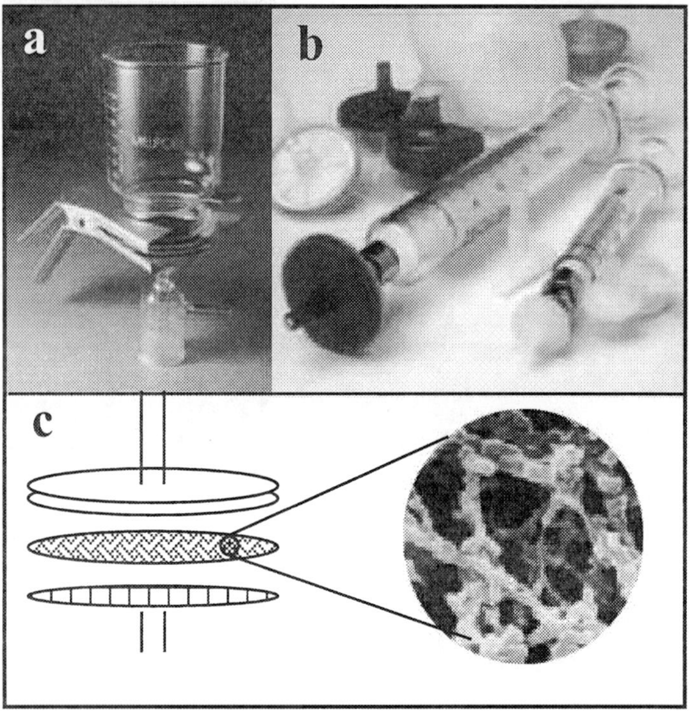

Figure 8.3: Filtration devices (a) Membrane filter (b) Syringe filter (c) Set up of a syringe filter and cross-section showing pores of a membrane filter.

Osmotic Pressure

Osmosis can be defined as movement of water through semi-permeable membrane from its high concentration to its lower concentration. Like most living organisms, microorganisms also contain water more than 70% of their weight. Another important concept in this regard is that of water activity level (a_w). The a_w of any medium is defined as amount of available water and it is different from the total 'water content'. For example, if we add salt, sugar or glycerol to any microbiological media, it's a_w is reduced while water content remains same. Fungi have better ability to survive at low water availability, say up to 0.85 a_w while bacteria require higher a_w level, i.e. more than 0.95. This is the reason why it's easy to limit bacteria by adding salts and sugar but fungus can still colonize on such foods. High concentrations of salts and sugars create a hypertonic environment that causes water to leave the microbial cell. As water leaves the cell, plasma membrane shrinks away from the cell wall (plasmolysis) and microbial cells die. This knowledge is used for preservation of food including mainly meats, fish, fruits and vegetables.

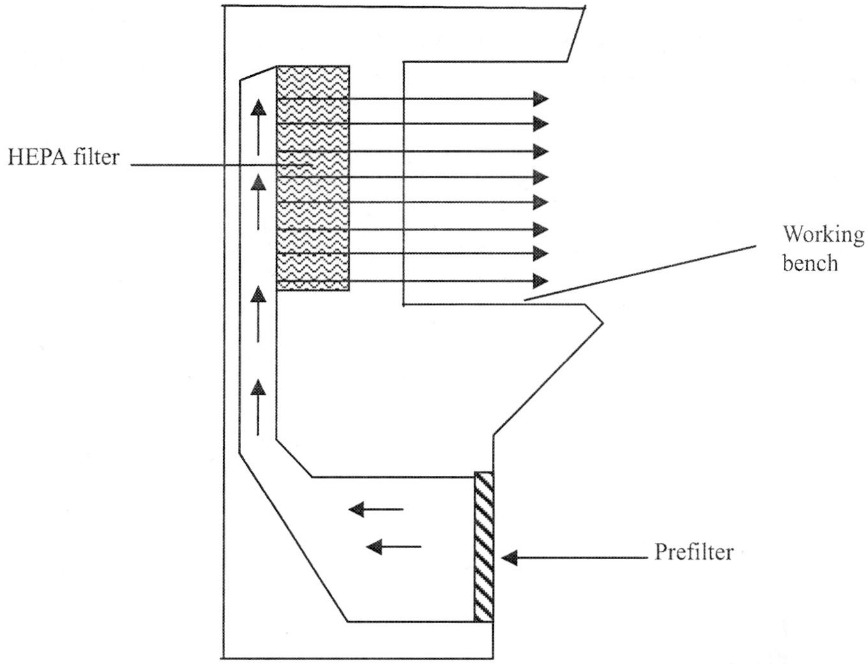

Figure 8.4: Working of a horizontal laminar flow.

Radiation

Radiations produce many effects on microbial cells, depending on its intensity, wavelength and time of duration. There are mainly two types of radiations that are employed in sterilization, namely, ionizing radiation and non-ionizing radiation.

Ionizing Radiation

X-rays, gamma rays and high-energy electron beams are included in ionizing radiation. These rays have small wavelengths (less than 1 nm) and high energy, and thus are more penetrating. Gamma rays are generated by radioactive cobalt and electron beams are generated by accelerating electrons to high energy in special machine. X-rays are produced by machine similar to electron beams. Gamma rays penetrate deeply but may require more time to sterilize large masses. High-energy electron beams have more penetrating power and require few seconds of exposure. The ionizing radiations are lethal because these ionize water to form highly reactive hydroxyl radicals (OH^-). These radicals react with DNA of the cell and alter it undesirably leading to death of the cell. These radiations can directly hit DNA to cause non-lethal mutations, sufficient enough to kill the microorganism. Sterilization by ionizing radiation is used in food industry for food preservation, and sterilization of syringes, surgical gloves and catheters in medical field.

Non-ionizing Radiation

Non-ionizing radiations have longer wavelength and less energy than ionizing radiations. UV light is most commonly used for sterilization and killing microorganisms. UV light damages the DNA of exposed cells by forming bonds between adjacent thymines (thymine dimer) in DNA. These thymine dimers inhibit correct replication of DNA during reproduction of the cell. The UV light of wavelength 260 nm is more effective in killing the microorganisms. UV radiation is used for sterilization of inoculation rooms, operating rooms, and laminar air flow. UV radiation has less penetration power and hence microorganism to be killed must be directly exposed to the rays. UV radiation can damage the eyes and may cause skin cancer in human and hence should be used cautiously.

CHEMICAL METHODS OF STERILIZATION

Many chemicals are known to control the growth of microorganisms on both living beings and non living objects. Most chemical agents reduce microorganism population to safer level or kill vegetative forms of pathogen. Over the time some chemicals have found a prominent place in sterilization or disinfections. Some of the important chemical agents are described below.

Phenol and Phenolics

Phenol is one of the first chemical agents used by Lister for disinfection in operation theatre. At concentration above 1% phenol has sufficient antibacterial effect. Its derivatives called Phenolics, contain a molecule of phenol that has been chemically changed to make it less irritating and also increase its antibacterial activity. Mode of action of phenolics proceeds through injuring plasma membrane, denaturing protein and inactivating enzymes. Phenolics are used in soap and detergents and antibacterial agents.

One of the most useful phenolic is bisphenol which is used in soap and lotions used for surgical scrubs. It shows bacteriostatic effect towards Gram-positive bacteria.

Phenol is also used as the standard to measure the efficacy of other disinfectants of similar chemical structure or compounds with antimicrobial activity. The procedure used is the **phenol-coefficient technique** and is based on the procedure outlined in AOAC phenol-coefficient method. This method is used to test disinfectants soluble in water and having similar mode of action as phenol. Usually, strains of *Salmonella typhi* or *Staphylococcus aureus* pathogenic bacteria are used as assay organisms. In this method, a series of dilutions of the test disinfectant (5 ml per tube) is prepared and 0.5 ml of 24 hours old broth culture is added to the tubes. Simultaneously, similar assay is also performed with phenol. All tubes (disinfectant + organism and phenol + organism) are incubated at 20°C. At time intervals of 5, 10, and 15 minutes, subcultures are made with a loop into sterile medium. The inoculated subculture tubes are incubated and subsequently examined for growth. The maximum dilution of the disinfectant killing the test organism in 10 minutes but not in 5 minutes is divided by the greatest dilution of phenol showing the same result. This fraction obtained out of the division is called as the phenol coefficient of the substance tested.

Halogens

Halogens mainly, iodine and chlorine, are effective antibacterial agents. **Iodine (I_2)** is one the oldest known and most effective antiseptics, effective on all kinds of bacteria, many endospores, many fungi and some viruses. Iodine acts by combining with the tyrosine, a common amino acid of many enzymes and cellular proteins, resulting in loss of protein function. Iodine also oxidizes the sulphahydryl (–SH) groups of sulphur containing amino acids and distorts the structure of protein. Widely used skin antiseptic, tincture of Iodine is solution of iodine in alcohol. Organo-iodo compounds or iodophores also exhibit antimicrobial activity.

Chlorine (Cl_2), as a gas or in combination with other chemicals is used as a sterilization agent and disinfectant. Germicidal action is caused due to hypochlorous acid (HOCl) formed by following reaction with water:

$$Cl_2 + H_2O \rightarrow H^+ + Cl^- + HOCl$$

$$HOCl \rightarrow H^+ + OCl^-$$

Hypochlorous acid is a strong neutral oxidizing agent that prevents the cellular enzyme system from functioning. In India, chlorine gas is widely used for disinfecting municipal water.

Alcohols

Alcohols are much effective in killing bacteria and fungi but not endospores and non-enveloped viruses. The mode of action of alcohol is protein denaturation. Alcohol can disrupt the cell membrane and dissolve many lipids. Ethanol and isopropanol are two widely used alcohols as sterilizing agents. 60% and 90% alcohol is more effective than pure ethanol as protein denaturation requires water.

Heavy Metals and Their Compounds

Many heavy metals such as silver, mercury and copper are known as germicidal and antiseptic. Their antimicrobial activity is also referred to as **oligodynamic action.** This action can be demonstrated by putting a silver or copper coin in an inoculated Petri plate. Very little amount of metal diffuses from coin and inhibits the growth of bacteria forming a zone of inhibition around the coin. 1% silver nitrate solution is used in disinfection of eyes. Compounds of mercury, e.g. mercuric chloride are used as disinfectants due to their broad spectrum bacteriostatic activity. Copper in the form of copper sulphate is a common agent used to kill green algae in water bodies. Copper sulphate is also added to lime in our country in white washing of houses. A mixture of copper sulphate and hydrated lime (calcium hydroxide) is also used as fungicide (Bordeaux's mixture). Heavy metals are also included in antibacterial paints and emulsions for similar reasons.

Quaternary Ammonium Compounds (Quats)

Quaternary ammonium compounds are strongly bactericidal (Gram-positive bacteria), virucidal, fungicidal and amoebacidal. Thes are cations with general structure NR_4^+ with R being alkyl group. They alter the permeability of cell membrane and cause the loss of essential cytoplasmic compounds such as potassium and disturb the polarity. Due to these properties they are widely used surface acting antimicrobial agents.

Aldehydes

Formaldehyde and Glutaraldehyde are the examples of very effective antimicrobials. They denature protein by forming covalent cross–links with several organic functional group on protein molecule including ($-NH_2$, –OH, –COOH and –SH). Formaldehyde gas is used as disinfectant for inoculation rooms and operation theatres. Glutaraldehyde is less irritating and more effective than formaldehyde. Glutaraldehyde is used to disinfect hospital instruments and respiratory therapy equipment. A 2% solution of glutaraldehyde is bactericidal, tuberculocidal and produces virucidal effect in 10 minutes.

Gas

Gases by virtue of their property of quick diffusion have a better reach as compared to liquids. This is why several gases such as ethylene oxide, propylene oxide, α-propiolactone and formaldehyde are used for sterilizing large rooms, operation theatres, and inoculation chambers by filling the fumes of these gases in these spaces for hours. Ethylene oxide denatures microbial proteins by replacing their hydrogen in $-NH_2$, –COOH, –SH and –OH group by alkine group ($-CH_2-CH_2-OH$). Ethylene oxide (C_2H_4O), because of its high penetrating power, is used to sterilize large packets, bundles of clothes and plastics.

Oxides and Peroxides

These compounds affect microorganisms by oxidizing their cellular compounds. **Ozone** is a common sterilizing agent used to sterilize water and air in industries. However, it is a toxic and unstable gas and needs to be produced on-site. **Hydrogen peroxide (H_2O_2)** is another commonly used chemical sterilizing agent. It is relatively non-toxic once diluted to low concentrations (although a dangerous oxidizer at high concentrations) and leaves no residue. Peracetic acid is used as liquid chemical sporicide.

Antibiotics

Antibiotics are substances produced by one microorganism which inhibit or kill another microorganisms. There are many antibiotics which follow different modes of action to inhibit or kill microbes by interrupting cell wall synthesis, inhibition of protein synthesis, injury of plasma membrane, inhibition of nucleic acid synthesis and inhibition of the synthesis of essential metabolites. Penicillin, streptomycin, cephalosporins, monobactams, vancomycin, etc. are used as antibacterial antibiotics. Polyenes, imidazoles and griseofulvin are used as antifungal antibiotics. Antibiotics are used to avoid or control the growth of sensitive microorganisms at various levels of microbiological practices. Antibiotics are added to exert a selection pressure during isolation of resistant microorganisms. These are used to avoid contamination during growth. However, the most common application of antibiotics remains in curing microbial diseases in human being. Most antibiotics are thermolabile (heat-sensitive) and must be sterilized by filtration methods only.

SUGGESTED READING

Cappuccino, J. G. and Sherman, N. (2002). *Techniques for isolation of pure culture In. Microbiology: a laboratory manual*. (VI Ed.) Pearson Education, Inc. Singapore.

Collee, J.G., Duguid, J.P., Frasee, A.G. and Marmion, B.P. 1989. *Practical medical microbiology*, 13th ed. Longman, U.K.

Collins, C.H. and Lyne, R.M. 1976. *Microbiological Methods*. Butterworth, Mass, U.S.A.

Pelczar M., Chan E.C.S. and Krieg, N.R. 1993. *Microbiology*, Tata McGraw Hill Publishing Co. Ltd., New Delhi.

Ronald M. Atlas 1993 *Handbook Of Microbiological Media*, CRC Press.

REVIEW QUESTIONS

1. What is meant by sterilization? Discuss various methods of sterilization used in microbiology.
2. Discuss the death pattern of microorganisms upon exposure to high temperatures.
3. Discuss following in detail:
   a) Autoclaving b) Filter sterilization
   c) Tyndallization d) Lyophilization
   e) Phenol coefficient e) Thermal death time (TDT)
4. Differentiate between disinfectant, antiseptic and antibiotics giving suitable examples.
5. Discuss the variety of materials to be sterilized in routine microbiological practice and criteria for selection for suitable method for their sterilization.

UNIT-IX

Instrumentation

Recent trends in microbiology and biotechnology involve a great deal of use of sophisticated instruments for determination of a variety of biological properties or parameters. Instrumentation in the field of biology involves application of a variety of instruments or devices for measurement of biological activities, maintenance of optimum conditions, monitoring, control and analysis, etc. It uses measurement principles and underlying laws of thermodynamics, stoichiometry, optics, electronics and a variety of disciplines of science and engineering. One of the aims of bioinstrumentation, particularly in the field of microbiology, is to develop devices that can directly measure biological parameters, e.g. biomass or enzymes in an operational bioreactor. A number of microprocessor based handy, reliable and compact instruments have been developed for biological estimations. Computers are becoming increasingly important in bioinstrumentation from point of view of automation, storage of data and more importantly, quick analysis of the data using graphics and statistical software. Rapidly expanding fields of molecular biology, genomics, proteomics, and metagenomics are centered on the use of state-of-the-art instrumentation for sequencing and analysis of macromolecules. In the present chapter principle and applications of some of the routinely used instruments are discussed.

SPECTROPHOTOMETER

Coloured solutions absorb light in visible region and the colour exhibited by the solution is the colour transmitted by it in the visible range. When a beam of light passes through a glass cell or cuvette containing a clear solution, the emergent light has less power due to absorption of radiant energy by the liquid.

Compound	Absorbs	Transmitted	Appearance
$NiCl_2$	Red	Green	Green
$CuSO_4$	Yellow, Red	Blue	Blue
NaCl	Nil	Nil	Colourless

Appearance of colour is due to reflected light and is related to the complementary spectra absorbed. The apparent colour will be always complementary to the colour of the absorbed light. Coloured solutions play a special role in quantitative estimation since these can be quantified on the basis of absorption of characteristic wavelength of light. Spectrophotometry is about quantification of substances based on their characteristic absorption and not emission. Direct measurement of a colourless substance cannot be made in visible region. However, if there is appreciable absorption in the ultraviolet range, it can be measured by UV-spectrophotometer in the range of 220 nm – 400 nm. Spectrophotometry makes an important part of such absorption measurement typically in the visible and ultraviolet region of spectra. The measurement in visible light region is also called as Colourimetry. Various electromagnetic regions of ordinary light may be divided into three parts:

(a) UV-light 1000–4000 Å
(b) Visible light 4000–7500 Å
(c) Infrared light 7500–10000 Å

Radiations below 4000 Å and above 8000 Å are invisible.

The Lambert's law states the relation between the light absorptive capacity and thickness of path as- "when monochromatic light passes through a transparent medium, the homogeneous medium absorbs incident radiation proportional to its thickness. Beer's law states that the fraction of the monochromatic radiant energy absorbed upon passing through a solution is directly proportional to the concentration of absorber. By combining these two, we obtain **Beer-Lambert's Law,** which states that the absorbance of a solution is directly proportional to the concentration of absorbing species when the path of light is fixed and directly proportional to the light path when concentration is fixed. A quantitative spectrometric analysis based on absorption of radiation by solution makes use of either of these relationships. The instrument used to measure how much light is absorbed and transmitted by substance in solution at different wavelength of radiation energy is called *spectrophotometer* and this process is known as spectrophotometry.

Spectrophotometry is a method of chemical analysis based on the absorption or attenuation of electromagnetic radiation of a specified wavelength or frequency by solutions. This is because of interaction of radiation with specific features of the molecular species being determined, such as the vibrational or rotational motions of the chemical bonds. The radiation can also interact with specific atoms or the whole molecule, for example, by causing the molecule to change its ionization state.

A spectrophotometer consists of two instruments, namely, a *spectrometer* for producing light of any selected colour (wavelength), and a *photometer* for measuring the intensity of light. The instruments are arranged such so that liquid in a cuvette can be placed between the spectrometer beam and the photometer. The amount of light passing through a tube is measured by the photometer. The photometer delivers a voltage signal to a display device, normally a galvanometer. The signal changes as the amount of light absorbed by the liquid changes.

PRINCIPLE

The Beer-Lambert law is a combination of two laws related to the concentration of the absorber and the path length or thickness of the path. The ratio of intensities of transmitted and incident light gives transmittance, **T**

$T = I_1/I_o$

I_o = intensity of incident light

I_1 = intensity of transmitted light

% Transmittance, or %T = 100 T

A 100% value of ***T*** represents a totally transparent substance, with no radiation being absorbed, whereas a zero value of ***T*** represents a totally opaque substance showing complete absorption.

We can define the absorbance (A), which is given by the logarithm (base 10) of the transmittance (T) as:

$A = \log_{10} I_0 / I_1$

$A = \log_{10} 1 / T$

$A = \log_{10} 100 / \%T$ or $A = 2 - \log_{10} \%T$

The absorption A is directly proportional to the path length (l) and concentration (C) of the substance (**Figure 9.1**).

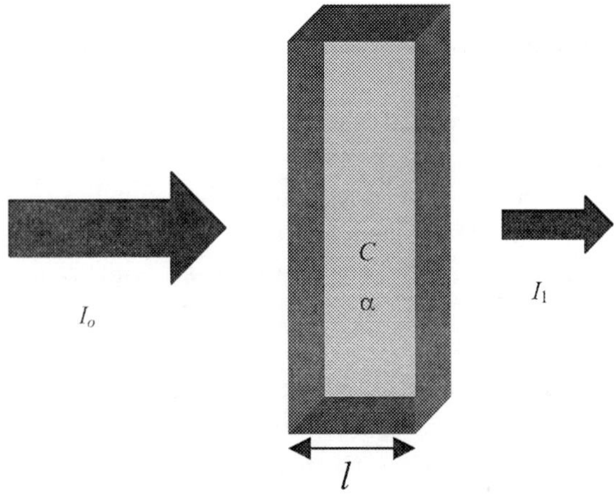

Figure 9.1: Schematic showing absorption of light as a beam travels through a cuvette of path length *l*.

thus, $A = \alpha lC$

where, α is the molar absorptivity expresed as $mol^{-1}cm^{-2}$.

Instrumentation

The material used in the optical parts of the instrument depends on the wavelength used. In the ultraviolet region, it is necessary to use prism, grafting reflecter and cuvettes made of silica. In the visible region, optical parts and cuvettes made up of borosilicate glass may be used. A spectrophotometer has the following basic components:

(i) Light source

A continuous source of radiant energy covering the region of spectrum in which instrument is designed to work. The wavelength range of visible light lies between 4000–8000 Å and in this region a tungsten filament lamp is most widely used. The wavelength range for ultraviolet light is between 2000–4000 Å. In this region hydrogen lamp or a Nernst glower is used.

(ii) Filter's and monochromators

A source generally emits continuous spectra. Therefore, a device is required to select a narrow band from wavelengths of continuous spectra. For this selection, filters or monochromator or both are used.

A light filter is a device that allows light of the required wavelength to pass but absorbs light of other wavelengths. Filters are of two types – **(a) Absorption filters** and **(b) Interference filters.** Absorption filters work by selective absorption of unwanted wavelength. An absorption filter is a solid sheet of glass that has been coloured by pigment which is dissolved or dispersed in the glass. The interference filter is a semitransparent metal film deposited on a plate of glass and then it is coated with a thin layer of semi-dielectric material, followed by another coating of a thin film of metal. When a ray of light is incident upon an interference filter, a part of light reflects back whereas the remaining light (unwanted) is transmitted.

A monochromator successfully isolates band of wavelength usually much more than a narrow filter. The essential element (prism or grating) of monochromators include an entrance slit, a dispersing and an exit slit. The function of prism or grating is to disperse the heterochromatic radiation into its different wavelengths.

(iii) Slits

Usually, two slits are used in spectrophotometer—(a) entrance slit (b) exit slit. The main function of entrance slit is to allow a narrow band of light so that there is no overlapping of monochromating image. Exit slit selects a narrow band of dispersed spectrum for observation by the detector.

(iv) Cell or cuvette

The cell holding the sample should be transparent to the wavelength region being recorded. A glass cell can work nicely for coloured range of spectrum, however, it can not be used in the case of UV spectrum since it is impervious to UV-light. A quartz cuvette is a must for analysis in the UV range, while it can also be used for coloured range as well.

(v) Detectors

Photosensitive devices such as photovoltaic cells are used as detector to detect and sense emergent radiation. The photovoltaic cell operates without any battery. It consists of metal base plate like iron, silver or aluminium which acts as electrode and other surface of a semiconductor metal like selenium covered with silver. Radiation when falls upon the surface of selenium, generates electrons at the selenium-silver interface. These electrons are collected by silver plate and the accumulation of electrons on the silver surface creates an electric voltage difference between the silver surface and base of the cell. Thus, a photocurrent flows which is directly proportional to the intensity of radiation beam. If this cell is connected to galvanometer, a current flow will vary with the intensity of incident light. The sensitivity of these signals is enhanced by using phototubes and photomultipliers as detectors.

Based on optical arrangements spectrophotometers can be classified into two types (i) Single beam spectrophotometer and (ii) Double beam spectrophotometer.

Single Beam Spectrophotometer

In this instrument, all of the light traverses only one path directly through the sample cell. In this case, the control or baseline is set once using the blank and then it is removed. After setting up the base line, sample is placed in the same holder for measurement of absorption. It is used for both research and routine quantitative determination of a variety of compounds (**Figure 9.2a**).

Double Beam Spectrophotometer

In this instrument, the light is split into two beams and traverses two paths. One beam going through the blank or reference cell is used as the reference and the other beam passes through the sample. Some double beam spectrophotometers have two detectors, one measures sample and other measures reference at the same time. In other instrument, two beams pass through a beam chopper, which blocks one beam at a time. The detector alternates between measuring the sample beam and the reference beam (**Figure 9.2b**). In double beam spectrophotometer, λ_{max} for any compound can be determined by scanning it against the complete spectrum.

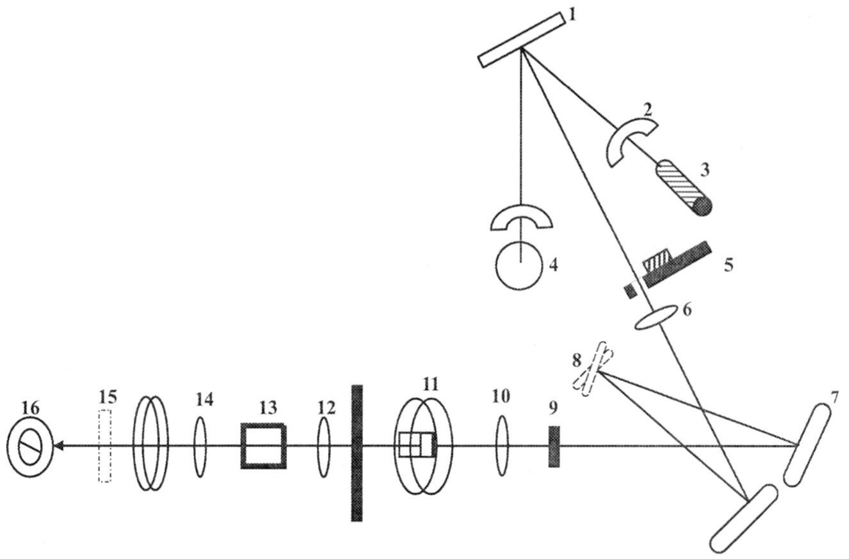

Figure 9.2(a): Single beam spectrophotometer
1- Toroidal Mirror, 2- Lens, 3- Quartz Halogen Lamp, 4- Deuterium lamp, 5- Filter wheel, 6- Lens/ entrance slit, 7- Spherical Mirror, 8-Blazed holographic grating, 9- Aperture, 10- Lens/ exit slit, 11- Beam Splitter, 12- Lens, 13- Sample, 14- Lens, 15-silicon sensor, 16- PMT (photomultiplier tube)

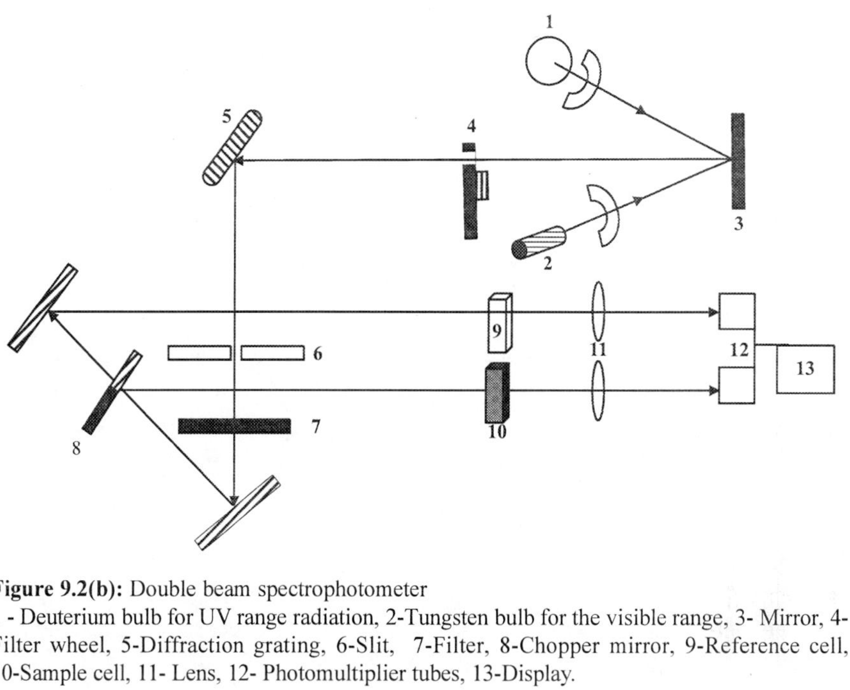

Figure 9.2(b): Double beam spectrophotometer
1 - Deuterium bulb for UV range radiation, 2-Tungsten bulb for the visible range, 3- Mirror, 4- Filter wheel, 5-Diffraction grating, 6-Slit, 7-Filter, 8-Chopper mirror, 9-Reference cell, 10-Sample cell, 11- Lens, 12- Photomultiplier tubes, 13-Display.

POWER SUPPLY

The power supply has typical functions including matching the line voltage to the instrument's operating level via a transformer and conversion of A.C. to D.C. with a rectifier, if direct current is required by the instrument.

APPLICATIONS OF SPECTROPHOTOMETRY

(i) Nucleic Acid Quantification

With the recent increase in interest in molecular biology, determination of nucleic acid purity and quantification at various stages of the gene cloning process have become routine. The purity of DNA and RNA extractions from cells can be readily measured using spectrophotometry because the absorbance ratio at 260 / 280 nm (both wavelengths are in the UV region) is a constant for the pure preparation (values are 1.8 and 2.0 for DNA and RNA, respectively); any deviation from these values is a result the presence of protein or phenol impurity.

(ii) Protein Determination

Determination of protein in solution can be conveniently performed using colourimetric methods in conjunction with a UV/Visible spectrophotometer as well as directly by absorbance at 280 nm

(iii) Enzyme Kinetics and Quantification

The investigation of enzyme kinetics allows us to find out parameters such as K_m, Vmax, optimum pH and temperature, etc. Absorbance versus time plots as a function of concentration at particular temperature are necessary to calculate kinetic parameters.

pH METER

pH is a quantitative expression of the acidity or alkalinity of a solution. Aqueous solutions having pH less than 7.0 are acidic and those with pH more than 7.0 are basic (alkaline). The pH value of 7.0 is neutral because at this pH the concentration of H^+ (or H_3O^+) equals the concentration of OH^- in pure water. The value of pH varies with temperature and hence by convention 25°C is considered as a standard temperature for taking pH. pH is formally dependent upon the activity of hydronium ions (H_3O^+), but for very dilute solutions, the molarity of H^+ may be used as a substitute (H^+ is often used as a synonym for H_3O^+).

Max Cremer in 1906 during his studies of liquid interfaces discovered that an electric potential is created between liquids separated by a thin layer of glass. Haber and Klemsiewicz discovered the glass electrode to measure hydrogen ion activity and logarithmic behaviour. Based on this, Soren Sorensen invented the pH scale in 1909. The first commercial pH meters were built by Dr. Arnold Beckman of United States in 1936 while working as a professor of chemistry at the California Institute of Technology.

The ultimate pH value of any solution depends upon the ionic activity of various compounds present in it. This property is difficult to predict or measure theoretically. Thus, the direct measurement of this value is not possible. pH of a solution is estimated by comparing its ionic activity with those of known (standard) solutions. Generally, the measurement of pH is based on electrochemical measurement. Sørensen (1909) proposed pH as a convenient notation for the hydrogen ion concentration of a solution. He defined the negative log of the hydrogen ion concentration as pH, expressed in molarity:

$$pH = -\log [H^+]$$

Therefore, for a 0.01M solution of HCl

$$pH = -\log [10^{-2}]$$

$$= 2.0$$

A pH meter works on the principle of electrochemistry. Electrons move as per the voltage or potential difference and when a metal is brought in contact with a solution of salts or acids, a similar electric potential is created. An electric potential is developed when one liquid is brought in contact with another one separated by a thin membrane. The thin glass bulb allows mainly the agile and small hydrogen ions to interact with the glass, the glass electrode measures the electrochemical potential of hydrogen ions $[H^+]$.

A pH meter uses the same principle and measures the electrochemical potential difference between a standard solution inside the glass electrode and an unknown liquid outside. A reference electrode is needed for completing the circuit. The change in potential is calibrated with the solutions of known pH and then this calibrated setup is used to measure pH of unknown solutions.

Water is a unique universal solvent being both an acid (H^+ ions) and a base (OH^- ions). It is thus both acidic and basic (alkaline) at the same time, making it neutral as the number of H^+ ions equals that of the OH^- ions. Because of its strong cohesion, only few water molecules dissociate (split) in their constituent ions: hydrogen ions (H^+) and hydroxyl ions (OH^-).

At any given moment only 0.1 millionth (10^{-7}) mol are dissociated in one litre of water (pH = 7). The potential difference between the inside of the glass electrode and the outside is caused by the oxides of silicon inside the glass as follows:

$$Si.O^- + H_3.O^+ \rightarrow Si.O.H^+ + H_2.O$$

The relation by which the **emf** (electromotive force) of a suitable pH measuring electrode is related to the hydrogen ion concentration was developed by Nernst as:

$$E = E^\circ - (RT/nF)\log[H^+]$$

where E^0 is potential dependent on the electrode system used, R is the gas constant, T is absolute temperature, F is Faraday constant equal to 96,485 coulombs/mole and n is the number of electrons involved in the equilibrium.

For $n = 1$, the factor $RT/nF = 0.0591$ pH

Thus, $E = E^\circ - 0.0591 \log[H^+]$

This equation establishes the relationship of 59.1 mV per pH unit at 25°C for any electrode system that follows the Nernst equation. Most pH measuring electrode employ ion-selective electrodes where ion accumulation occurs at the interface without any electron transfer. Such glass electrodes are widely used for pH measurements.

The hydroxyl ion concentration can be related to the hydrogen ion (H^+) or more accurately to hydronium ion H_3O^+ concentration by considering the dissociation of water:

$$H_2O \rightleftarrows H^+ + OH^-$$

The equilibrium constant equation for this reaction is

$$K_{eq} = [H^+] \times [OH^-] / [H_2O]$$

Thus,

$$K_{eq}[H_2O] = [H^+] \times [OH^-]$$

Since the concentration of water remains about constant throughout the ionization process, this term can be combined with the equilibrium constant, generating a new constant, **Kw**

$$K_w = [H^+] \times [0H^-]$$

For pure water at 25°C, both hydronium and hydroxyl ion concentration are equal to 1×10^{-7} M, therefore

$$K_w = (1 \times 10^{-7})(1 \times 10^{-7})$$
$$= 1 \times 10^{-14}$$

Once the ionic equilibrium is established, the potential difference between the glass wall and the solution is given by the equation:

$$\mathbf{E} = \mathbf{R} \times \mathbf{T} / [F \text{ x } \ln(a)]$$

where **E** = electron potential (Volt), **R**= molar gas constant 8.314 J/mol/°K, **F**= Faraday constant 96485.3 °C, **T**= temperature in °Kelvin and **a**= the activity of the hydrogen ions (hydronium ions). ln (**a**) = the natural logarithm, which converts to the decimal logarithm = **2.303** x log(**a**). The formula **R** x **T** / (2.303 x **F**) yields that there will be 0.0591 V (59.1 mV) per tenfold increase in hydrogen ions or one pH unit. The pH range of 0 – 14 accounts for hydronium activities from 10 to

10^{-14} mol/litre. A pH = 7 corresponds to **hydronium activity** of $1x10^{-7}$mol/litre. Since log (10^{-7}) = –7, the pH scale leaves the minus sign out.

Glass electrodes have problems with some substances low in H^+ ions, like alkali hydroxides (NaOH and KOH), pure distilled water, etching substances like fluoride, adsorbing substances like heavy metals and proteins.

At neutral pH (pH 7) the voltage at the probe's output is zero volts (0 * 17 + 7 = 7). At alkaline pH, the voltage at the probe's output ranges from **> 0 to +0.41 volts** (7 * 0.059 = 0.41). So for a sample of pH 10 (3 pH units more than neutral), 3 * 0.059 = 0.18 volts), the output of the meter's amplifier display will be 0.18 * 17 + 7 = 10. At acidic pH, the voltage at the probe's output ranges from **–0.41 volts to < 0**. So for a sample of pH 4 (3 pH units below neutral), –3 * 0.059 = –0.18 volts, the output of the meter's display will be –0.18 * 17 + 7 = 4. Calibration with preferably three standard buffer solutions (pH 4.0, 7.0 and 9.0) is usually performed every time when a pH meter is used. Recent instruments based on microprocessor retain their calibration for longer time in their memory.

A conventional pH meter has **three** main parts:

1. Reference electrode
2. Glass electrode
3. Electrometer

The use of contemporary electrode design and solid state technology permits a resolution of 0.005 pH units by most high quality instruments.

1. Reference electrode: A reference electrode maintains a constant electrical potential against which deviations due to pH change may be measured. The two most widely used reference electrodes are the calomel electrode and silver-silver chloride electrode.

(a) Calomel electrode: The most important electrode used for pH measurement is calomel electrode. Various types of vessels have been used for the purpose of setting up calomel electrode. It consists of glass tube having a side tube at the end. A small amount of pure mercury is placed at the bottom of the glass tube and covered by a paste of mercury chloride and potassium chloride solution. The vessel is then filled with potassium chloride solution by introducing the latter from the small side tube. The solution in the tube on the left helps to make connections through a salt bridge, with the other electrode or half cell, the potential of which is to be determined. The electrical connection with mercury is made by means of a platinum wire sealed in a glass tube. The potential of calomel electrode depends upon the concentration of KCl solution used. Electrodes having lowest temperature coefficient, such as that having 0.1 N KCl, are preferred. Half-cell of reference electrode can be represented as follows:

$$Hg \mid Hg_2Cl_2 \parallel KCl \mid \text{Test solution}$$

where the double line (||) represents the salt bridge linking half-cell to the indicator half cell.

(b) Silver-silver chloride electrode: The silver-silver chloride electrode is also used as a reference electrode, especially in connection with accurate measurements in chloride solutions. It consists of silver wire coated with silver chloride immersed in a solution containing chloride ion (NaCl) represented as

$$Ag \mid AgCl\,(s),\, NaCl\,(aq.)$$

The silver chloride electrode can be prepared by depositing silver electrolytically on a platinum wire and then coating with finely divided silver by the decomposition of silver oxide. The silvered wire is then used as an anode in a dilute solution of sodium chloride (**Figure 9.3**). The standard potential of this electrode is obtained by its combination with standard hydrogen electrode. The e.m.f. of the cell:

$$AgCl\,(s),\, NaCl\,(aq.) \mid H_2\,(1atm);\, Pt$$

is found to be –0.224 volts at 25°C. Hence, the standard potential of the electrode, E° is 0.224 at 25°C.

2. Glass electrode: The function of glass electrode is to create potential which responds to changes in the hydrogen ion activity of the solution being tested. The basic assembly of glass electrode is shown in **Figure 9.4**. It consists of a high resistance glass tube with a thin, low resistance glass bulb at the bottom. This bulb responds to change in pH. The tube is filled with dilute aqueous solution of hydrochloric acid. In contact with this solution is a silver-silver chloride electrode. When a pH sensitive electrode and a reference electrode are placed in solution, a galvanic cell is set up. The complete cell is as follows:

$$E_{cell} = E_{ref} + E_{glass}$$

More recently biocatalytic membrane electrodes have been developed for measurement of pH. In such electrodes, biocatalysts or enzymes immobilized onto a gel layer coating are employed. Urease enzyme is fixed in a layer of acrylamide gel held in place around the glass electrode bulb by a nylon mesh for developing ammonium selective electrode. The urease acts on urea present in sample to liberate ammonium ions which diffuse through the gel layer and are sensed by the internal electrode. A pH responsive electrode can be used as internal electrode since ammonium ions affect hydrogen ion concentration. The glass membrane of a pH electrode is fragile and easily damaged. It is important that its surface remains hydrated and so it should be kept immersed in water when not in use.

3. Electrometer: This is a meter that shows an output of the potential difference developed between the glass electrode and the reference electrode. The output is calibrated in terms of pH using known solutions and hence pH of the unknown

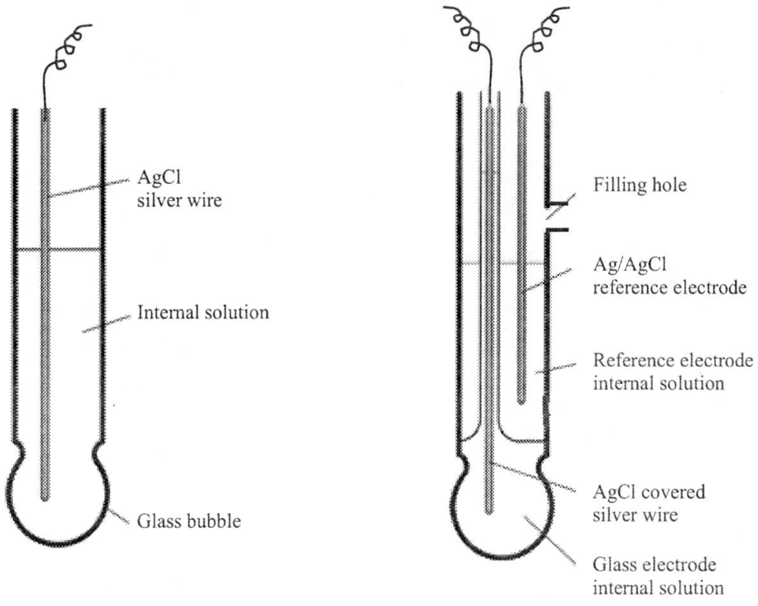

Figure 9.3: Single and combined electrodes used in pH meters.

solutions can be directly read from the display as an analogue signal or in form of digits.

Ion Selective Electrodes

The other method of pH measurement is based on ion-selective electrodes. Ion-selective electrodes measure ion activities and provide an ideal measurement because of their ability to measure selectively the activity of certain ions including Li^+, K^+, Na^+, Ca^{2+}, Cl^- and NO_3^- in solutions. In case of dilute solutions, ion activity usually approaches the ion concentration. There are three types of membrane sensors used in ion-sensing electrodes: glass, solid state and solid matrix sensors. Such glass membrane electrodes have active sites on the surface or in the hydrated layer of the glass. The electrode potential is built by a combination of cation exchange and cation mobility leading to charge accumulation at the glass solution interface.

The **solid-state sensors** of such ion-selective electrodes are more commonly used today in handy digital pH meters. Their body is fabricated using plastic or epoxy coating and sensing membrane made up of single pure and non-porous material bonded to the electrode which in turn is bonded to a micro porous surface for avoiding sample retention, e.g. Fluoride electrode. The active membrane is a single crystal of LaF_3 doped with europium (II) to facilitate ionic transport. The internal solution is 0.1M NaF and NaCl. The fluoride ion activity controls the

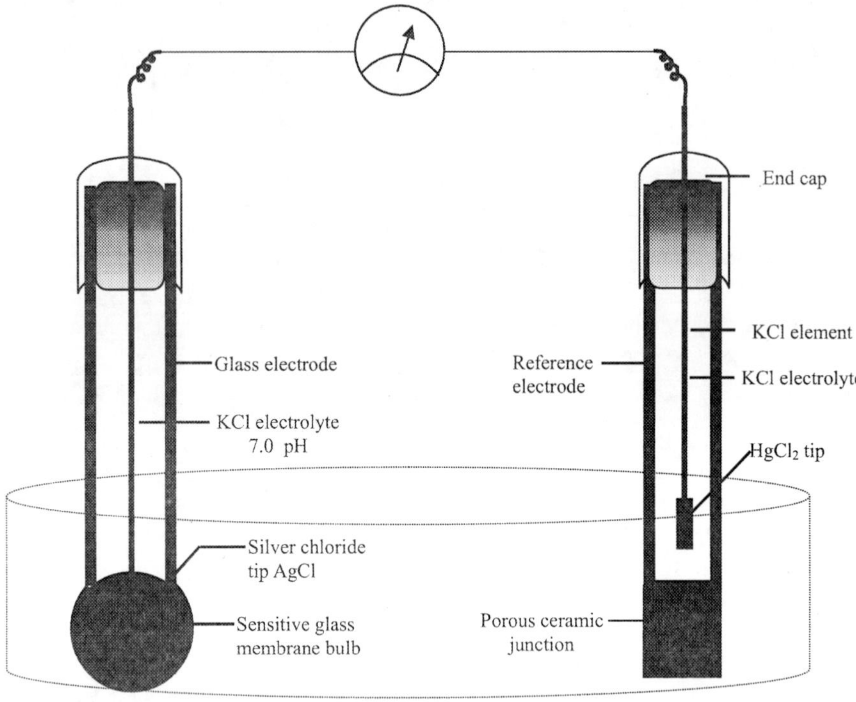

Figure 9.4: Diagram of pH meter showing the setup of glass and reference electrode.

potential of the inner surface of the LaF_3 and the chloride ion activity fixes the potential of the silver /silver chloride reference electrode. The external surface of the membrane responds to contact with any sample based on the fluoride ion activity. As described earlier any pH meter has three parts: a sensing electrode, a pH measuring system (high input impendence operational amplifier for measuring and displaying cell potential) and a reference electrode. The pH meter is a voltmeter that carries out some critical measurements and conversions. It measures the potential difference between reference and glass electrode and converts it into pH terms at a specified temperature. There is also a provision of standardization and correction of the measuring unit. Standardization of pH meter involves immersing the pH electrode into standard buffers of known pH and calibrating the pH meter with at least two reference points. The microcomputer based handy digital pH meters use appropriate algorithms that perform required standardization and calibration.

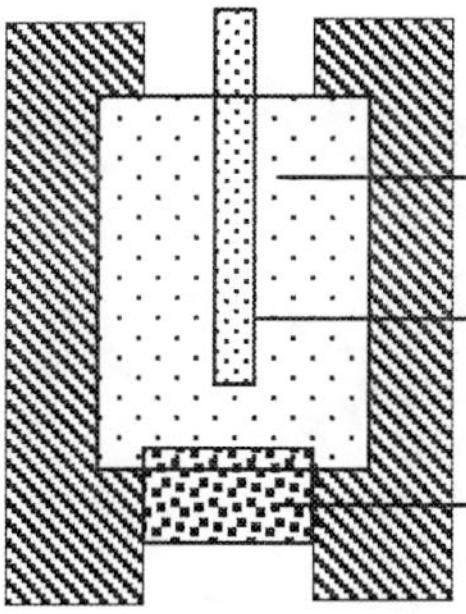

Figure 9.5: Cross-sectional view of a solid state pH sensor.

CENTRIFUGATION

Centrifugation is one of the most important and widely applied techniques in microbiology, biochemistry, cellular and molecular biology and medicine. Current research and many clinical applications rely on isolation of cells, subcellular organelles and macromolecules. A centrifuge uses centrifugal force (generated due to rotation of the particle around an axis) to isolate suspended particles from their surrounding medium. Applications of centrifuge include separation of microorganisms, sub-cellular organelles and isolation of macromolecules such as DNA, RNA, proteins, and lipids.

Centrifugation is a separation technique based upon the behaviour of particle in a centrifugal field. Molecules separate on the basis of their relative molecular mass, shape and density. Centrifuge is an instrument that makes the separation mixture rotate at high speed resulting in separation of particles according to mass and density under the influence of centrifugal force. The centrifugal force acts upon the mixture in the same manner as a gravitational field, but it can be increased by increasing rotational speed. Centrifugation increases the gravitational potential energy of the particles in a solution in a vessel rotating at high speed. The particles in the solution will then sediment faster because a centrifugal force is acting upon them in addition to gravitational force. This technique was discovered in 1923 by **Svedberg and Nicholas.**

Particles suspended in a liquid medium are held in tubes or bottles which are located in a rotor. The rotor is positioned centrally on the drive shaft of the centrifuge. Particles that differ in density, shape or size can be separated because they sediment at different rates in the centrifugal field.

PRINCIPLE OF CENTRIFUGATION

Particles in a liquid suspension eventually settle down at the bottom due to gravitational force ($1 \times g$). However, the length of time required for such

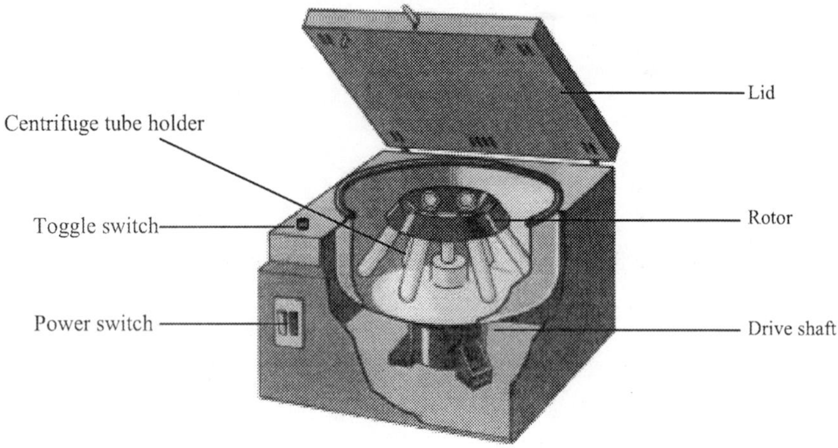

Figure 9.6: Centrifuge.

sedimentation is too long for experimentation. Also, extremely small particles, e.g. proteins, may not settle at all in solution, unless subjected to high centrifugal force. When a suspension is rotated at certain revolutions per minute (RPM), centrifugal force causes the particles to move radially away from the axis of rotation.

The rate of sedimentation is dependent upon the applied field (G) being directed radially outward and is determined by the square of the angular velocity of particle from the axis of rotation according to the equation.

$$G = \omega^2 r$$

where ω (omega) = angular velocity of rotor = 2π rev sec^{-1} = 2π rev min^{-1} / 60

Thus, centrifugal field, $G = \omega^2 r = 4\pi^2 (\text{rev min}^{-1})^2 r /3600$

It is generally expressed as relative centrifugal force (RCF), a multiple of gravitational field (g). Relative centrifugal force is ratio of the weight of the particle in centrifugal field to the ratio of weight of same particle when acted on by gravity alone, and is then referred to as the relative centrifugal field (RCF). For example, an RCF of 1000 x g indicates that the centrifugal force applied on the particle is 1000 times greater than gravitational force.

$$\text{Relative centrifugal force (g)} = 4\pi^2 (\text{rev min}^{-1})^2 r/ 3600 * 981$$

$$= 1.118 \times 10^{-5} (\text{rev min}^{-1})^2 r$$

$$g = RCF = 0.00001118\ rN^2$$

where

g = relative centrifugal force

r = rotational radius (centimeters, cm)

N = rotating speed (revolutions per minute, r/min)

From above equation it is clear that RCF is directly proportional to (i) rev min^{-1} and (ii) radial distance (r). Therefore, to make a separation under higher RCF either or both of these should be increased. For a centrifuge with constant 'r' for rotor, rpm can only be increased to achieve higher RCF. It should be noted that for a specific particle, the distance r increases with the progress in centrifugation. As centrifugation progresses the particle will sediment progressively toward the bottom of the sample tube with continuously increasing distance (r) from the rotor axis resulting in respective increase in RCF.

The other important consideration in centrifugation is the density of the medium in which separation is being carried out. When a particle is centrifuged then density and viscosity of medium in which it is sedimenting and extent to which its shape deviates from spherical also play a role in sedimentatation. The net outward force experienced when centrifuged at an angular velocity of ω radians s^{-1} is given by

$$F = 4/3\ \pi\ r^3_p\ (\rho_p - \rho_m)\ \omega^2 r$$

where

$4/3\ \pi\ r^3_p$ = the volume of a sphere of radius (r)

ρ_p = density of particle

ρ_m = density of suspending medium

If a particle is rigid and spherical and moving at a known velocity, then the frictional force (F_o) opposing motion is given by

$$F_o = \nu f$$

ν = velocity

f = friction coefficient of particle in solvent = $6\pi\eta r_p$

η = viscosity coefficient of medium

$$F_0 = 6\pi\eta r_p\ \nu$$

A spherical particle of known volume and density present in the medium of constant density accelerates in centrifugal field and its velocity increases until net force of sedimentation is equal to the frictional force resisting its motion through the medium that is

$$F = F_o$$

$$4/3\ \pi\ r^3_p(\rho_p - \rho_m)\ \omega^2 r = 6\pi\eta r_p\ \nu$$

Under this condition the net force acting on the particle is zero hence, the particle no longer accelerates but achieves maximum velocity, as a result it sediments at a constant rate.

TYPES OF CENTRIFUGES

Various kinds of centrifuges are available and can be selected depending upon the scale of operations and the kind of separations to be achieved. Some examples of types of centrifuges are:

(1) Small bench-top centrifuge
(2) Large capacity refrigerated centrifuge
(3) High-speed refrigerated centrifuge
(4) Ultra centrifuge

(1) Small Bench Top Centrifuge

These are simple, cheap centrifuges which are used to collect small amount of sediments such as microbial cells, crude precipitate. These have maximum speed upto 5,000–15,000 rpm and maximum RCF 4,000–10,000 g. These are operated at room or ambient temperature but if desired can be put in a refrigerating system for achieving sedimentation at low temperature. These are easy to use machines used for small working volumes (15–50 ml) resulting in 250 mm^3–1.5 cm^3 of precipitates in few minutes (5–10 minute). These centrifuges are used for rapid sedimentation of larger macromolecules and microbial cells (bacteria, yeast), blood cells, etc.

(2) Large Capacity Refrigerated Centrifuge

They have refrigerated rotor chamber and vary mainly in their maximum volume capacity from above. They utilize variety of interchangeable rotors such as swinging bucket and fixed angle rotor. Interchangeable rotors varying in volumes and number of tubes are also used. Usually, large capacity tubes and vials of 50 and 100 cm^3 capicity are used. These are used for collection / compaction of sediments from large volumes of liquids at low temperatures, e.g. fermentation broth, RBC, yeast cells, nucleus, chloroplast and crude protein precipitate.

(3) High-Speed Refrigerated Centrifuge

These have maximum speed 25,000–30,000 rpm and produce maximum RCF 100,000 g. They utilize variety of interchangeable swinging bucket and fixed angle rotors. These are used to collect microorganisms, cellular debris, large cellular organelle (chloroplast), and protein precipitate but can not be effective for separation of virus and small organelles and ribosome. Continuous flow centrifuge is a simple modified high speed centrifuge in which particle suspended in medium flow continuously through the rotor. As medium enters rotating rotor, particles sediment against its wall and excess clarified medium overflows through the outlet. This centrifuge finds application in harvesting of bacteria and yeast from large volume of fermented broth during downstream processing.

(4) Ultra Centrifuge

They are capable of operating at very high speeds approaching 70,000 – 1, 00,000 rpm (maximum RCF 800,000 g). These are relatively sophisticated instruments consisting of motor and rotor contained in a protective armored chamber that is refrigerated and evacuated and have an optical system to observe the sedimenting material during centrifugation to determine concentration and distribution in the sample at any given point of time during centrifugation. The temperature is maintained between 0–4°C during the operation of the instrument so that heat generated due to rotor moving at high speed is negated. The shaft is made of aluminium or titanium alloy of high tensile strength to withstand great force generated during centrifugation. The rotor is prevented from operating at speed above its maximum rate speed by an over speed device (OSD). Based on application, ultracentrifuge is mainly of two types (i) Preparative ultracentrifuge (ii) Analytical ultracentrifuge.

(i) Preparative Ultracentrifuge

They are capable of operating at speeds approaching 70,000 rev min^{-1} (maximum RCF 60,000 g). Rotor chamber is refrigerated, sealed and evacuated to minimize any excessive frictional resistance between air and spinning. They have IR – temperature sensor to continuously monitor rotor temperature and control the refrigeration system and over speed system and electronic circuit to detect rotor imbalance. This centrifuge is used for assay and separation of the major lipoprotein fraction from plasma and deprotenization of physiological fluid for amino acid analysis. For safety reasons, rotor chamber region of high-speed centrifuge is always enclosed in heavy armor plating. This form of centrifugation is called preparative and the objective is to isolate specific particles in bulk that can be reused, e.g. sedimentation of proteins, enzymes, whole cells (*Rhizobium*, Yeast), etc.

(ii) Analytical ultracentrifuge

The purpose of analytical centrifugation involves measurement of the physical properties of the sedimenting particles such as sedimentation coefficient or molecular weight. Molecules are observed by optical system in a manner similar to spectrophotometer during centrifugation. This allows observation of movement of macromolecules in solution under centrifugal field.

The samples are centrifuged in special tubes or cells having quartz windows that lie parallel to the plane of rotation of the rotor head. As the rotor turns, the light passes through the cell and produces the images/information of macromolecules. This information is projected by an optical system on to film or a computer. The changes occurring in concentration of the solution at various points in the cell are determined by absorption of a light of the appropriate wavelength as per the principles of spectrophotometer.

ROTORS

Rotors can be classified into three common types, viz. (i) Fixed angle rotor, (ii) Swinging bucket rotor, (iii) Vertical tube rotor. Each type of rotor has strengths and limitations leading to its specific application (**Figure 9.7**).

Fixed Angle Rotor

In fixed-angle rotors, the sample tubes are held fixed at the angle of the rotor cavity. It has tubes located in holes in the rotor body set at fixed angle between 14–40°. Under the influence of centrifugal field, which is exerted at an angle to the tube wall, particle moves radially outward and has only a short distance to travel before colliding with and precipitating on the outer wall of the centrifuge tube. When the rotor begins to rotate, the solution in the tubes reorients. This rotor type is most commonly used for pelleting applications. Examples include pelleting of bacteria, yeast, proteins and other mammalian cells. It is also useful for isopycnic separations of macromolecules such as nucleic acids. A region of higher concentration is formed that has a density greater than that of surrounding medium

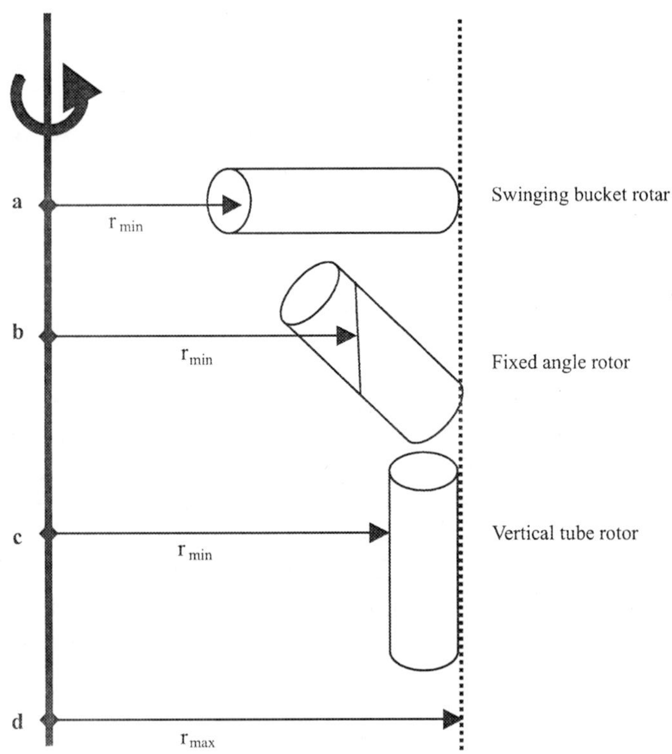

Figure 9.7: Type of rotors. R_{min} and R_{max} are minimum and maximum radius from the axis of rotation in centimetres, respectively.

with the result that the precipitate sinks and collects as small compact pellets at the outer most point of tube. Fixed angle rotor is very useful for differential separation of particles whose sedimentation rate differ significantly, e.g. sub-cellular fractionation **(Figure 9.8).**

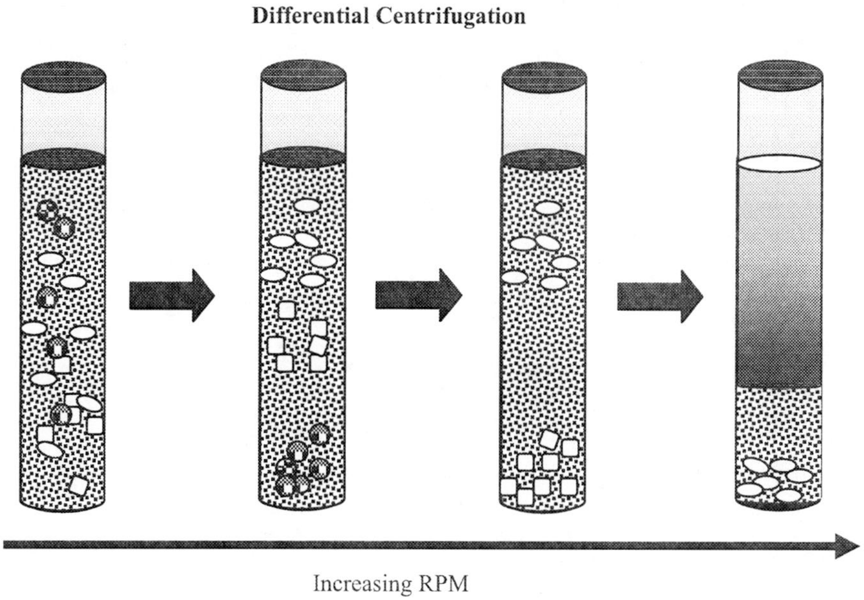

Figure 9.8: Cellular fractionation using differential centrifugation.

Swinging Bucket Rotor

In swinging bucket rotors, the sample tubes are loaded into individual buckets that hang vertically while the rotor is at rest. When the rotor begins to rotate the buckets swing out to a horizontal position. Tubes in the bucket also swing out to a horizontal position so that during centrifugation the tube and solution in tube is angled at 90° to the axis of rotation and parallel to applied centrifugal field. The tube returns to its original position when the rotor comes to rest. This rotor is especially useful for density gradient centrifugation. The longer path length permits better separation of individual particle types from a mixture. However, this rotor is relatively inefficient for pelleting.

Vertical Tube Rotor

Vertical tube rotor can be regarded as zero angle rotor in which tubes are aligned vertically in the body of the rotor at all times during operation. Solution in tube reorient through 90° acceleration to the axis of rotation. Sedimentation occurs across the diameter of the tube. Sedimentation is achieved more quickly than

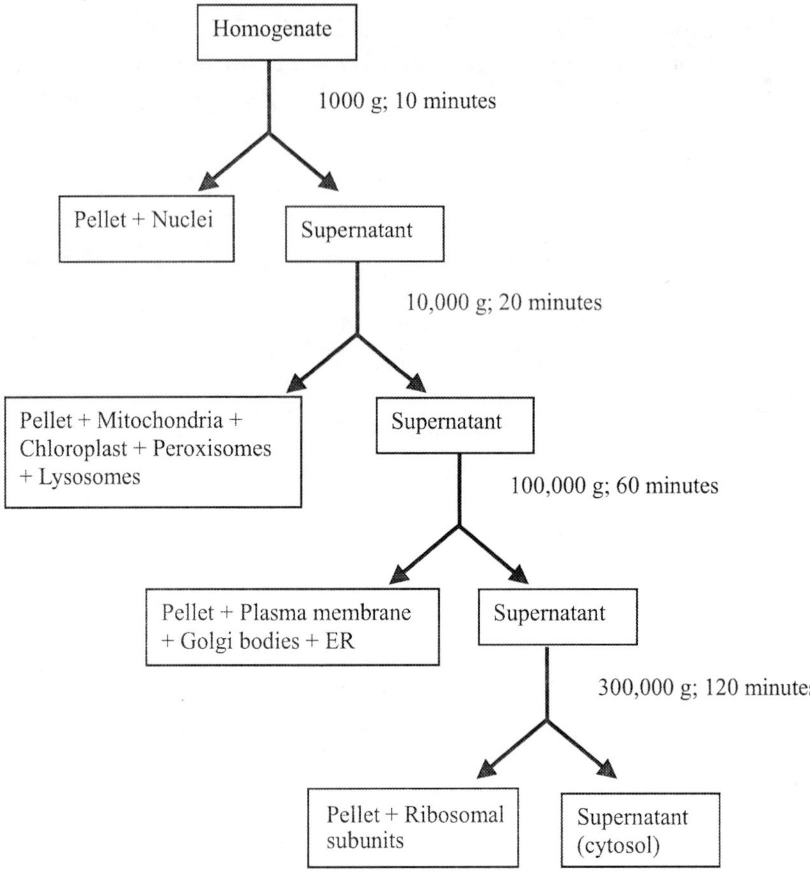

Figure 9.9: Separation of subcellular material using differential centrifugation.

swinging bucket or fixed angle rotor. However, this type of rotor is not suitable for pelleting applications but is most efficient for isopycnic (density) separations due to the short path length. Applications include isolation of plasmid DNA, RNA and proteins.

TYPES OF CENTRIFUGAL SEPARATION METHODS

Differential Centrifugation

Introduced by Albert Claude, this method is based upon the difference in the sedimentation rate of particles of different size. Centrifugation will initially sediment the largest particles. Separation is achieved primarily based on the size of the particles. This type of separation is commonly used in simple fractionation and in obtaining partially-pure preparation of sub-cellular organelles. For the study of sub-cellular organelles, tissue or cells are first lysed open to release their contents. This crude cell mixture is called as homogenate. During centrifugation of a cell homogenate, larger particles sediment faster than smaller ones and this provides the basis for obtaining crude organelle fractions by differential centrifugation. A cell homogenate can be centrifuged at a series of progressively higher rpm or **g** forces and time intervals to generate pellets of partially-purified organelles.

This is also called as sub-cellular fractionation as shown in **Figure 9.9**. When a cell homogenate is centrifuged at 1000 x g for 10 minutes, unbroken cells and heavy nuclei pellet to the bottom of the tube. The supernatant can be further centrifuged at 10,000 x g for 20 minutes to pellet sub-cellular organelles of intermediate size such as mitochondria, lysosomes, and microbodies. Some of these sedimenting organelles can be obtained in partial purity and are typically contaminated with other particles. Repeated washing of the pellets by resuspending in isotonic solvents and re-pelleting may result in removal of contaminants that are smaller in size. Obtaining partially-purified organelles by differential centrifugation serves as the preliminary step for further purification using other types of centrifugal separation techniques.

In differential centrifugation, the material to be separated (e.g. tissue homogenate) is divided centrifugally into fractions by increasing the applied centrifugal field. However, in this method bigger particles when moving down carry the smaller particles with them, leading to contamination.

Density Gradient Centrifugation

Density gradient centrifugation is used to obtain much higher degree of separation with purity, e.g. to obtain only one kind or similar sub-cellular organelles and macromolecules from a tissue homogenate. Density gradient centrifugation is also employed for quantitative separation of all components of a mixture and for determination of buoyant densities and estimation of sedimentation coefficients.

Density gradients can be generated by placing layer after layer of gradient media such as sucrose in a tube with the heaviest layer at the bottom and the lightest at the top in either a discontinuous or continuous mode. The cell fraction to be separated is placed on top of the layer and centrifuged. Density gradient separation can be classified into two types:

(i) Rate-zonal separation (based on size)

(ii) Isopycnic separation (based on density)

Rate zonal (size) separation

Rate-zonal separation is based upon particle size and mass instead of particle density for sedimentation. **Figure 9.10** illustrates a rate-zonal separation process based on size of particles. Examples include separation of cellular organelles such as endosomes or separation of proteins, such as antibodies. For instance, antibody classes have very similar densities, but different masses. Thus, separation based

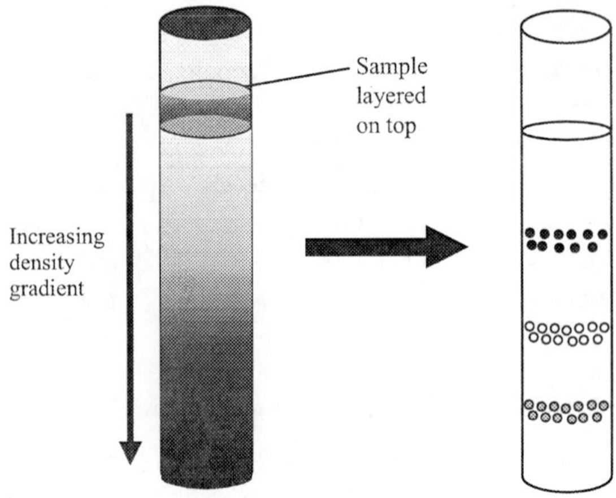

Figure 9.10 Rate zonal centrifugation.

on mass will separate the different classes, whereas separation based on density will not be able to resolve these antibody classes. Certain types of rotors are more applicable for this type of separation than others (**Table 9.1**).

Table 9.1 Types of rotors and their uses

Type of Rotor	Sedimentation		Pelleting
	Rate Zonal	Isopycnic	
a. Fixed Angle	Limited	Variable	Very good
b. Swinging Bucket	Good	Better	Inefficient
c. Vertical	Good	Excellent	Not suitable
d. Zonal	Excellent	Good	Not suitable

First careful layering of sample solution on top of preformed liquid density gradient is done. Highest density of the gradient should not exceed that of densest particle to be separated. Gradient fraction is to stabilize the liquid column in tube against movement resulting from convection currents and helps to improve resolution. The loaded sample is centrifuged until desired degree of separation is effected that is for sufficient time for particles to travel through the gradient to form discrete zones or bands which are spaced according to relative velocities of particles. Rate zonal technique is used for separation of enzymes, hormone, RNA-DNA hybrid, ribosomes, lipoprotein fractionation and bulk preparation but cellular organelle (mitochondria, lysosome, and peroxisome) cannot separated.

Isopycnic technique

Isopycnic technique depends entirely upon the buoyant density of the particle and not on its shape or size and is independent of time. The size of particle affects only the rate at which it reaches isopycnic position in the gradient. It is used to separate particles of similar size but of differing density, e.g. separation of subcellular organelles on the basis of difference in density (Fig. 9.11). Separations as sharp as between heavier DNA (containing isotopes 13C or 15N) and normal DNA can be achieved using caesium chloride (CsCl) isopycnic centrifugation. In this procedure, sample is layered on top of the density gradient that covers whole range of particle densities that are to be separated. Maximum density of the medium always exceeds density of most dense particle. A particle continues to sediment until it reaches to a region where its buoyant density becomes equal to the surrounding density. At this point of iso-density the particle stops and no further sedimentation occurs. Different particles set to form a band or zone w.r.t. their own buoyant density. Caesium chloride (CsCl) density gradient is used to purify viruses, proteins, nucleic acids, analysis of human plasma, and separation of liposomes of differing buoyant density. After running the centrifuge at 100,000 × g for an hour, we can find separate disks of cellular organelles at the point of change in density between layers. Caesium chloride gradient is commonly used to achieve precision in isopycnic separation.

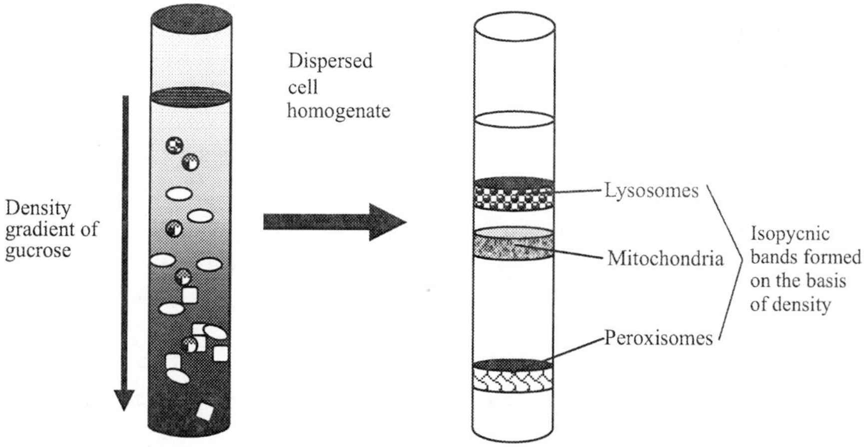

Figure 9.11: Isopycnic (Density) separation.

FERMENTERS

The term fermentation is derived from the Latin verb *fervere* (meaning to boil) for describing the appearance of effervescence of CO_2 due to activity and growth of yeast on fruit juices or malted extracts. The boiling appearance is due to the release of CO_2 from anaerobic breakdown of sugars by yeast under the microaerophilic conditions.

Fermentation of alcohol is an ancient art. Fermentation has gained several novel meanings while retaining the old ones. The term fermentation has been routinely applied to any process employing microorganisms for production of useful commodities including industrial products. However, biochemically, fermentation stands for a process that leads to generation of energy from organic compound under anaerobic conditions.

Thus, one must note that the meaning of the term 'fermentation' is altogether changed with reference to the context. With regards to physiology or metabolism, fermentation refers to an energy generating process where organic compound acts as both electron donor and acceptor under anaerobic conditions. In context of industrial microbiology or biotechnology, fermentation is defined as the process by which large quantities of living cells are grown under aerobic or anaerobic conditions for producing some useful commodity. Mostly, aerobic cultures are used to produce industrial high value products and thus most fermentation processes are aerobic as well. Again a simple definition of the term can be a process in which industrial microorganisms are grown under defined and controlled conditions with an aim of optimizing the growth of the organism or yield of the desired microbial product. In today's biotechnological context, the most generalized definition encompasses any process which is carried out in vessels known as fermenter. **Figure 9.12** shows general steps involved in any fermentation process. Any bioprocess involves mainly (i) Upstream processing and (ii) Downstream processing.

DESIGN OF A FERMENTER

Fermenter is a specially designed vessel in which favourable culture conditions are maintained to provide best possible yield of the product and growth. It also allows easy manipulation of important factors and parameters along with the monitoring and control of conditions. Grossly, fermentation can be classified on the basis of different products such as biomass, microbial enzymes, microbial metabolites and biotransformation of precursors.

The process development for the production of acetone during the first world war (1914–1918) by submerged fermentation using *Clostridium acetobutylicum* led to the development and use of first ever large vessels. The first large aerobic fermenters were used in Europe in 1930s for production of baker's yeast. Prior to discovery of antibiotics during 1940s, ethanol, glycerol, organic acid, enzymes, and sorbose were the chief products of fermentation. Submerged culture techniques were developed quite rapidly for the purpose of penicillin production using aerated and agitated vessels for aerobic mould *Penicillium notatum.* Since then many fermenter types ranging from simple tank to complex integrated system of automated microprocessor based monitoring and control have been developed.

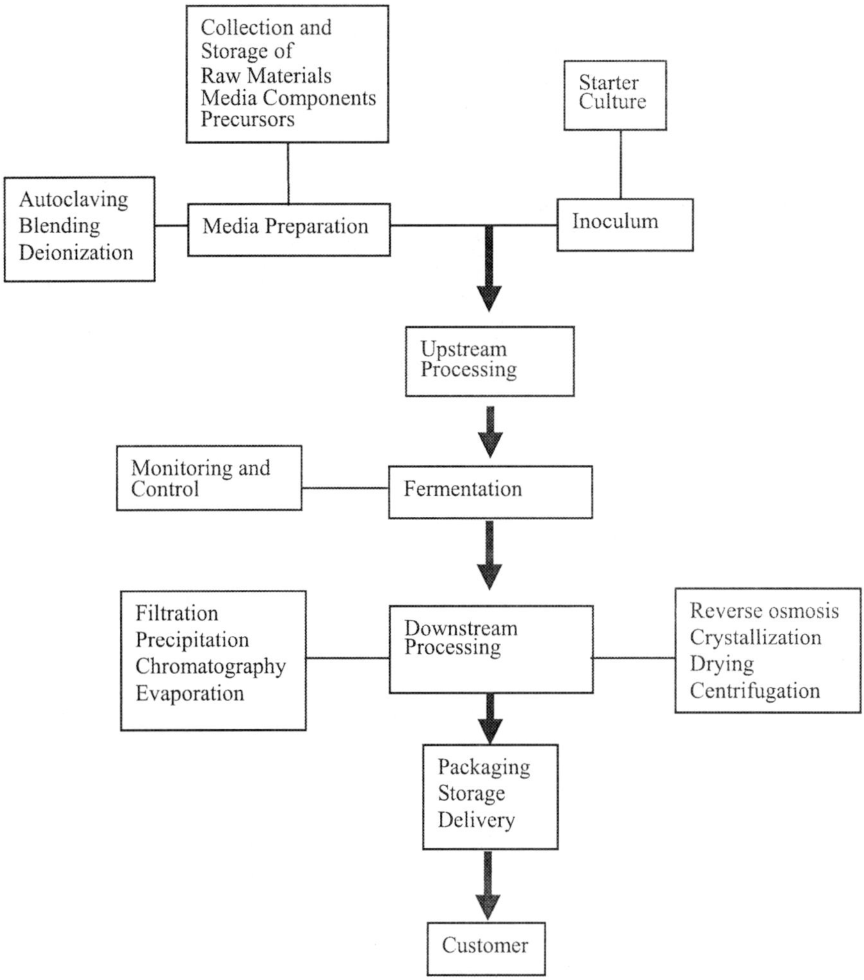

Figure 9.12: Flow chart of events in an industrial fermentation process.

Construction Material

Vessels of most fermenters are made up of either glass or stainless steel. Glass is most suitable for small size fermenter because it is smooth, non-toxic, inert, transparent and corrosion proof. A round bottom glass vessel with stainless steel top plate or a glass cylinder with stainless steel top and bottom plates are in common use. However, for larger volumes (more than 100 litres) vessels made up of stainless steel are used. This is because it is not advisable to make use of a glass fermenter of large volume since glass is fragile.

Agitator

Agitators are required to achieve various mixing objectives including bulk liquid and gas mixing, air dispersion, oxygen transfer, heat transfer and overall uniformity in the constituents and conditions (**Figure 9.13a**). Various types of agitators are used as per the suitability to the specific fermentation process. Suitability of impeller is particularly concerned with the ability of the microorganism to bear shear forces due to turbulence caused by agitation and rotation of blades. Some commonly used impellers use Disc turbine, Vaned disc, Open turbine, Marine propeller, Rushton turbine etc. (**Figure 9.13b**).

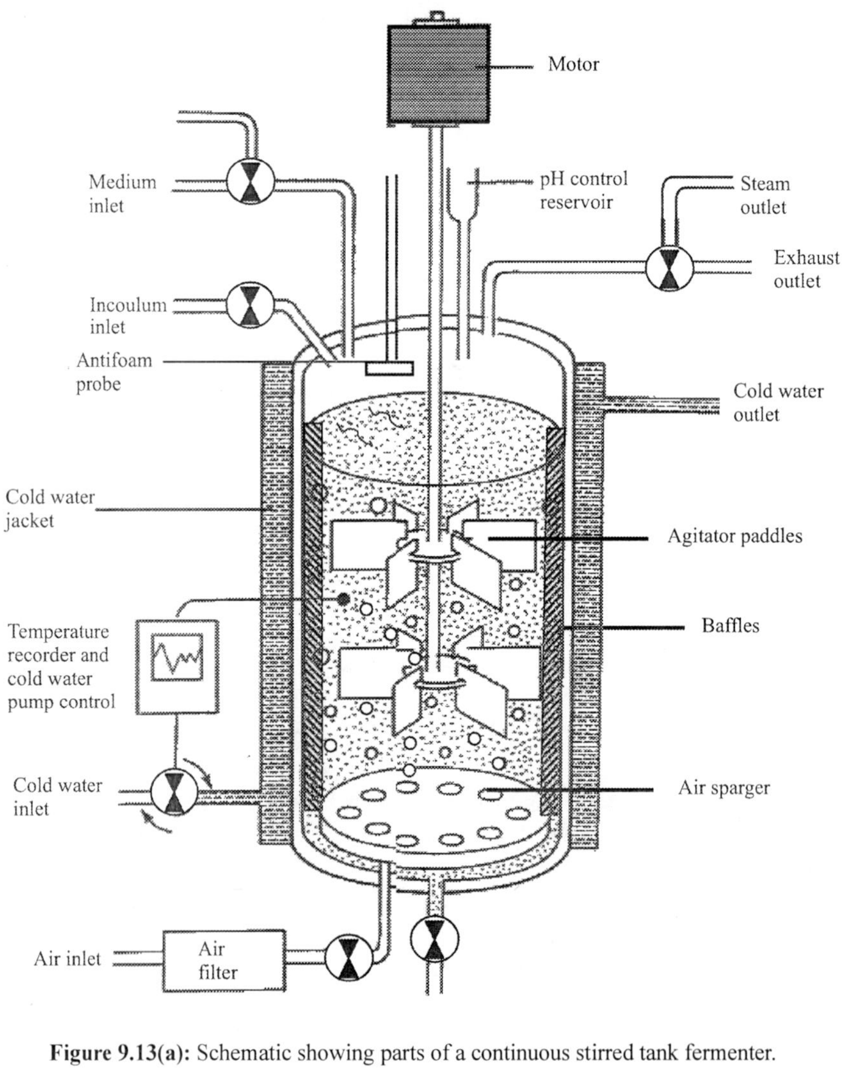

Figure 9.13(a): Schematic showing parts of a continuous stirred tank fermenter.

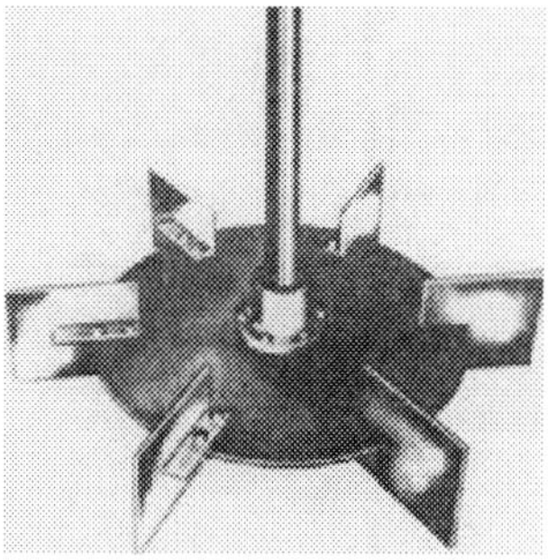

Radial flow flat blade disc turbine

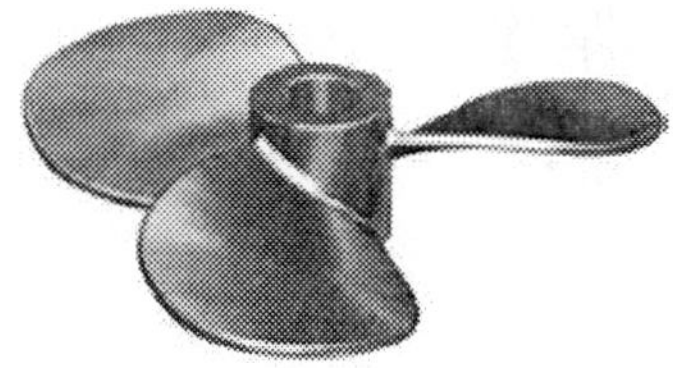

Square-pitch marine type impeller

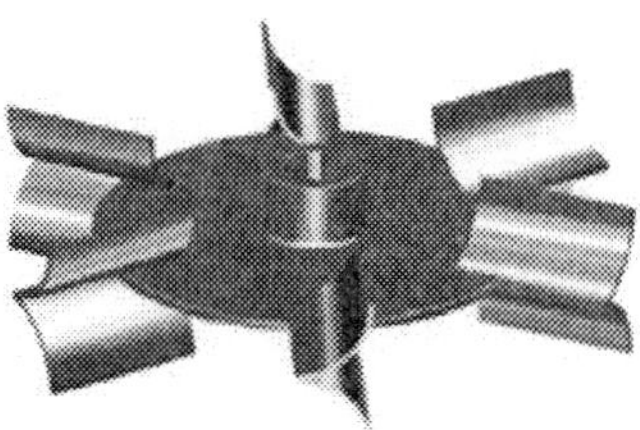

Double pitched Rushton turbine

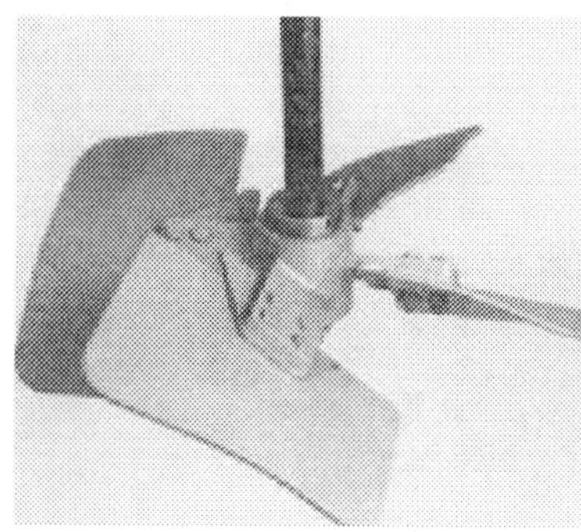

315 Fluidfoil impeller

Figure 9.13(b): Types of impellers commonly used in fermenters.

Stirrer Glands and Bearings

Maintenance of asepsis is of utmost importance for success of any fermentation process. At the same time various ports, glands and seals are used on the head plate of the fermenter to accommodate the important probes, stirrer shaft, ducts, etc. These are brought inside through the ports which should remain hermetically sealed during operation. Various kinds of bearings and mechanical seals have been introduced in conjunction with the agitator, shaft and motor. Ports and vents including those used for inoculation, sampling and entrance of probes also require to be aseptically sealed. Reliable aseptic seals are required between glass and glass, glass and metal and metal and glass joints. Between glass and metal, a seal can be made using compressible gasket while between metal to metal only O-ring in groove is suitable. Usually, nitryl or butyl rubbers that can withstand steam sterilization and fermentation conditions are used for this purpose.

Baffles

Any circulating fluid will produce a vortex at the origin of its motion. Continuous circular movement of the fluid due to agitator leads to the formation of vortex and as a result particulate or dense matter, such as microbial cells, may settle down at the centre of the vessel. To avoid this, four metal strips, about $1/10^{th}$ of the vessel diameter, are fixed along the length of the fermenter. These strips prevent vortex formation and also improve aeration efficiency by increasing turbulence.

Sparger

Most industrial fermentations are aerobic, e.g. production of penicillin and many other fungal enzymes. However, they are carried out under submerged conditions to achieve maximum production and best utilization of the medium in aerated vessels. In submerged condition, aerobic organisms take up oxygen from the small gas bubbles coming out of sparger. Oxygen is 6000 times less soluble than CO_2 in water and thus a continuous supply of oxygen is necessary in such cases. An adequate provision of oxygen supply and transfer is necessary for optimal growth of microorganisms. Sparger is an aeration device that introduces sterile air into the fermenter. The air bubbles coming up from sparger are shattered to smaller bubbles by the blades of the agitation device. These small bubbles travel through the liquid medium to the microorganism.

A variety of spargers suiting to the needs of different fermenters are available:

(i) Porous sparger
(ii) Orifice sparger
(iii) Nozzle sparger
(iv) Combined sparger agitator

Sensors and Probes

The success of any fermentation process depends upon the maintenance of well-defined culture conditions for the growth of microorganism and product formation. To achieve this objective it is most important to monitor the fermentation process and control any undesirable change rapidly. All such parameters can be classified into physical, chemical and biological types. Temperature, pH, agitation speed, concentration of dissolved oxygen are some of the most important parameters. Apart from these, several other conditions (such as biomass, enzyme and glucose, etc.) are also to be analyzed all over the period of operation. Monitoring of these conditions is achieved by means of specific probes and data obtained is analyzed continuously by microprocessor based units or modules (**Table 9.2**). Deviation from the specified optimum condition can be corrected by a manual or automated control system.

Table 9.2 List of important physical, chemical and biological parameters and sensors / probes used for monitoring

(i) Physical parameters	
Temperature	Autoclavable mercury glass thermometer Thermistor based thermometer, Thermocouple
Pressure	Stainless steel diaphragm / gauge, Burdon tube pressure gauge
Flow measurement	Thermal mass flow meter, Magnetic flow meter
Viscosity	On-line viscometer, Off-line viscometer, Shaft strain
Liquid level	Conductance probes, Capacitance probes, γ-ray sensors, Camera
Airflow	Orifice plate, Thermal mass flow meter, Vortex shedding meter
Agitator speed	Inductive proximity sensor, Optical shaft position encoder
(ii) Chemical parameters	
pH	Steam sterilizable (autoclavable) pH probe
Dissolved oxygen	Polarographic / galvanic probe
Vent oxygen	Polarographic sensor, Dual channel paramagnetic device, Mass spectrometer
Vent CO_2	Infrared analyzer, Mass spectrometer
Other vent gases	Catalytic degradation, Mass spectrometer
Dissolved CO_2	Gas-permeable membrane plus pH electrode (diffusion tube with mass spectrometer)
Ions in solution	Specific ion electrodes, Dialysis probes, Off-line analysis
Redox potential	Specific ion electrodes
(iii) Biological measurements	
Substrates (e.g. glucose, phosphate)	Enzyme-coupled electrodes, Off-line biochemical estimation
Products (e.g. penicillin, ethanol)	Enzyme-coupled electrodes, Off-line biochemical estimation
Biochemicals (e.g. RNA, ATP, NADH, cytochromes)	Off-line analysis, Spectral analysis
Biomass	Fluorimeter, Turbidometer, Mass balance, Off-line dry weight, Vent gas analysis, Packed cell volume (PCV)
Heat generation	Microcalorimetry, Heat balance

PURPOSE OF A FERMENTER

Essentially a vessel as small and simple as an Erlenmeyer flask can be considered as a fermenter. However, when we plan to produce any microbial product at large scale the scale of process has to be enlarged many times. Conditions in such a large vessel cannot be maintained easily and necessary provisions are to be made within the vessel. Looking at this, the main features of any fermenter can be underlined as follows:

(1) A fermenter can be operated continuously with stringent maintenance of aseptic conditions and total containment of the microorganisms.
(2) There must be provision of aeration and agitation suiting to the requirement of microorganisms.
(3) The operational power consumption should be less.
(4) There should be provision of measurement and manipulation of important parameters, e.g. pH, temperature and foam, etc.
(5) Vessels should be versatile and must allow multiple applications by making minor changes.
(6) The fermenter should allow easy harvesting, cleaning and operation.

Thus, it is apparent that the design of a fermenter will need technical input from a variety of professionals including microbiologist, bioprocess engineer, mechanical engineer, fabricators, electronic and microprocessor (software) engineers, etc. Apart from the design, a fermenter needs a number of accessory units including machines providing sterile compressed air, chilled and hot water, steam, control and monitoring equipment, etc. An uninterrupted power supply is very essential since any discontinuity in aeration may lead to undesirable consequences.

Industrial fermenters are designed to provide the best possible conditions for the growth and production of microbial product at large scale. The methodology of a bioprocess development involves detection and assay of the desired microbial product by screening a lot of microbial industrial strains. Comparison of the productivity leads to selection of strain. The selected strain is then grown in small size laboratory fermenters (10–15 litres). The process of production is optimized by manipulating conditions, nutrients, incubation time, inoculum level, etc. The process is then scaled up to the pilot plant (50–100 litres) and then if the result is satisfactory, the process is further scaled up to industrial scale (1,000–10, 0000 litres or more capacity).

Any industrial fermenter must be able to bear pressure of large volumes of aqueous medium and must have all the provisions for manipulation of culture conditions as desired. Continuous fermentations are run continuously without emptying the vessel at each harvest of product or cells. Fresh and sterile medium is added intermittently or continuously, equal to the amount of spent medium withdrawn with cells and product. By controlling the amount of limiting component in nutrient medium, the growth of the organism is maintained in the desired phase of growth cycle. Continuous fermentation set up requires monitoring of growth, amount of product formed, regulation of addition of fresh medium, and withdrawal of spent medium. Provision of media reservoir, periastaltic pumps, valves and piping for addition and withdrawal is also necessary. Major problems with this kind of fermentation are that of contamination and genetic stability of the microbial strain over the long period of incubation.

Main Products of Fermentation

- Alcohol
- Vinegar
- Baker's yeast
- Glycerol
- Organic acids: citric acid, lactic acid
- Organic solvents: acetone, butanol
- Antibiotics: penicillin, streptomycin
- Amino acids: L-lysine, glutamic acid
- Microbial enzymes: amylase, cellulase, protease, xylanase and inulinase
- Single cell protein: *Spirulina*, mycoprotein
- Heterologous proteins: recombinant DNA vaccine, insulin
- Monoclonal antibodies.

Any industrial fermentation process can be divided into different stages including:

(i) Collection and storage of raw materials, e.g. sugar cane molasses, malt, barley, etc.

(ii) Upstream processing which involves the processing of raw materials for the fermentation, preparation of inoculum.

(iii) The fermentation process where microorganism grows and brings about conversion of raw materials.

(iv) Downstream processing for separation, purification and concentration of the product.

(v) Packaging, storage and delivery of the product.

Among these stages, upstream and downstream processing make the two most important stages of any industrial fermentation. The upstream processing involves all factors and processes that lead to fermentation and also includes the fermentation process. The major components of upstream include maintenance of microbial strain, preparation of inoculum or seed culture, preparation of fermentation media and running or operation of fermenter. On the other hand, the downstream processing includes cell harvesting, product purification, cell disruption (for intracellular products) and packaging of the product.

The submerged liquid cultures are mainly managed in two formats or operated in two modes:

(i) Batch fermentation

(ii) Continuous fermentation

In the batch mode, the reactor is filled with medium, sterilized, inoculated and fermentation is allowed to proceed. After a due incubation period, when fermentation is over, the contents are emptied for downstream processing. The reactor is then cleaned, refilled, reinoculated and the process is repeated again.

Batch processes are dynamic processes that are never in a steady state. Often in batch processes the critical parameter is gas exchange or balance between respiration rate and oxygen transfer. In batch operation, the sterilized media components are supplied at the beginning of the fermentation with no additional feed after inoculation. When cells are grown in a batch reactor, they go through a series of stages following a typical growth curve.

In case of continuous fermentation, fresh media is continuously added and spent medium containing product is continuously removed. As a result, cells continuously receive fresh medium while product, waste and spent medium is continuously harvested. The reactor can thus be operated for long periods of time without any gap or shut down. Continuous reactors are very cost effective, easy and more productive than batch reactors (**Figure 9.14a**). The underlying reasons behind greater productivity are that the fermentation does not need to be shut

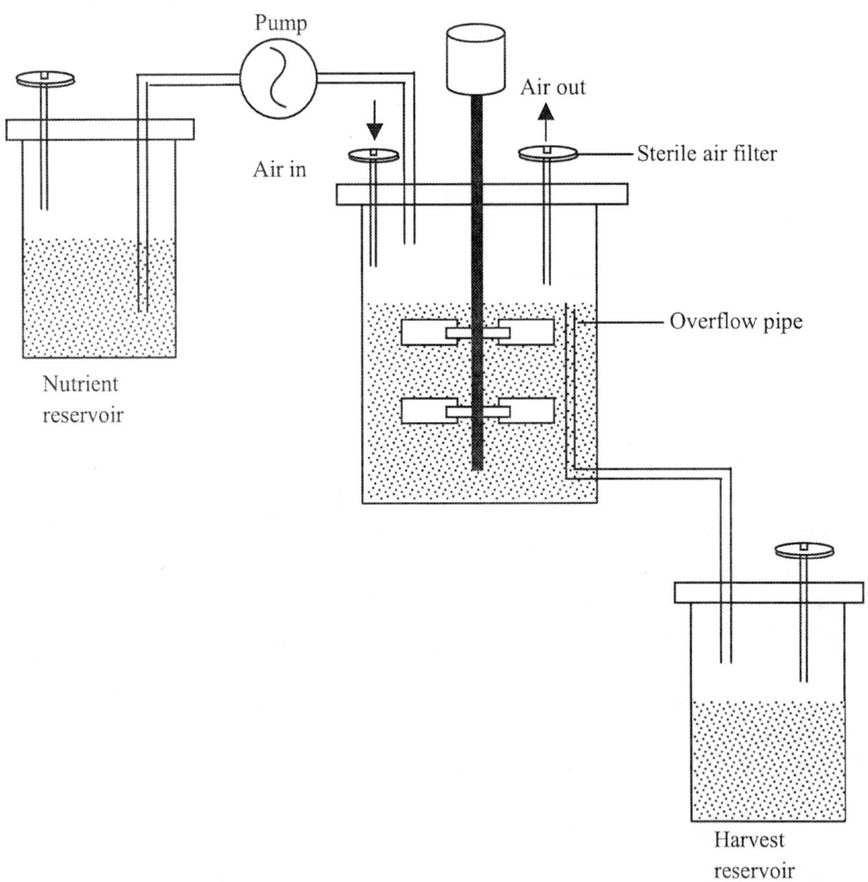

Figure 9.14(a): Set up for a continuous fermentation.

down as in batch cultures and the growth of the microorganism is maintained in the desired phase for longer time. This avoids repetitions of several steps. An in-between alternative of these two formats is **fed batch fermentation** and it is the most common type of method used in industrial sectors. In this kind of fermentation, fresh media is added periodically or continuously to the bioreactor but unlike a continuous reactor, there is no continuous removal. The fermenter is emptied or partially emptied when reactor is full or fermentation is finished. As with the continuous reactor, it is possible to achieve high productivities in fed batch also due to the fact that the growth rate of the cells can be optimized by controlling the flow rate of the feed entering the reactor.

Bioreactors can be classified on the basis of different criteria and parameters including:

(i) Mode of operation of bioreactor: batch, continuous, or fed-batch.
(ii) Configuration and design of the fermenter: tank dimensions (height: diameter <3), column dimensions (height: diameter > 5).
(iii) Physical form of microorganism or biocatalyst: free cells in submerged cultures; bound or immobilized cells / enzymes
(iv) Aeration: liquid phase; gas phase; combined
(v) Agitation: mechanical stirring, airlift, bubble column
(vi) Hydrodynamics: homogeneous mixing; incomplete mixing; no mixing.

Table 9.3 lists some of the major types of fermenters and their distinguishing features.

Table 9.3 Basic types of bioreactors and their characteristic features

Type of Bioreactor	Characteristic feature
Continuously Stirred Tank Reactors (CSTR)	These reactors employ impellers for homogeneous distribution of oxygen, cells, heat, media and other materials.
Bubble column reactors	Tall reactors having very large height : width ratio (more than 10) and use forced air alone for mixing of the contents
Airlift reactors	These reactors are similar to bubble column reactors, but have an internal or external draft tube which improves circulation and oxygen transfer and balances shear forces in the fermenter.
Fluidized bed reactors	In fluidized bed reactors, cells are "immobilized" as small particles which move with the fluid. The small particles create a large surface area for cells to stick to and enable a high rate of transfer of oxygen and nutrients to the cells.
Packed bed reactors	In packed bed reactors, cells are immobilized on large particles but these particles do not move with the liquid. Packed bed reactors are simple to construct and operate but can suffer from blockages, channelling and poor oxygen transfer.
Solid state fermenter	Fermenters that employ particulate organic substrate in absence or near absence of water.

Continuously Stirred Tank Reactors (CSTR)

Continuously stirred tank reactor (CSTR) is the choice for many fermentation processes because it offers several advantages. The idea of continuous stirring is to achieve complete homogenization, even dispersion of gas-liquid mixtures, thorough aeration of liquid and ease of heat and mass transfer. The CSTR has a rotating agitator with two to four impellers attached on the shaft. The agitator can either be electrically driven or more recently, magnetically driven agitators are also being used. In fermenters with electrically driven motor, the port from where agitator shaft enters the vessel through head plate, is often a potential route of contamination. Contamination is avoided by using double seals. The effectiveness of agitation depends upon the design of the impeller blades, speed of agitation and the rheology of medium. Most CSTRs have height-diameter aspect ratios of 3:1 or 4:1. Casters must create high turbulence to maintain high transfer rates, but this also generates considerable shear force that is detrimental to sensitive cells, e.g. animal cells may be disrupted.

CSTRs, therefore, are not appropriate in case of highly sensitive cells (without cell walls) and other alternatives such as airlift or supported biofilm reactors may be used. Baffles are always used in CSTR so as to avoid formation of any vortex. These are usually flat vertical plates whose width is about one-tenth of the vessel diameter. Four such baffles are fitted along the vessel walls to promote efficient mixing and mass transfer by increasing turbulence, preventing vortex formation and eliminating 'dead spaces'. Most important parameters in CSTRs include kind, position and the number of impellers and size of the paddles.

These parameters determine the hydrodynamic transfer patterns in the reactor, which in turn influence mixing times, mass and heat transfer coefficients, shear rates etc. The conventional fermentation is carried out in a batch mode. Most industries utilize the advantages of CSTRs in producing important commodities in batch fermentations. The batch stirred tanks have demerit of their low volumetric productivity. Also the transitional state of fermentation imposes stress upon the microbial cells due to depletion of nutrients. The fed batch fermentation, in which limiting nutrients are added periodically, successfully eliminates this limitation. The operational costs are not very high and the reactors can be used with a variety of microbial species. CSTRs require provision of *ex-situ* or *in-situ* sterilization, steam, water supply as coolant, sensors and computer aided modules for real-time monitoring of important process parameters such as pH, temperature and foam, etc.

Airlift Fermenter (ALF)

These are pneumatic reactors without any provision for mechanical stirring and the expansion of compressed gas is utilized to bring about the mixing. The turbulence caused by movement of gas causes the fluid flow and results in mixing of the liquid. This movement is accomplished with the help of a draft tube provided in

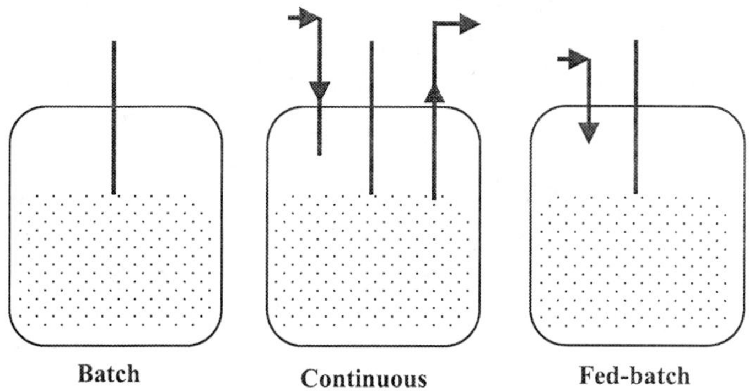

Different modes of fermentation: batch, continuous and fed-batch

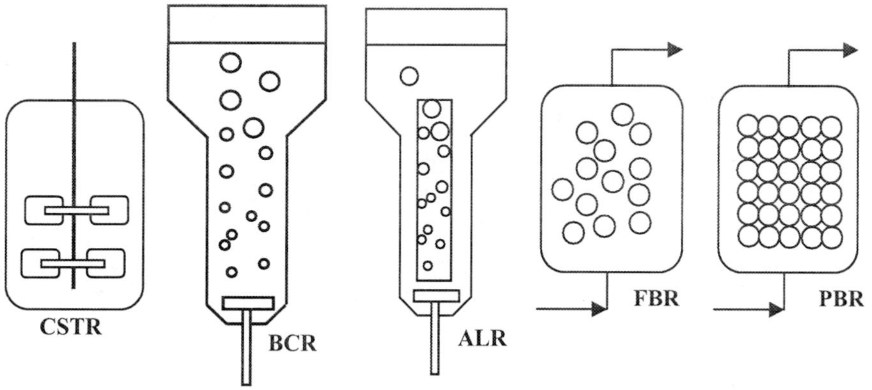

Various types of fermenters: CSTR (continuously stirred tank reactor), BCR (bubble column reactor), ALR (airlift reactor), FBR (fluidized bed reactor), PBR (packed bed reactor).

Figure 9.14(b): Different modes and types of fermentations.

the central section of the fermenter. The introduction of the fluid (air/liquid) causes upward motion and results in circulatory flow in the entire reactor. The movement of fluid is fast enough to allow cooling by means of an external jacket only and there is no need of internal cooling coils. The airlift fermenters have very low operational and maintenance costs and due to low liquid velocities, shear forces are also negligible. Therefore, these can be used for both free and immobilized cells. Airlift reactors eliminate shear effects generally encountered in mechanically agitated reactors. These reactors are suitable for aerobic cultures since oxygen mass transfer coefficient are quite high. ALFs are also cost-effective because of simple instrumentation and no input of power for mechanical agitation and aeration.

Fluidized Bed Bioreactor (FBR)

These reactors are used for systems involving use of cells as biocatalysts in different phase systems including all the three phases, viz. solid, liquid & gas. These are commonly operated in co-current up-flow with liquid as continuous phase. The bed is loosely packed and fluidization is obtained either by circulation of liquid or by feeding gas inside the reactor. Immobilized enzymes FBR involve two-phase systems, i.e. solid and liquid but the use of aerobic cells requires provision of gas (air). FBR and the airlift bioreactor can be distinguished on the basis that the latter have a draft tube for effecting agitation. The circulatory motion of the liquid in ALF is induced due to the draft tube. The particles used in fluidized bed can be of different types including cells / biocatalyst attached to inert carriers, or entrapped in porous material or self-immobilized as floccules. The advantages of FBRs include very low abrasion or shear of cells or solid particles as compared to CSTRs. There is full possibility of reuse of immobilized catalyst in FBRs. These reactors employ smaller size particles thus facilitate higher mass transfer rates and better mixing. The productivity obtained in FBRs is usually higher than that of CSTRs.

Packed or Fixed Bed Bioreactors

In these reactors immobilized live cells or biocatalysts are packed in the column without any free space for movement and fed with nutrients either from top or bottom. Trickling filter used for bioremediation of waste water can be a good example of packed bed or fixed bed bioreactors where an attached biofilm is used for purifying water. Many packed beds employ alginates, carrageenan and other matrices for immobilization. The major drawback with the packed bed bioreactors is the alteration in their hydrodynamics with the change in flow during the operation. Use of soft gels (e.g. alginate) usually leads to bed compaction which generally occurs during fermentation resulting in high pressure drop across the bed. The bed compaction may also disrupt the gel integrity and channeling may occur due to turbulence in the bed.

Solid State Fermenters (SSF)

Solid state fermentation can be defined as the fermentation where there is absence or near absence of free flowing water in the substrate. Solid state fermentation involves the growth of microorganisms on particulate organic material moistened with the minerals or salt solutions or plain water. The substrates used are often agricultural produce or low-value agricultural residues or lignocelluloses, such as cereals, fruit peels, wheat bran, wheat straw, wood chips, husk, etc. Many traditional processes such as food fermentations, composting and silage making are basically solid state fermentations. Commercially, several important microbiological products including enzymes, organic acids and alcohols are now produced by solid state fermentations. Solid state fermentation does not require any kind of sophisticated instrumentation or equipments or control mechanisms as used in submerged fermentations. It does not require any costly instrumentation for continuous agitation, aeration, etc. SSF is very cost-effective and generally the productivity is much higher than submerged fermentation since in the absence of water there is hardly any repression of enzymes and feed back inhibition of metabolic processes. The downstream processing is also easier and comparatively less amount of effluent or waste water is generated. However, it has some demerits also. It is very difficult to monitor and control SSF processes due to physical hindrances and lack of knowledge of kinetics of microbial growth under these operating conditions. Maintenance of desired temperature and moisture simultaneously is again a tough job. In case of sporulating moulds, containment of spores is also very difficult. However, it still remains to be a method of choice especially in case of fungal strains that do not perform in submerged fermentations. A routine SSF involves pretreatment of a substrate such as chopping, cutting, washing, etc., moistening the substrate with appropriate solution and sterilization followed by inoculation with spores. During incubation, the medium can be shaken or mixed intermittently to somewhat homogenize the contents. **Figure 9.15** shows some of the devices used in solid state fermentations. After completion of the due incubation period, product is extracted out of fermented bran using appropriate solvent and purified by downstream processing.

ELECTROPHORESIS

Electrophoresis is defined as the migration of individual ions, molecules or macromolecules (e.g. proteins) through a solution or medium under the influence of electrical field. This movement is due to the Lorentz force, which may be related to fundamental electrical properties of the substance under study and the ambient electrical conditions by the equation given below. Lorentz force (**F**) in the electrical field (***E***) and the charge (**q**) of the substance can be related as

$$F = qE$$

Application of potential difference in a solution makes some ions or molecules to behave as charged species and hence move towards either electrode depending upon their charge. The reason being that a charged particle suspended between the poles of electrical field tends to move towards the pole bearing opposite charge. However, the rate of movement of particle will depend upon several factors including: (i) characteristics of the molecule or particle, i.e. charge, mass, chemical composition (ii) electric field (iii) temperature (iv) medium. Thus, when a mixture of ionic species, in a particular medium and temperature, is subjected to an electric field, its components migrate as per their charge: mass ratio. Any technique employing such a migration of components under the influence of electric field for separation or purification is referred to as **electrophoresis**.

Electrophoresis can be carried out in a free solution or through a supporting matrix or medium (e.g. paper or gel). The latter is also referred as electrochromatography or zone electrophoresis. In free solution electrophoresis,

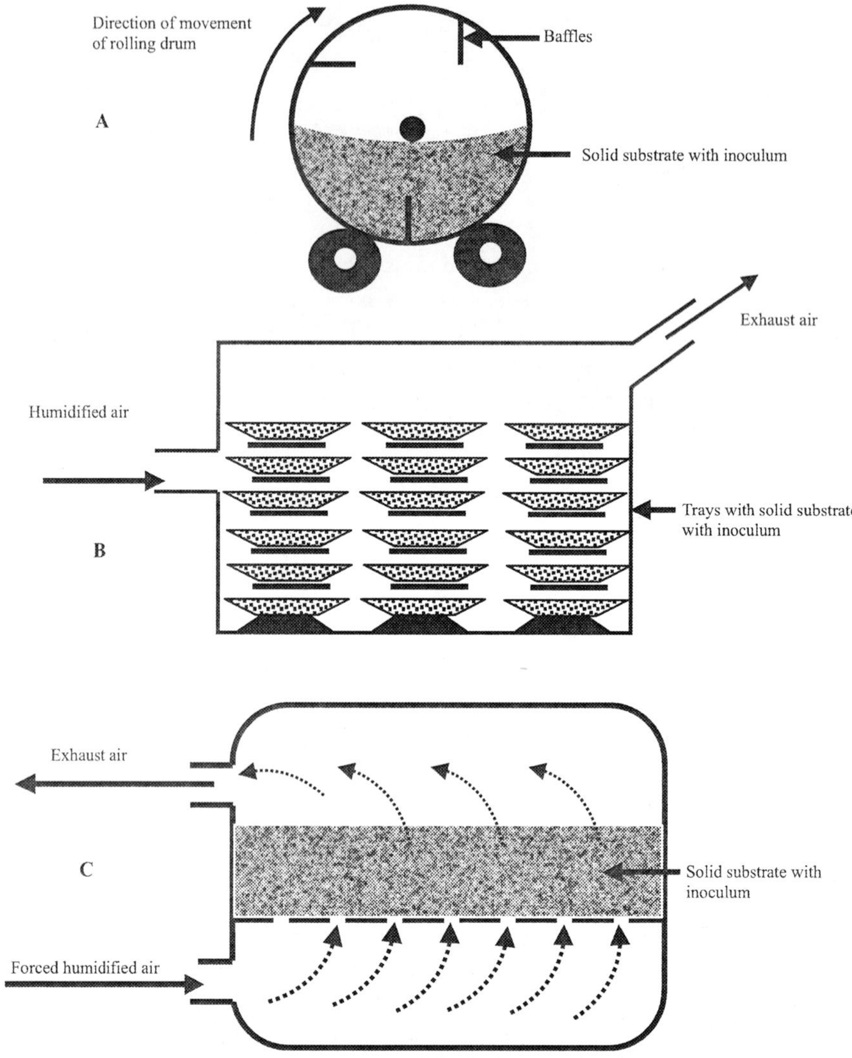

Figure 9.15: Types of solid state fermenters: (A) Rotating drum fermenter (B) Tray fermenter (C) Bed system.

the sample solution (sol) is injected at the bottom of a U-tube filled with buffer solution and an electric field is applied by means of electrodes located at the ends of the U-tube. Charged particles separate as per their differential movement on the basis of their mobility which is directly proportional to their charge: mass ratio. The free solution method was perfected by Tiselius for separation of proteins. He was awarded Nobel Prize for this. He perfected the separation of serum proteins using an apparatus consisting of a U-shaped glass cell of rectangular cross-section having two parts, one of which can be made to slide on the other (**Tiselius method**).

The lower part of the tube was filled with protein sample as a buffered solution, while the upper part contained only buffer. The electromotive force was applied at the two ends by electrodes placed at the ends of the U cell. Temperature was maintained at 4°C since the density of water at this temperature is maximum and hence density differences and convection currents are minimized. Migration of proteins was recorded by very sensitive photographic devices. However, the method has several limitations such as back mixing of the separated components due to thermal convection or mechanical jerks and need of an optical system for location and measurement of fractions.

In **moving boundary electrophoresis**, the protein sample, on the basis of their differential mobility, separate as distinct boundaries with clear demarcations. Each fraction of proteins is separated as a cylinder of solute migrating independently. Fractions are detected by scanning tube with a refractometer and concentration of different components is calculated with respect to the change in refractive index. Another modification of free solution electrophoresis is **density gradient electrophoresis** which employs a vertical cylinder filled with an electrolyte containing inert solute, e.g. sucrose with graded density throughout the tube. The sample is forced to move through the density gradient. Back mixing of sample components is avoided due to absence of convection currents and complete separation is achieved.

Introduction of an anti-convection medium, such as paper or gel, to stabilize separated components as discrete bands lead to development of zone electrophoresis or electro-chromatography. The technique resembles chromatography with an additional parameter of applied electric field which quickens the rate of migration of sample components. The separations are a result of electrophoretic mobility of various components in matrix or fine meshes of the stabilizing medium.

The stabilizing media are numerous and range from a filter paper to agarose and polyacrylamide gels, etc. The choice depends on the sample to be separated and underlying detection methods. Paper chromatography is widely used as paper is cheap and readily available. However, it can carry a very small amount of sample and hence many samples are difficult to detect. On the other hand, gels can be cast as thick as desired with sample capacities up to 0.1 ml and offer great deal of anticonvection stabilization on behalf of their fine mesh network. Depending upon the physical nature of gels they are cast in columns in vertical cylinders or glass plates or as horizontal thin layers supported by glass or plastic trays. The agarose and polyacrylamide gels are used routinely for electrophoretic separations of DNA and proteins, respectively.

Requirement for Conducting Electrophoresis

Gel electrophoresis unit: The unit consists of buffer chambers and a support for stabilizing media (paper or gel). It is horizontal in case of paper and vertical in case of gel.

Paper or gel: Various types of papers such as (Whatman 1, Eaton-Dikemann 301) can be used for separation of proteins samples upto 20 ml of sample. For PAGE, gel is cast in between pair of glass plates separated by plastic spacers and held vertically by clamps.

Electrodes: These are made up of graphite rods, stainless steel, silver chloride or most commonly platinum wires in the form of fine wires or sheets.

Source of current: Usually referred as Power Supply, it supplies direct current between 100–300 volts by rectifying the alternating current of our local power supplies.

TYPES OF ELECTROPHORESIS

Paper Electrophoresis

A paper strip is wetted in buffer and the excess of buffer is removed by absorbent filter papers. This strip is positioned horizontally in the electrophoresis apparatus consisting of two buffer chambers filled with buffer solution. The sample is introduced at the centre of the wet strip and a direct current (D.C.) power supply is tuned on that applies 100–300V across the electrodes. As a result of electric field, charged components of the mixture will migrate towards the electrode having opposite polarity. After the run, paper is dried and developed using suitable reagents for localization of components (as in case of paper chromatography).

Agarose Gel Electrophoresis

Agarose is linear polysaccharide made up of alternating unit of galactose and 3, 6-anhydrogalactose. Agar is the mixture of polysaccharides isolated from certain see weeds. Gels are formed by suspending dry agarose in aqueous buffer, then boiling the mixture until a clear solution forms. This molten solution is then poured and allowed to cool at room temperature to form a rigid gel. Gelling properties are attributed to both inter and intra-molecular H-bonding within and between the long agarose chains. The pore size of gel is controlled by the initial concentration of agarose. Agarose gels are used for the electrophoresis of nucleic acids, proteins and in techniques such as immunoelectrophoresis or flat-bed isoelectric focusing.

Polyacrylamide Gel Electrophoresis (PAGE)

Proteins are one of the most important biomolecules from structural and functional point of view. Analysis and identification of proteins is a decisive factor in most areas of research in biology. Electrophoresis is specially useful for analysis of most functional proteins, e.g. immunoglobulins and enzymes. The benefit of this method is that we can separate as well as visualize proteins and hence these can be characterized in terms of number, molecular weight, activity, isoelectric point, etc.

The first step is their separation using gel electrophoresis. When placed in an electric field, protein molecules with a net charge will move towards either of the electrodes. The greater the net charge, the faster will be the movement. Proteins in their native state are separated on the basis of their mass (size) and their net charge. Smaller molecules move faster as compared to larger molecules through the fine mesh of the gel. Generally, the buffer used in reservoir has a pH around 9.0–9.5. Most proteins at this pH have a net negative charge and thus migrate towards anode placed in lower reservoir. If the electrical potential is E, the electrophoretic mobility of the molecule, μ is the ratio of the velocity of the particle V to the electric potential. It can also be defined as the ratio of the net charge of the molecule, Z to its frictional coefficient, f:

$$\mu = \mathbf{V/E = Z/f}$$

Gels used for protein separation are usually made up of polymerized acrylamide that provides an inert support. Hence, the process is named as poly acrylamide gel electrophoresis (PAGE). The pore size of the gel is maintained by using appropriate concentration of acrylamide and the cross-linking agent methylene bis-acrylamide. Higher concentration of acrylamide results in small pore size. The gel is cast between glass plates separated by a distance of 0.5–1.0 mm by plastic or rubber spacers and plastic combs of desired size is put in the gel before it gets set. A blue dye, bromophenol blue, is used for loading the protein samples in the well. Since its molecular weight is small it travels ahead of the protein sample and the frontline of the migrating gel can be located with the help of this dye. An electric current of 100–300 mA is applied through the reservoir buffers and gel through which the protein moves. The gel is then removed from glass slabs and stained with Coomassie blue or silver stain to visualize different proteins as distinct bands.

In case of denaturing gel electrophoresis or SDS-PAGE the proteins are heat denatured and coated with negative charge of sodium dodecyl sulphate (SDS) molecules and the basis of their separation is mass only. First the protein samples are completely denatured by boiling and treating with a reducing agent, e.g. 2-mercaptoethanol or dithiothreitol to break intramolecular bonds in the protein sample. SDS, a strong anionic detergent disrupts nearly all the non covalent bonds in the protein and unfolds it completely. One molecule of SDS binds via its hydrophobic

Table 9.4 Change in protein conformation and charge after SDS treatment in denaturating gel electrophoresis

	Mass	Charge	Shape	After SDS treatment		
				Shape	Charge	Mass
Protein A	30kD	+3	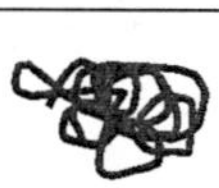		–300	30kD
Protein B	42kD	–4		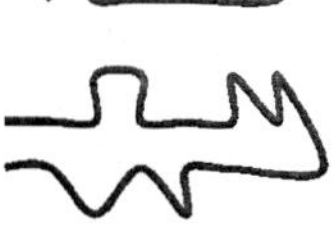	–420	42kD

alkyl chain to the polypeptide backbone for every two amino acid residues, which gives the denatured protein a large net negative charge that is proportional to its mass (**Figure 9.16**). This sample is then applied to the gel and electrophoresed.

Proteins are visible as blue colour bands and their molecular weight is ascertained with the help of standard markers.

Gel electrophoresis can be used to identify protein of interest using immunoblot, Western blot and zymography. For western blotting, the proteins are transferred to a nitrocellulose membrane by overlaying the gel with nitrocellulose and blotting

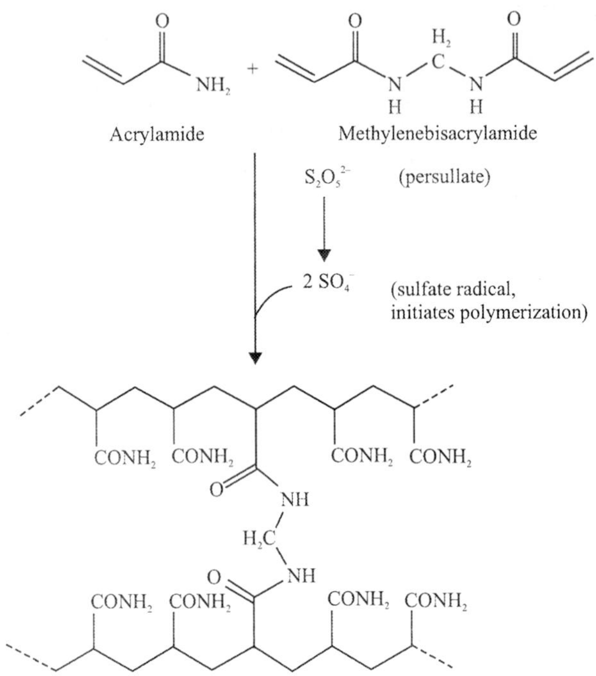

Figure 9.16: Formation of polyacrylamide gel by cross-linking between acrylamide and bis-acrylamide.

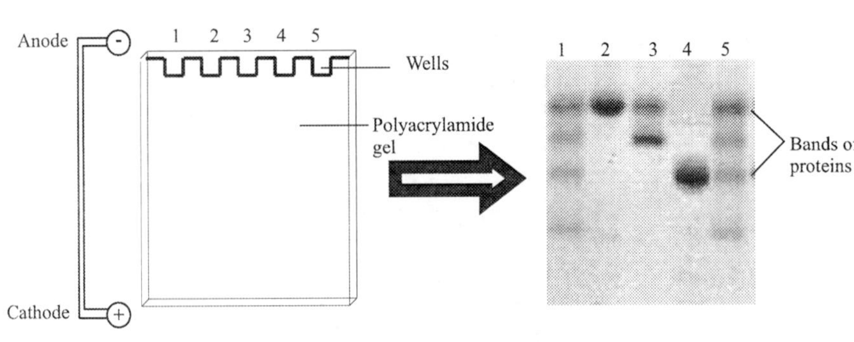

Figure 9.17: Polyacrylamide gel electrophoresis (PAGE) for separation of proteins.

the protein up by suction or electric current. The nitrocellulose membrane binds the proteins and becomes an exact image of the gel, i.e. the bands of proteins are held in the similar positions as in the gel. The remaining binding sites on the nitrocellulose paper are blocked by skimmed milk and then the membrane is placed in a solution containing homologous antibody to the desired protein. Unbound antibody is removed by washing and desired protein is localized using radio-labelled secondary antibody and placing the nitrocellulose on an X-ray film. The secondary antibody can also be coupled to an enzyme as in the case of ELISA.

DENSITOMETER

While the spectrophotometer is used for measuring absorbance of clear solutions to interpret the concentration of solute on the basis of absorption at λ_{max}, a densitometer quantifies the amount of a substance in form of a spot on a film, paper, glass or a solid opaque surface. It provides an estimate of the substance in

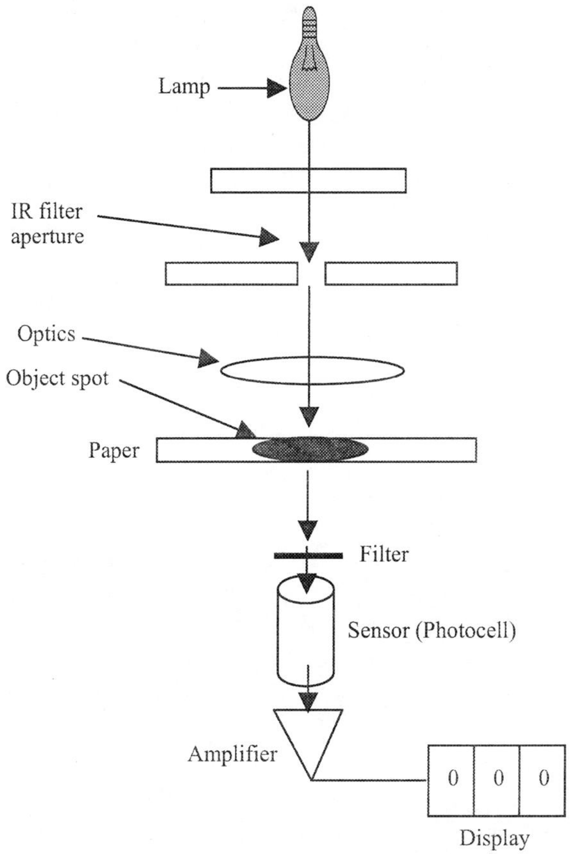

Figure 9.18: Line diagram of a transmission densitometer.

terms of the degree of darkness or the optical density. The densitometer has a light source directed at a photoelectric cell or sensor that receives light coming through the object sample or spot and determines the density of the sample based on the differences in the absorption. Based on the optical principle and the test material used, there are two types of densitometer:

(i) Transmission densitometers that measure transparent materials: this is used for measuring the thickness of films, colour on a transparent film, chemicals or compounds forming a transparent spot on a film (**Figure 9.18**)

(ii) Reflection densitometers that measure light reflected from a surface: these are used to measure the density of spots formed on opaque sheets, e.g. Aluminum sheets (**Figure 9.19**).

Commonly these are used to measure thickness of transparent films of varying densities in photography. A common application of reflection densitiometry is the measurement of density of spots on the developed TLC plates. In such a case, the concentration of a particular compound of interest resolved from a mixture on TLC plate can be estimated based on absorption.

A densitometer consists of a light source that has a stable output, optical lenses to focus the light into a defined light spot on the sample, filters to define the spectral response of the unit and a photocell to detect the reflected light. The sample is usually illuminated at 90° and viewed at 45° to the surface. This view angle eliminates reflections due to glossy surface of the film and only the diffuse

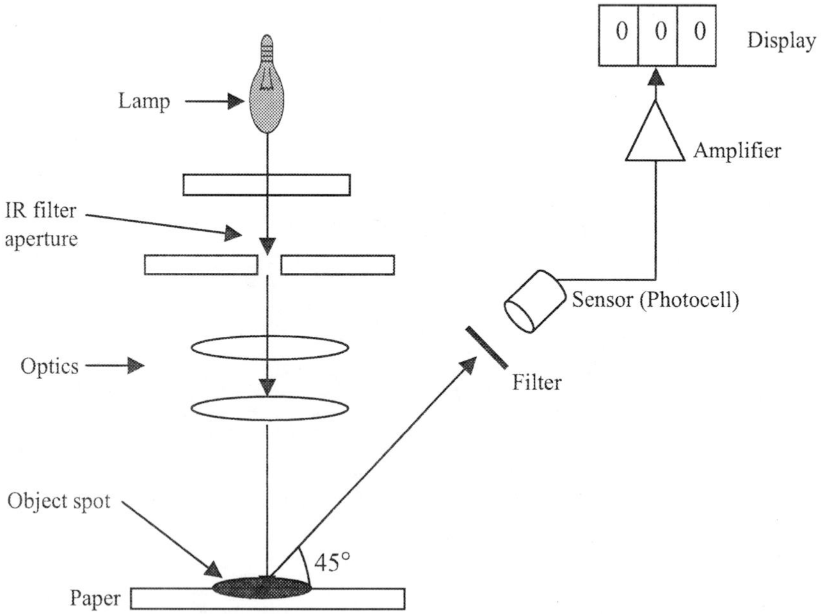

Figure 9.19: Line diagram of a reflection densitometer.

reflections are accounted by the detector. The density measurement scale is logarithmic, i.e. if a density equivalent to 1.00 indicates that 10% of the light has been reflected and then a density of 2.00 shows a 1% reflection meaning that the latter is twice as dark. Over a given narrow range, the density readings from a densitometer are approximately proportional to the thickness of the spot.

FLUORIMETER

Fluorescence is the emission of a longer wavelength radiation by a substance as a consequence of absorption of energy from a shorter wavelength radiation, continuing only as long as the stimulus is present. Fluorescence can be distinguished from phosphorescence as in the latter; the emission persists for a perceptible period of time after the stimulus has been removed. Fluorescence indicates that the absorbance of radiation results in emission of electromagnetic radiation (often in the visible range, ~300 – 800 nm, but may involve other ranges, e.g. x-ray). A great range of chemicals and biomolecules fluoresce naturally. These include proteins (heme proteins-haemoglobin, myoglobin, cytochrome C, hemocyanin; redox & photoproteins, phytochromes, phycoerythrin, phycocyanin, green fluorescent protein, phycobiliproteins, rhodopsin, ferredoxins), natural pigments, amino acids e.g. tryptophan, tyrosine and purine nucleotides. Several chemical dyes (ethidium bromide), minerals and metallic chelaters also fluoresce. Such compounds or molecules can be estimated quantitatively based on fluorescence and measured using **a spectrofluorimeter or simply a fluorimeter.** Thus, fluorimeter is an instrument used to determine the intensity and the wavelength of the light emitted as fluorescence from a molecule excited at a specific wavelength. Characteristic fluorescence bands may be used to identify compounds based on their fluorescence spectra.

Principle

Fluorescence is an emission incident and in Fluorimetry the energy transition from a higher to lower state of the molecule is measured. Consecutively, for the transition from higher to lower states to occur, a preceding excitation event, for example, caused by absorption of electromagnetic radiation, must have taken place. The wavelength(s) of absorbed radiation must be at lower value (higher energy) than the released (fluoresced) wavelength. The difference between these two wavelengths is known as the stock shift and in general the best results are founded from compounds involving large shifts. It is possible for a compound to absorb in the ultraviolet region and emit or fluoresce in the visible.

The emitted radiation appears as band spectra because there are many wavelengths dependent upon the final vibrational and rotational energy levels achieved. These band spectra are generally independent of the wavelength of the exciting radiation and have a mirror image association with the absorption peak with the greatest wavelength. Fluorescence spectra give information about events that occur in less than 10^{-8}s.

The quantum efficiency (Q) is usually independent of the exciting wavelength.

$$Q = \frac{\text{Quanta fluoresced}}{\text{Quanta absorbed}}$$

At low concentration, the intensity of fluorescence (I_f) is related to the intensity of the incident radiation (I_o) by:

$$I_F = 2.3 I_0 \varepsilon_\lambda c d Q$$

where c is the concentration of the fluorescing solution (molar), d is the light path in fluorescing solution (cm), ε_λ is the molar extinction coefficient for the absorbing material at wavelength λ ($dm^3mol^{-1}cm^{-1}$). The technique of spectrofluorimetry is most accurate at very low concentrations, whereas absorption spectrophotometry is least accurate at these concentrations.

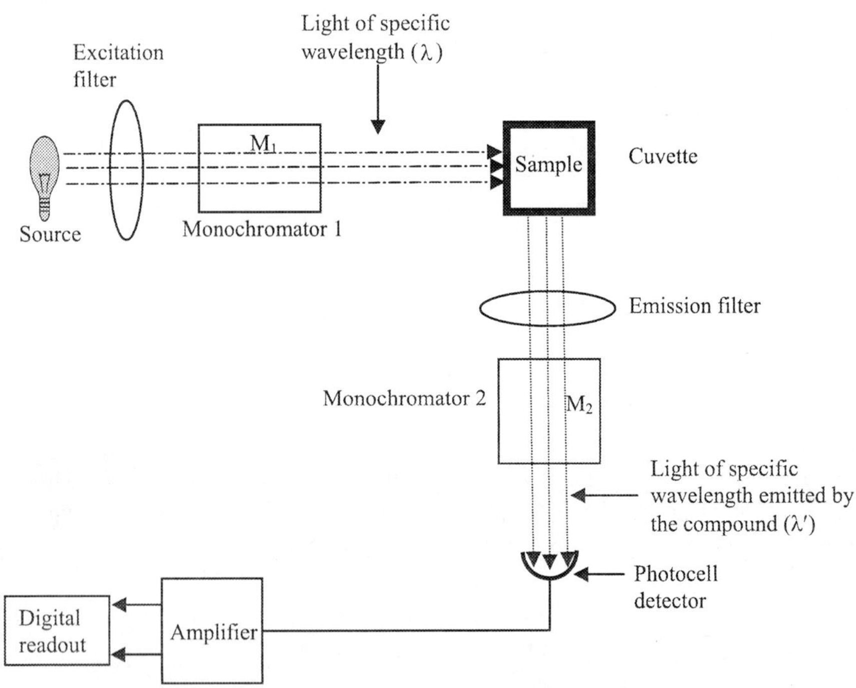

Figure 9.20: Schematic showing setup and working of a fluorimeter.

INSTRUMENTATION

There is a direct dependence between fluorescence intensity and concentration and hence the instrumentation in fluorimetry uses simple electronics and optics.

Two monochromators are used for specific purpose. The first (M_1) for selecting the excitation wavelength and the second monochromator (M_2) is used for selection and determination of the specific fluorescence spectrum emanating from the sample cuvette. The source of radiation is mercury lamp or xenon arc lamp that produces excitation wavelengths in the ultraviolet region. The detector is usually a sensitive photocell made up of metal plate coated with semiconductor. Fluorimetry should be done under standard temperature control since the intensity of fluorescence may vary between 10% and 50% for 10°C change.

SUGGESTED READING

Browen T.J. 1970. *An Introduction to Ultracentrifugation*. Wiley-Interscience, New York.

Campbell I.D. and Dwek R.A. 1984. *Biological Spectroscopy*. Benjamin / Cummings, Menlo Park.

Cooper T.G. 1977. *The Tools of Biochemistry*. Wiley, New York.

Schachman H.K. 1959. *Ultracentrifugation in Biochemistry*. Academic Press, Inc., New York.

Ribinson J.R. 1961. *Fundamentals of Acids-Base Balance*. Blackwell, Oxford 1961.

Willard, H., Merritt, L., Dean, J. and Settle, F. 1988. *Instrumental Methods of Analysis*, Wadsworth Publishing Company.

Wilson, K. and Walker, J. (Eds) 2000 *Principles and Techniques of Practical Biochemistry*, Cambridge University Press.

Fermentation Technology

Stanbury, P.F., Whitaker, A and Hall, S.J. 1995. *Principles of Fermentation Technology* 2nd edition. Pergamon Press.

Vandamme, E. J. 1984. *Biotechnology of Industrial Antibiotics*, Marcel Dekker, Inc., New York.

Wiseman, A. 1983. *Principles of Biotechnology*, Surrey University Press, London.

Wiseman, A. 1978. *Topics in enzyme and fermentation biotechnology*, 8, Ellis Horwood Ltd., Chichester.

Aiba, S., Humphrey, A., and Millis, N.F. 1965. *Biochemical Engineering*, Academic Press, New York.

Atkinson, B., and Mavituna, F. 1983. *Biochemical engineering and biotechnology Handbook*, Macmillian Publishers Ltd., New York.

Atkinson, B. 1974. *Biochemical reactors*, Pion Ltd., London.

Bailey, J. E., and Ollis, D. F. 1977. *Biochemical engineering fundamentals*, McGraw-Hill Book Company, London.

REVIEW QUESTIONS

1. Explain the principle and working of a spectrophotometer.
2. What is Beer-Lambert's law? How it is applicable in spectrophotometric estimations?
3. Describe the principle and working of a pH meter and explain the following:
   a) Ionic product of water b) Nernst equation
   c) Calomel electrode d) Reference electrode

4. Discuss various types of methods of centrifugation and their uses.
5. Write a note on important parts of a fermenter including:
   a) Construction material b) Agitators (Impellers)
   c) Aeration device (Spargers)
6. Describe various types of sensors used for measurement of physical and chemical parameters in a fermenter.
7. What do you mean by solid state fermentation (SSF)? Highlight its advantages and limitations.
8. Write a note on principle and objectives of polyacrylamide gel electrophoresis.
9. Write notes on the working and application of:
   a) Fluorimeter b) Densitometer

PART II

Principles of Biochemistry and Microbial Genetics

UNIT X

Carbohydrates

Carbohydrates are the most abundant substances of biological origin on the earth. The major share of carbohydrates is synthesized by photosynthetic plants by reduction of atmospheric CO_2. Carbohydrates such as sugar and starch make the major chunk of human diet while cellulose is the prominent feed for ruminant animals. All these complex polysaccharides are made up of repeating monomer, glucose. Glucose is readily oxidized into CO_2 and H_2O via glycolysis and Kreb's citric acid cycle by aerobic respiration.

Apart from being energy source, they also make structural components of plants and animals, e.g. cell walls of plant and bacterial cells. Carbohydrates show great diversity and distribution all over the earth. As the name suggests, carbohydrates are hydrates of Carbon ($C.H_2O$) where the ratio of C:H:O is 1:2:1. Most carbohydrates follow this formula, e.g. Glucose $C_6H_{12}O_6$ = 6.($C.H_2O$) or Ç6.6H_2O. However, some compounds, e.g. formaldehyde (CH_2O or HCHO) is a gas and is not considered to be a carbohydrate. Many a carbohydrates contain elements other than C, H, and O. Some carbohydrates such as methyl and uronic acids, although considered to be carbohydrates, do not conform to the $C.H_2O$ formula.

Carbohydrates are polyhydroxy aldehydes or ketones or polymeric substances which yield polyhydroxy aldehydes or ketones upon hydrolysis. Glucose is a polyhydroxy aldose. Starch and cellulose yield glucose on hydrolysis while inulin yields fructose. Based on difference in their size and degree of polymerization (DP), carbohydrates can be classified as monosaccharides, oligosaccharides and polysaccharides. D-glucose or dextrose is the most abundant monosaccharide on the earth.

Oligosaccharides consist of few units of monosaccharides joined covalently together by glycosidic bonds. Disaccharides, such as sucrose, are the most abundant among oligosaccharides. Oligosaccharides with more than two monomeric units are often found to occur in conjugation with proteins (glycopeptides or peptidoglycan, as in bacterial cell walls) and lipopolysaccharides. Polymers containing more than 10 monosaccharides are referred as polysaccharide units.

MONOSACCHARIDES

These are regarded as simplest carbohydrates. These are crystalline, colourless, readily soluble in water and sweet in taste due to presence of hydroxyl groups. Monosaccharides have varying members of linearly linked C-atoms. One of these carbons is bonded to oxygen atom by double bond. If this carbonyl (C=O) group is terminal, then it makes an aldehyde group (CHO) and the sugar is referred as *Aldose*. If the C=O is stituated anywhere else, except terminus, the sugar is referred as *Ketose.*

Monosaccharides containing 4, 5, 6 or 7 carbon in their backbone are referred as tetroses, pentoses, hexoses and heptoses, respectively (**Table 10.1** and **Figure 10.1**). Both aldose and ketose variants of these chain lengths are observed, e.g. aldo-tetrose, keto-tetrose. Hexoses such as glucose and fructose are most abundant in nature. Pentoses D-ribose and 2-deoxy-D-ribose are an essential component of nucleic acid backbone. All the monosaccharides, excepting dihydroxyacetone contain one or more asymmetric (chiral) carbon atom and occur as optically active enantiomers in nature.

Three dimensional representations of sugar structures are made using Fischer projection formulas. Such stereoisomers of each sugar can be divided into two groups. The chiral carbon which is farthest from the carbonyl carbon is the reference carbon. Sugars are named as D- or L- isomer with reference to chiral carbon of glyceraldehyde (**Figure 10.2**). By convention, if hydroxyl (–OH) group is on the right in the projection formula, the sugar is D-isomer.

D-glucose has two different ring structures, viz. α and β having different optical properties. D-glucose crystallized from aqueous solution acquires α-D-glucopyranose form while D-glucose crystallized from pyridine solution will acquire β-D-glucopyranose structure. The solution structure resembles pyran (a six membered ring structure) and hence these are referred as pyranoses. These forms interchange in an aqueous solution by a process called mutarotation to establish an equilibrium where the solution contains 1/3 α-D-glucopyranose and 2/3 β-D-glucopyranose with small traces of linear glucose molecule.

Table 10.1 Classes of monosaccharides based on number of carbons

No. of carbons	Category	Examples
3	Triose	Glyceraldehyde, Dihydroxyacetone
4	Tetrose	Erythrose, Threose
5	Pentose	Arabinose, Ribose, Ribulose, Xylose, Xylulose, Lyxose
6	Hexose	Allose, Altrose, Fructose, Galactose, Glucose, Gulose, Idose, Mannose, Sorbose, Talose, Tagatose
7	Heptose	Sedoheptulose

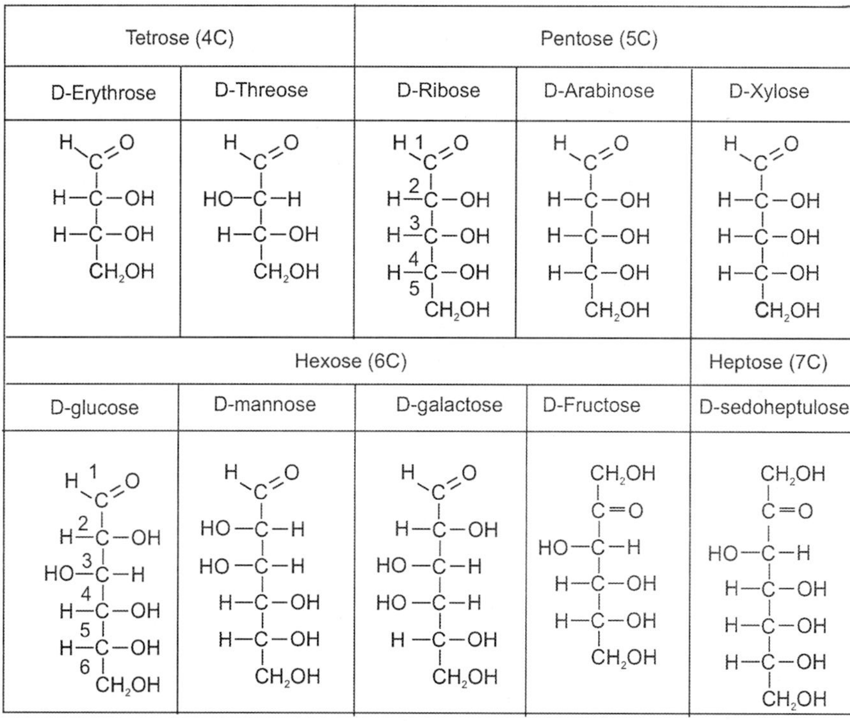

Figure 10.1: Structure of some monosaccharide of varying chain length.

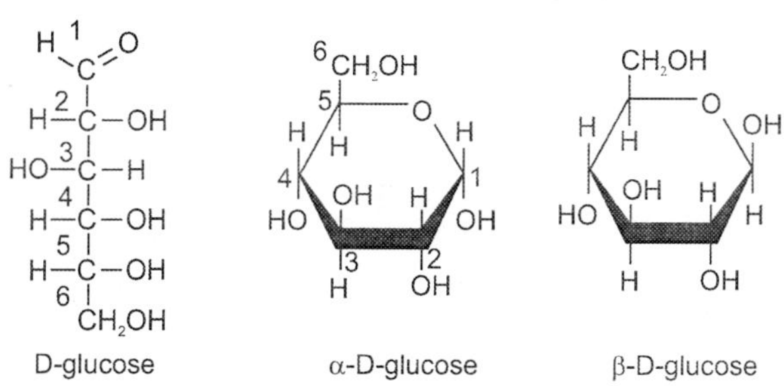

Figure 10.2: Sterioisomers of glucose.

Formation of pyranose ring results from condensation of aldehyde and alcohol group resulting in formation of an intramolecular hemiacetal. After condensation, the carbonyl carbon becomes asymmetric to take up one more stereoisomeric (optically active) form. Such forms which differ in configuration only with regard to hemiacetal carbon atom are referred as *anomeric* isomers. Only aldoses having five or more C-atoms can make pyranose ring structure. Ketohexoses upon cyclisation produce hemiketals. Such less stable cyclic forms of 5 member rings are referred as furanoses (similar to furan). Among the two forms of fructose, β-D-fructofuranose is more common.

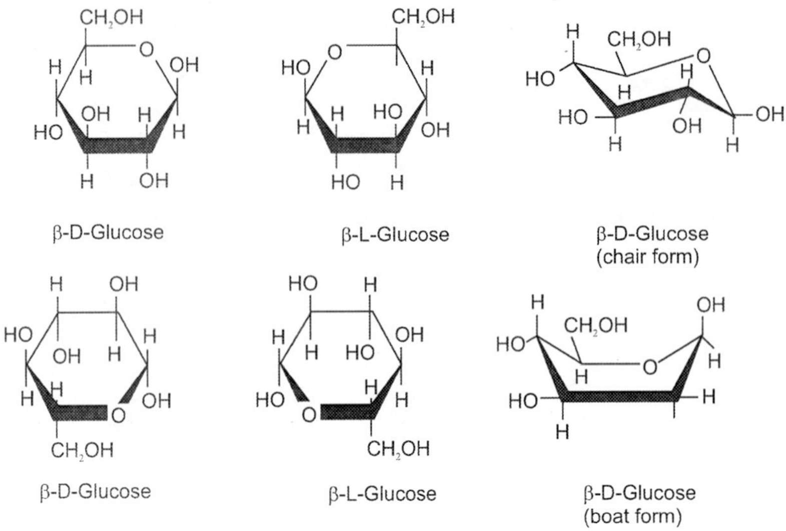

Figure 10.3: Conformations of glucose.

These rings are not planar but assume either **boat or chair** conformations (**Figure 10.3**). Conformations can be altered without breaking any bond (boat - chair) but any change in configuration requires breakage of covalent bond ($\alpha - \beta$). However, enantiomers cannot be made to correspond to their mirror images by flipping and rotating. Glucose is sometimes illustrated as a "chair form" because it is a more accurate representation of the bond angles of the molecule. The "boat" form of glucose is unstable.

Most of the biologically functional hexoses are D-isomers (e.g. D-glucose, D-fructose). Sugars differing in configuration around only one carbon atom are called epimers of each other, e.g. D-glucose and D-mannose. Similarly, D-glucose and D-galactose are epimers with reference to C-4 carbon. Monosaccharides containing 5 or more C atoms usually acquire shape of a cyclic ring in aqueous solutions. The carbonyl group forms a covalent bond with the oxygen of hydroxyl group.

Monosaccharide Derivatives (Alcohols, Acids and Amines)

Biological world is full of derivatives of hexoses where the hydroxyl group is substituted by another group or the carbon atom is oxidized to carboxylic acid. Aldoses upon oxidation of their aldehydic carbon yield aldonic acids. Glucose yields gluconic acid upon oxidation of its carbonyl carbon while oxidation of its carbonyl atom present at the other end forms corresponding glucuronic acid.

Such acids participate in formation of stable intramolecular esters, called lactones. Monosaccharides are often found as phosphorylated derivatives in various metabolic reactions and structural components. Sugar phosphates formed by condensation of phosphoric acid and sugar molecule, such as ribose phosphates and deoxyribose phosphate make the backbone of nucleic acids. Sugar phosphates are negatively charged due to ionization of phosphoric acid. Such charged moieties are confined within the cells as they can not move across the cell membrane.

During some metabolic reactions some monosaccharides are modified as per the specific requirement of the cell. However, these retain the basic configuration of a saccharide, but due to addition of different functional groups their properties are altered. Glycols or sugar alcohols, also known as polyols, polyhydric alcohols, or polyalcohols, are formed after hydrogenation of the aldoses or ketoses. Sorbitol (or glucitol) is derived from glucose and has the same linear structure as the chain form of glucose (**Figure 10.4**), but the carbonyl carbon of aldehyde (**–CHO**) group is hydrogenated resulting in a **–CH_2OH** group. Some other common glycols include erythritol and xylitol. Sugar alcohols are sweet in taste and have about half the calories of sugars. These are used in low-calorie or "sugar-free" products.

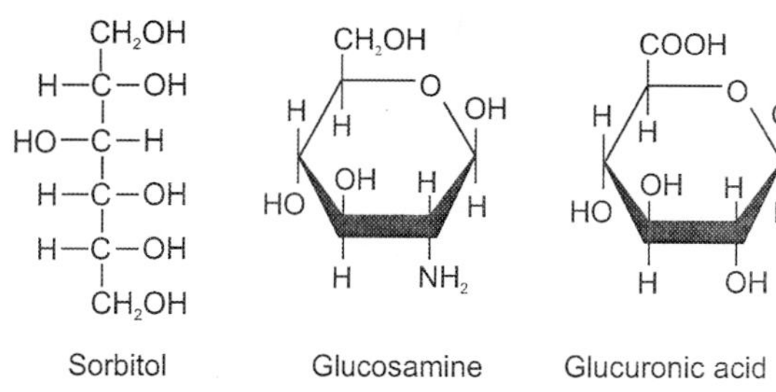

Figure 10.4: Derivatives of glucose.

Xylose, a pentose, yields Xylitol upon hydrogenation. It is a very common ingredient in "sugar-free" candies and gums because it is approximately as sweet as sucrose but is 40% low calories. In amino sugars or aminosaccharides the hydroxyl group is replaced with an amino (**–NH_2**) group.

Addition of NH_2 group to a monosaccharides results in formation of amines. N-acetyl glucosamine is a major component of bacterial cell wall. Oxidation of aldehyde group yields uronic acids. Uronic acids have a carboxyl group (**–COOH**) on the carbon that is not part of the ring. Their names retain the root of the monosaccharides, but the *-ose* sugar suffix is changed to *-uronic acid*. For example, galacturonic acid has the same configuration as galactose, and the structure of glucuronic acid corresponds to glucose.

DISACCHARIDES

A disaccharide is made up of two monosaccharides joined covalently by an O-glycosidic bond. The bond may be α or β bond depending on the configuration of the anomeric carbon atom involved in the bond formation. Usually, the anomeric carbon atom of only one of the two monosaccharides is involved in the bond so

Disaccharide	Constituent monosaccharides and nature of glycosidic bond
Maltose (Starch hydrolysate)	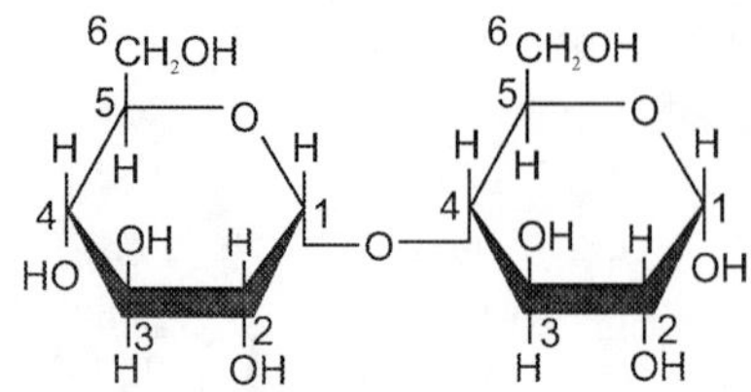 glucose α, 1 → 4 glucose
Lactose (Milk sugar)	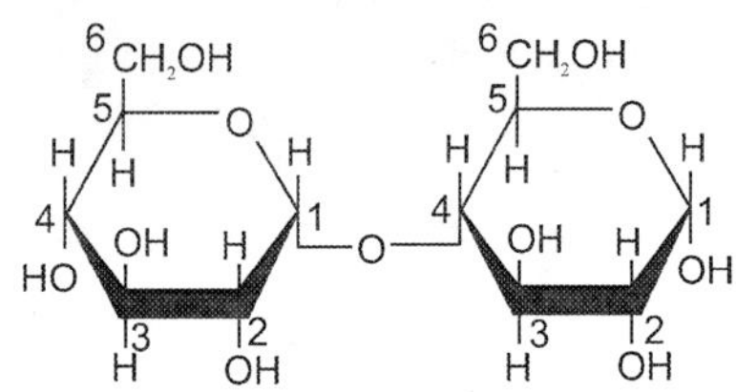 galactose β, 1 → 4 glucose
Sucrose (Table sugar)	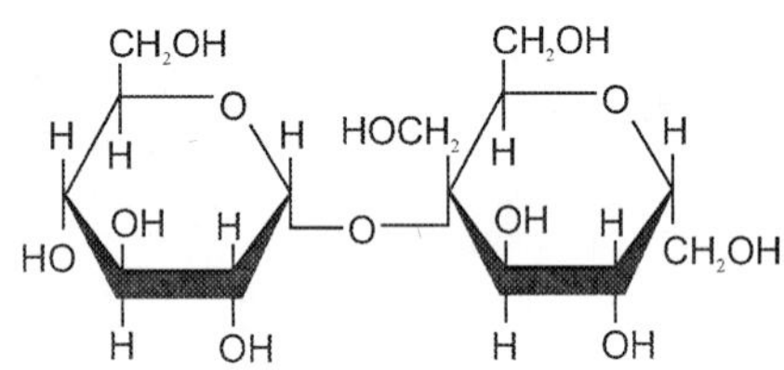 glucose α, 1 → 2 fructose

Figure 10.5: Structures of some disaccharides.

that the disaccharide still has one free reducing aldehyde or ketone group (e.g. maltose, fructose) (**Figure 10.5**). In sucrose both anomeric carbon atoms are bonded together making surcrose a non-reducing disaccharide. These glycosidic bonds can be readily hydrolysed by acid to liberate free monosaccharide units from disaccharides.

Trehalose, a disaccharide found in fungi shows a unique α 1 → 1 glucose to glucose bonding pattern while melibiose, found in legumes is made up of galactose and glucose joined together by α, 1 → 6 glycosidic bonds. **Cellobiose** is a disaccharide consisting of two β-D-glucose molecules that have a β, 1 → 4 linkage as in cellulose. Cellobiose has no taste, whereas maltose and trehalose are about one-third as sweet as sucrose.

TRISACCHARIDES

Sugars having three monomeric units are also known to occur in nature. Such **trisaccharides** include raffinose (or melitose) (glu +fru+ gal), rabinose, rhamnose, gentianose and melzeitose (**Figure 10.6**). Raffinose, a trisaccharide, is very common in legumes and cruciferous vegetables. It is made up of galactose connected to sucrose via α, 1→6 glycosidic linkage. Human beings cannot metabolize saccharides with α, 1→6 linkage and these sugars are selectively digested in the large intestine by bacteria.

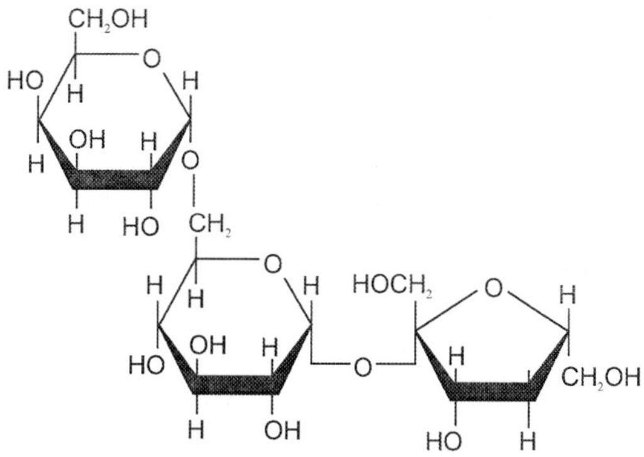

Figure 10.6: Raffinose- A trisaccharide.

POLYSACCHARIDES

Polysaccharides are among the most abundant naturally occurring polymers on the Earth and most commonly are seen as structural and storage polysaccharides of plants. Generally, polysaccharides are named by replacing *–ose* of their monomeric unit by suffix *an.* A common name for carbohydrate monomers is glycose and polymers is glycan, e.g. glucose: glucan, mannose: mannan, fructose: fructan, xylose: xylan. However, some popular names are in more use, e.g. starch, cellulose, pectin, inulin, chitin and heparin. Polysaccharides vary with respect to the monomer type, chain length, bond linkage, branching, etc. Broadly, polysaccharides can be classified into (i) homopolysaccharides, containing only one kind of repeating monomeric unit and (ii) heteropolysaccharides, containing two or more different kinds of monomeric units. The general purpose of polysaccharides is storage (starch and glycogen) or providing mechanical strength in form of structural components, e.g. cellulose fibrils in plant cell wall. However, it is very difficult to find a truly pure natural polysaccharide and thus, the term homopolysaccharide is used to refer when at least 95% of monosaccharides are of one kind.

Starch: It is a very common storage polysaccharide of plants, consisting of D-glucose units linked by α, 1 → 4 glycosidic linkages. It consists of soluble amylose and insoluble amylopectin components in the ratio of 1:4, respectively. Amylose is a linear macromolecule with about 200–500 monomeric units (mol. wt. 150,000) (**Figure 10.7 A**). It combines with iodine to give intense blue color. Amylopectin is branched and has over 1000 glucose units linked by similar α, 1 → 4 bonds with intermittent α, 1 → 6 glycosidic bonds in between leading to a branched structure having molecular weight in the range of 200,000–1,000,000 (**Figure 10.7B**). Enzymes that can hydrolyze starch to yield maltose and glucose are referred as amylases (α-amylase, glucoamylase). Incomplete hydrolysis of starch yields dextrins, e.g. erythrodextrins which give red color with iodine. Branching is more frequent and shorter in starch at every 25th glucose unit.

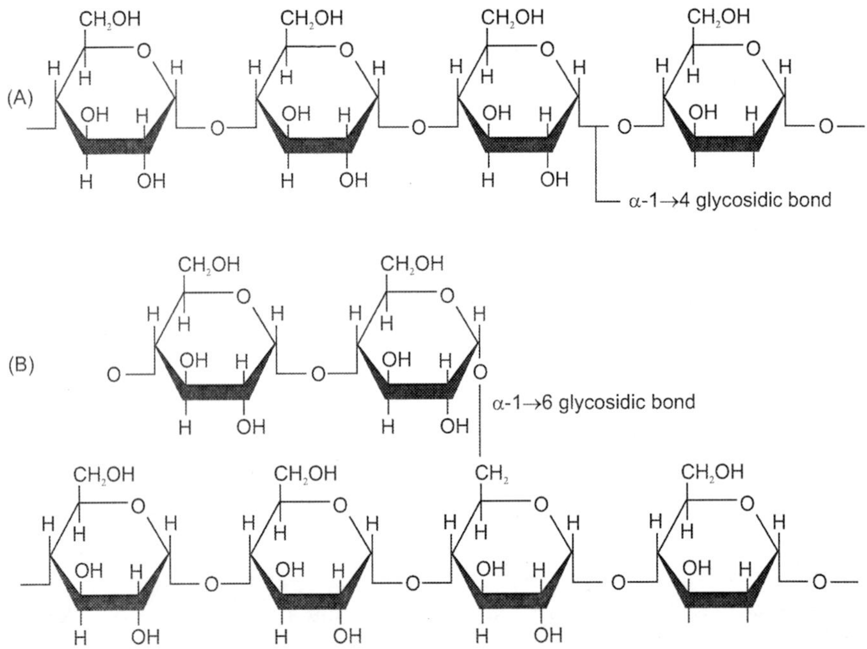

Figure 10.7: Structure of starch (A) amylose and (B) amylopectin.

Glycogen: It is the principal storage carbohydrate of animals found in muscles and liver. The liver glycogen supplies glucose to all the organs while muscle glycogen provides glucose during muscle action for the purpose of energy generation. It is composed of repeating monomeric unit, glucose linked by α, 1 → 4 bonds with some α, 1 → 6 glycosidic bonds giving rise to branches. The molecular weight is in between (Mr 300,000-more than 100 million) and degree of branching differs. Branching is more frequent and shorter in glycogen at every 12th glucose unit.

These branched glucose chains are arranged like branches of a tree originating from a pair of molecules of **glycogenin**, a protein that acts as a primer at the core of the structure.

Cellulose: Cellulose is the most abundant polysaccharide on Earth, because it is most common structural component of most plants. It is also found in the cell walls of fungi. It is insoluble in boiling water and does not give color with iodine. Enzymes that digest cellulose are referred as cellulases and are not found in man and animals. Therefore, cellulose cannot make a dietary component of humans. Ruminant animals digest it with the aid of commensal cellulolytic bacteria residing in their rumen. It is a linear polymer of β-D-glucose units linked via β, 1 → 4 glycosidic bonds. These linkages are much more resistant to acid hydrolysis as compared to α, 1 → 4 bonds of starch or glycogen. Several of these chains are held together by H-bonds to make microfibrils (**Figure 10.8 A**). Several microfibrils combine to make this macromolecule mechanically very strong. Dextran consists of glucose units linked by α, 1 → 6 glycosidic linkages and the side branches originate via α, 1 → 3 or α, 1 → 4 linkages (**Figure 10.8 B**). Low value- cellulosic raw material resulting out of agricultural practices is used as feed for cattles and ruminants. However, recently it is also being looked upon as a rapidly renewable resource for production of biofuel. This technology utilizes microbial enzymes which liberate glucose from the macromolecule and then it can be converted into bio-fuel ethanol by ethanolic fermentation.

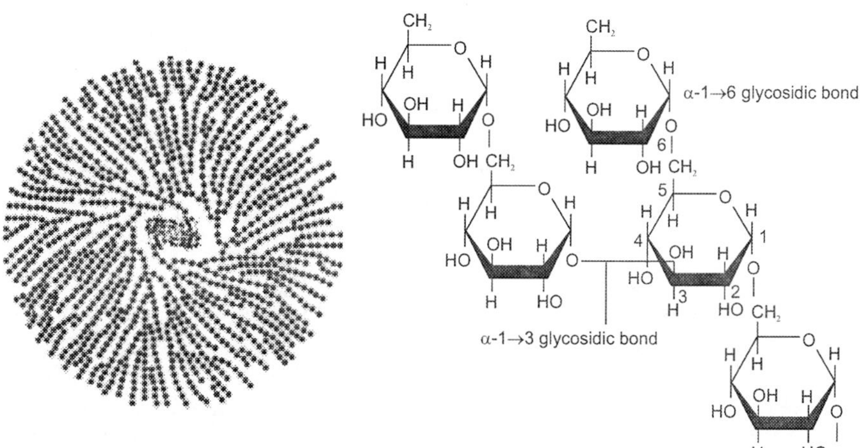

(A) (B)

Figure 10.8: Structure of (A) cellulose and (B) dextran.

Xylan: Xylans are major hemicelluloses of plant cell wall and similar to other polymers of plant origin display large polydiversity and polymolecularity. Xylans, having backbone of β-1→4 glycosidically linked xylose units with arabinose, acetyl and methyl glucuronic acid side groups are reported to be present in all terrestial plants including hardwoods, softwoods and grasses (**Figure 10.9**) while β-1→3 linked xylose backbones have been witnessed in marine algae. Xylans from annual plants are more heterogenous than those form woody plants. Xylans with both sorts of linkages are reported from some seaweed such as *Palmeria palmate.* Unsubstituted xylan, composed of only D-xylose residues is considered to be a homoxylan e.g. esparto grass (*Stipa tenacissima*) xylan and tobacco stalk xylan containing 95–96% xylose. Xylan content in hardwoods is 15–30% of total hemicelluloses while it is only 7–10% in softwoods. The xylan of dicots is essentially an acetyled methyl glucuronoxylan in which 4-O-methyl glucuronic acid residues are the main substitutions at C-2 in approximately one in ten of the β-D- xylopyranose units via β- 1,2-glycosidic linkage while about 10–70% of xylopyranose units are acetylated at C-2 or C-3 positions. The xylan of softwood is mainly arabino-4-O-methyl glucuronoxylan. In addition to 4-O- methyl glucuronic acid, it also contains α-arabinofuranosyl units which are α-1, 3 linked to the xylan backbone and the ratio of arabinose side groups to xylose residues is 1: 8. Graminaceous (*Poaceae*) xylans are arabinoxylans having predominantly arabinofuranosyl units at C-3 positions. Xylan forms intrachain hydrogen bonds which support a twofold ribbon-like extended structure.

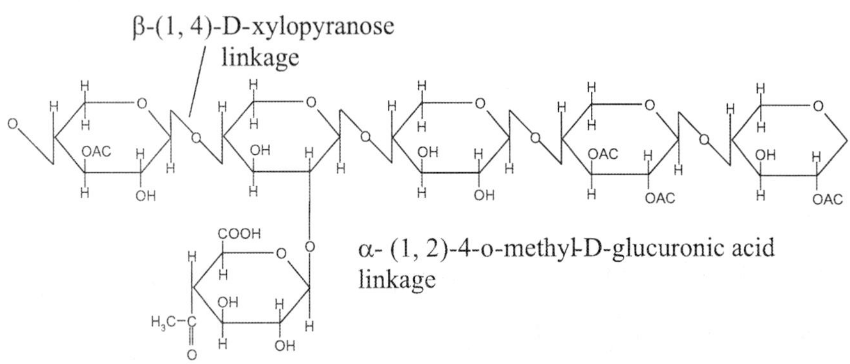

Figure 10.9: Methylglucuronoxylan (Hardwood).

Inulin: Inulin is a polydispersed fructan consisting mainly of β, 2-1 fructoysl-fructose links terminated by a sucrose residue (**Figure 10.10**). It serves as a storage polysaccharide in many plants including members of Liliaceae, Gramineae, Compositae etc. and is accumulated in the underground roots and tubers of several plants including Jerusalem artichoke (*Helianthus tuberosus*), chicory (*Cichorium intibus, Cichorium endivia*) and dahlia (*Dahlia pinnata*). Currently two members of compositae: Jerusalem artichoke and chicory are preferred for the industrial production of inulin, the latter being the most commonly used source. The degree

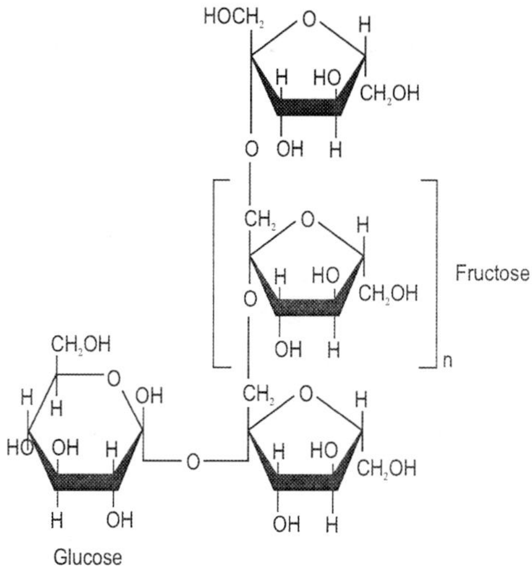

Figure 10.10: Structure of inulin.

of polymerization in inulin (M.W. 60,000) of plant origin is < 200 and varies according to plant species, weather conditions and age. Some well known plant sources of inulin are several vegetables and fruits, including onions, leeks, garlic, bananas, asparagus, etc. The fructans, inulin and oligofructose are known to possess many of the physiological properties of dietary fibers (DF) and their determination in food samples can be done by treating samples with enzymes (inulinase, amyloglucosidase, etc.) and analyzing the sugars released by ion-exchange chromatography.

Chitin: Chitin is a polymer of N-acetyl D-glucosamine units connected together by β, 1 → 4 glycosidic bonds. Many such parallel chains are held together by hydrogen bonds to make it a mechanically strong molecule. It is mainly found as cell wall component of fungi and outer covering of insects (**Figure 10.11**).

Mannans: Mannans are made up of repeating units of manose linked via α, 1 → 6 bonds having small side branches of one to three mannose residues joined by α,

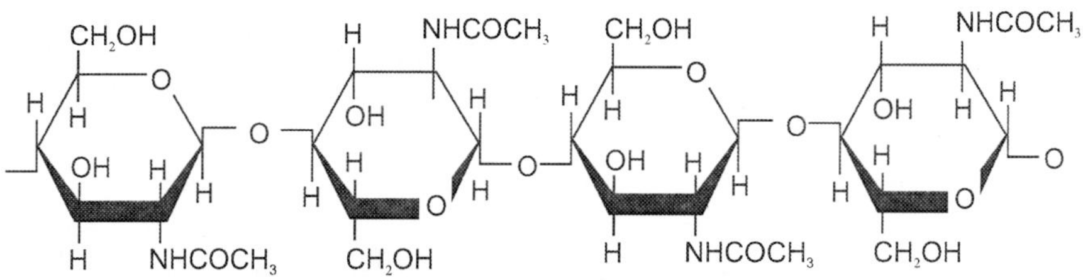

Figure 10.11: Structure of chitin.

$1 \rightarrow 5$ or α, $1 \rightarrow 3$ bonds. Mannans are commonly found in cell walls of unicellular ascosporous fungi, i.e. yeasts.

Heteropolysaccharides

Peptidoglycan: Bacterial cell wall is composed of a heteropolysaccharide of alternating β $1 \rightarrow 4$ linked N-acetylglucosamine (NAG) and N-acetylmuramic acid (NAM) units. Linear chains of these polysaccharides are cross-linked by short peptides. Glycosaminoglycans are composed of disaccharide units consisting of either N-acetylglucosamine or N-acetylgalactosamine and glucuronic acids. These are important structural polysaccharides found in animal cartilages, lubricating fluids of the joints and as synovial fluid, etc.

Hyaluronic Acid: Hyaluronic acid is a major component of cartilage and tendons and contributes to their tensile strength and elasticity. Glycosaminoglycan present around ovum, as an extracellular matrix is degraded by hyaluronidase enzyme of sperms for successful penetration. This enzyme is also secreted by some pathogenic microbes for invading host tissues.

The common gelling agent **Agar** is a polymer of agarobiose, a disaccharide made up of D-galactose and 3, 6-anhydro-L-galactose (**Figure 10.12**). Agar is extracted from seaweed *Gelidium* and is used in many media formulations as a solidifying agent. Purified agar or agarose is used in gel electrophoresis for separation of nucleic acids. Agar is successfully used in microbiological media since it can not be hydrolyzed or used up by most microorganisms and does not have any nutritional value. **Carrageenan** is a collective term for polysaccharides extracted from seaweeds. It differs from agar in having sulfate groups ($-OSO_3^-$) in place of some hydroxyl groups. Carrageenan is also used as thickening and solidifying agent for food products.

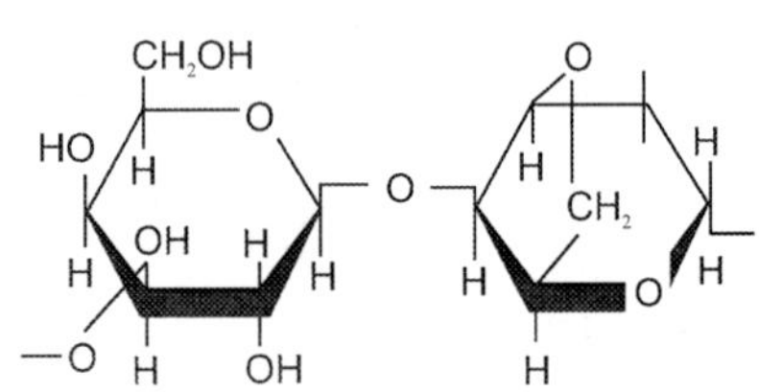

Figure 10.12: Structure of agarobiose.

Guar Gum: Guar is a legume that has been traditionally cultivated as livestock feed. Guar gum is the ground endosperm of the seeds. Majority of guar gum is **guaran**, (**Figure 10.13**) a water soluble polysaccharide consisting of linear chains of mannose with β, $1 \rightarrow 4$ linkages to which galactose units are attached with α, $1 \rightarrow 6$ linkages. The ratio of galactose to mannose is 1:2. Guar gum is endosperm of seeds of a leguminous plant, guar. It is a very efficient thickening agent and it is used as a stabilizer and a dietary fibre in food and pharmaceutical industries.

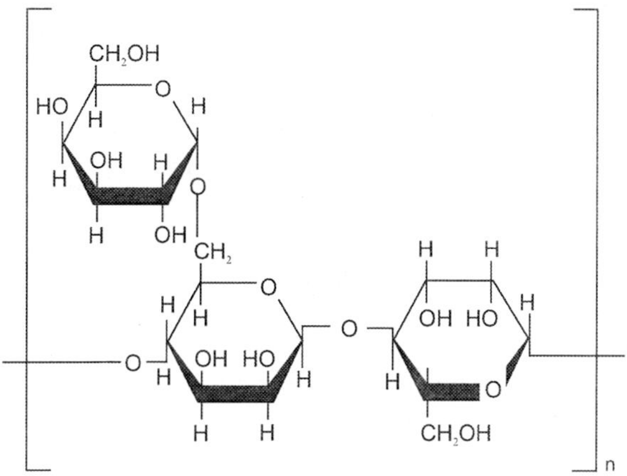

Figure 10.13: Structure of **guaran** a galactomannan.

Pectin: Pectin is methylated ester of polygalacturonic acid consisting of chains of 300 to 1000 galacturonic acid units joined with α, 1 → 4 linkages (**Figure 10.14**). The Degree of esterification (DE) affects the gelling properties of pectin. It is an important ingredient of fruit pulps, jellies, and jams. In plants pectin occurs as a cementing material in the cell walls. Most commonly, citrus fruits are used as a source of pectin and their white rind portion contains about 30% pectin.

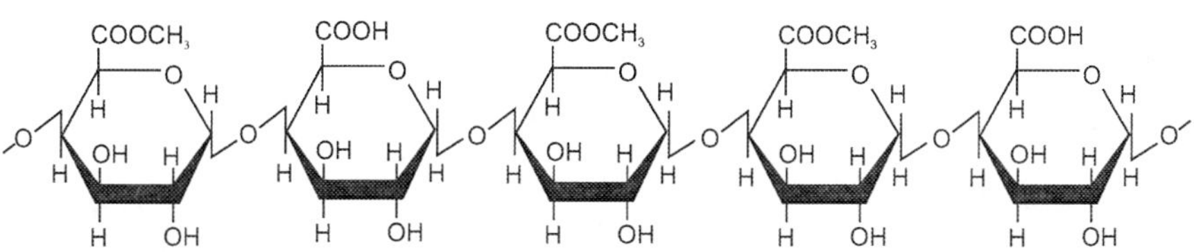

Figure 10.14: Structure of pectin: The structure shown here has three methyl ester forms (**–$COOCH_3$**) for every two carboxyl groups (**-COOH**), hence it is has a 60% degree of esterification, called as DE-60 pectin.

Xanthan Gum: Xanthan gum is a polysaccharide with a β-D-glucose backbone as cellulose (**Figure 10.15**), but alternate glucose unit is linked to a trisaccharide of mannose, glucuronic acid, and mannose. The mannose linked to the backbone bears an acetic acid ester on its 6th carbon, and the other mannose of the trisaccharide is linked through 6th and 4th carbon of pyruvic acid. The negatively charged carboxyl groups on the side chains because the molecules form very viscous fluids when mixed with water. Xanthan gum is used as a thickener for sauces, to prevent ice crystal formation in ice cream, and as a low-calorie substitute for fat. Xanthan gum is produced by the bacterium *Xanthomanas campestris* found on member of crucifereae such as cauliflower cabbage. It makes a very viscous

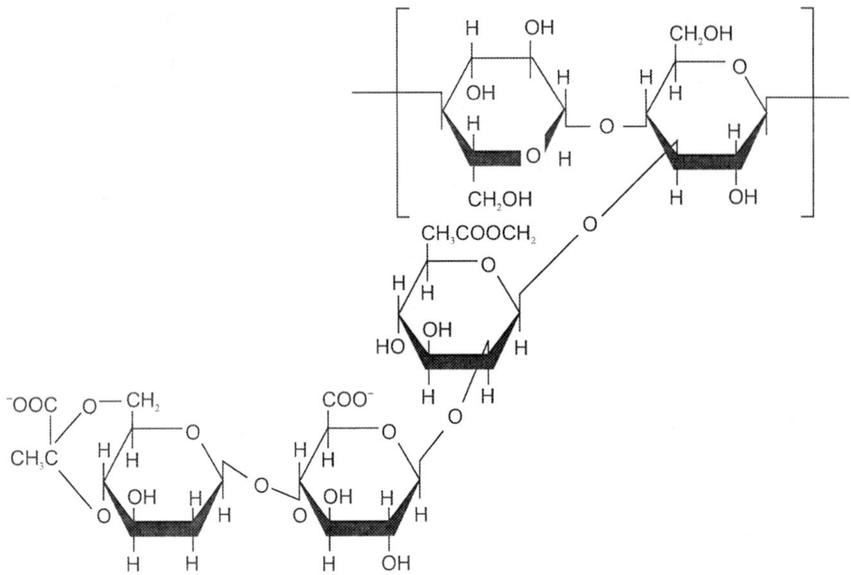

Figure 10.15 Structure of xanthan gum.

aqueous solution and therefore used as a thickener and fat substitute in food and ice creams; it is often used along with guar gum to achieve greater thickness

Glucomannan: It consists of glucose (G) and mannose (M) in a proportion of 5:8 joined by β,1 → 4 linkages with the polymeric repeating unit having sequence: "GGMMGMMMMMGGM". Short side chains of 11–16 monosaccharides are also found at intervals of 50–60 units of the main chain attached by β, 1 → 3 linkages (**Figure 10.16**). Acetate groups are present on 6th carbon at every 9–19 units of the main chain and their hydrolysis favors the formation of intermolecular hydrogen bonds that are responsible for the gelling action. It is a dietary fibre that accumulates in the tubers of *Amorphophallus konjac* and is also used as a hunger suppressant since its aqueous solution swells to give a feeling of fullness in the stomach.

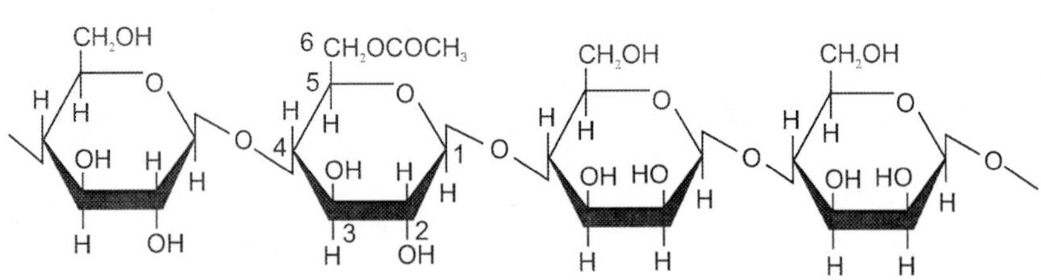

Figure 10.16: Structure of glucomannan.

Glycoproteins and Glycolipids: Carbohydrates occurring in conjunction with proteins and lipids are called as glycoproteins and glycolipids, respectively.

Hydrophilic carbohydrates alter the nature and solubility of hydrophobic proteins and lipids. This alteration in behaviour of biomolecules is meant for achieving: specific function related with transport or biological activity. Proteins conjugated with polysaccharides are also resistant to degradation by proteases.

SUGGESTED READING

Ashford, D.A. 1992. *Carbohydrate analysis.* Curr. Opin. Biotechnol. 3 (1): 45

Davidson E.A. 1967. *Carbohydrate Chemistry.* Hort, Rinehart and Winston, New York.

Florkin M. and Stotz E. 1963. *Comprehensive Chemistry; Carbohydrate* Sec. II, Vol. 5. Elsvier Pub. Co., New York.

Nelson, D.L. and Cox, M.M. 2000 *Lehninger Principles of Biochemistry*, (3rd edition), W.H. Freeman and Company.

Percival, E.G.V. and Perícival E. 1962. *Structural Carbohydrate Chemistry.* Prentice-Hall.

Sharon N. 1980. *Carbohydrate.* Scientific American. 243 (5): 90–116.

Stryer, L., 1995 *Biochemistry*, (5th edition) W.H. Freeman and Company.

REVIEW QUESTIONS

1. What are carbohydrates? Classify carbohydrates on the basis of complexity and composition.
2. Discuss the types and characteristics of important monosaccharides and disaccharides.
3. Give a comparative account of:
   a) Homopolysaccharides and heteropolysaccharides
   b) Storage and structural polysaccharides
4. Comment upon the following:
   a) Starch　　b) Cellulose
   c) Xylans　　d) Mannans

UNIT XI

Amino Acids and Proteins

Proteins of all living organisms are made up of amino acids. A total of 20 standard amino acids are the building blocks of proteins. The term amino acid refers to α-amino carboxylic acids which are repeating monomeric units of proteins. The general structure of an α-amino acid is given in **Figure 11.1**.

```
R – CH – COOH
     |
    NH2
```

Figure 11.1: General structure of α-amino acid.

The carbon atom bearing carboxylic group is called as α carbon. A typical amino acid has an amino group, a carboxyl group, a hydrogen atom and a side chain (R-group) attached to a central α-carbon atom (C_α). Proline (α-imino acid) is an exception having a secondary amino group. All of the 20 standard amino acids, except glycine, have four different groups arranged tetrahedrally around the C_α atom and hence show stereoisomerism. However, only L isomers of these amino acids are used in making of proteins. More than 200 such amino acids are known but only 20 are obtained from hydrolysis of different natural proteins.

CLASSIFICATION OF AMINO ACIDS

There are various ways of classifying amino acids into different groups. These can be classified on basis such as:

(i) Whether they can be synthesized in the body
(ii) Whether they contain any aromatic group
(iii) Whether they contain sulphur, extra –OH groups, dicarboxylic or amide groups
(iv) Whether they ionize as acidic, basic or neutral ion
(v) Their affinity for water: Hydrophilic or hydrophobic

Essential amino acids cannot be synthesized *de novo* in the body and thus should be incorporated in the diet itself. Such indispensable amino acids include valine, leucine, isolucine, threonine, methionine, phenylalanine, tryptophan and lysine. Their deficiency cannot be made up by the body itself and hence results in diseases such as nervous disorder, improper growth and can be fatal in some cases. Semi-essential amino acids (arginine and histidine) can be partially complemented by body but still an extraneous source is required to fulfil the total requirement. All proteins are made up from the same set of 20 standard amino acids (**Figure 11.2**). Proteins, whether as structural components or as functional biomolecules, occur in all living cells and parts thereof. A single cell contains a great variety of proteins ranging to thousands. Proteins are the ultimate molecules through which genetic information is expressed and executed. The enormous diversity of proteins ranging from oldest of life forms such as archaebacteria to higher plants and animals have a unique common feature that they are made out of a combination of covalently linked 20 amino acids.

In such manner, amino acids link up to produce proteins which are entirely different in structure and function, e.g. hormones, enzymes, antibodies, keratin (nails, hair, skin, hooves, horns), antibiotics to name a few.

The names of amino acids are trivial (without any scientific basis or have common names based on their source from which they were discovered or their taste, e.g. asparagine (first noticed in *Asparagus*), glutamate (first noticed in wheat gluten), tyrosine (first isolated from cheese or *Tyros* in Greek), glycine (glycos-meaning sweet). Amino acids differ from each other in their side chains or 'R' groups which vary in length, type and charge.

The 20 most commonly occurring amino acids are regarded as primary amino acids to distinguish them from their own derivatives or modifications and from other amino acids which are structurally similar but do not participate in protein structure. Each of 20 standard amino acids are assigned three letter abbreviations and one letter symbol for brief description of protein sequence.

All amino acids, except glycine contain a chiral center (α-carbon). All these amino acids exist in two optically active forms – stereoisomers or enantiomers and rotate the plane polarized light. The denotation of all stereoisomeric forms occurring in nature is based on a convention that uses smallest sugar with a chiral carbon, i.e. 3-glyceraldehyde as reference. Stereoiosmers having configuration similar to α glyceraldehyde where α stands for *levo* or left rotatory and the stereosiomers similar to D-glyceraldehyde are designated as D-type as in case of D-glucose (D-Dextrorotatory derived from dextro meaning 'right'). All the amino acids used in proteins are unexceptionally α-stereoisomers. D-amino acids are witnessed only in peptidoglycan of bacterial cell walls.

PHYSICAL PROPERTIES OF AMINO ACIDS

Regarding their physical characteristics, amino acids are colourless, crystalline substances. The crystal form may vary from slender needles (tyrosine) to thick

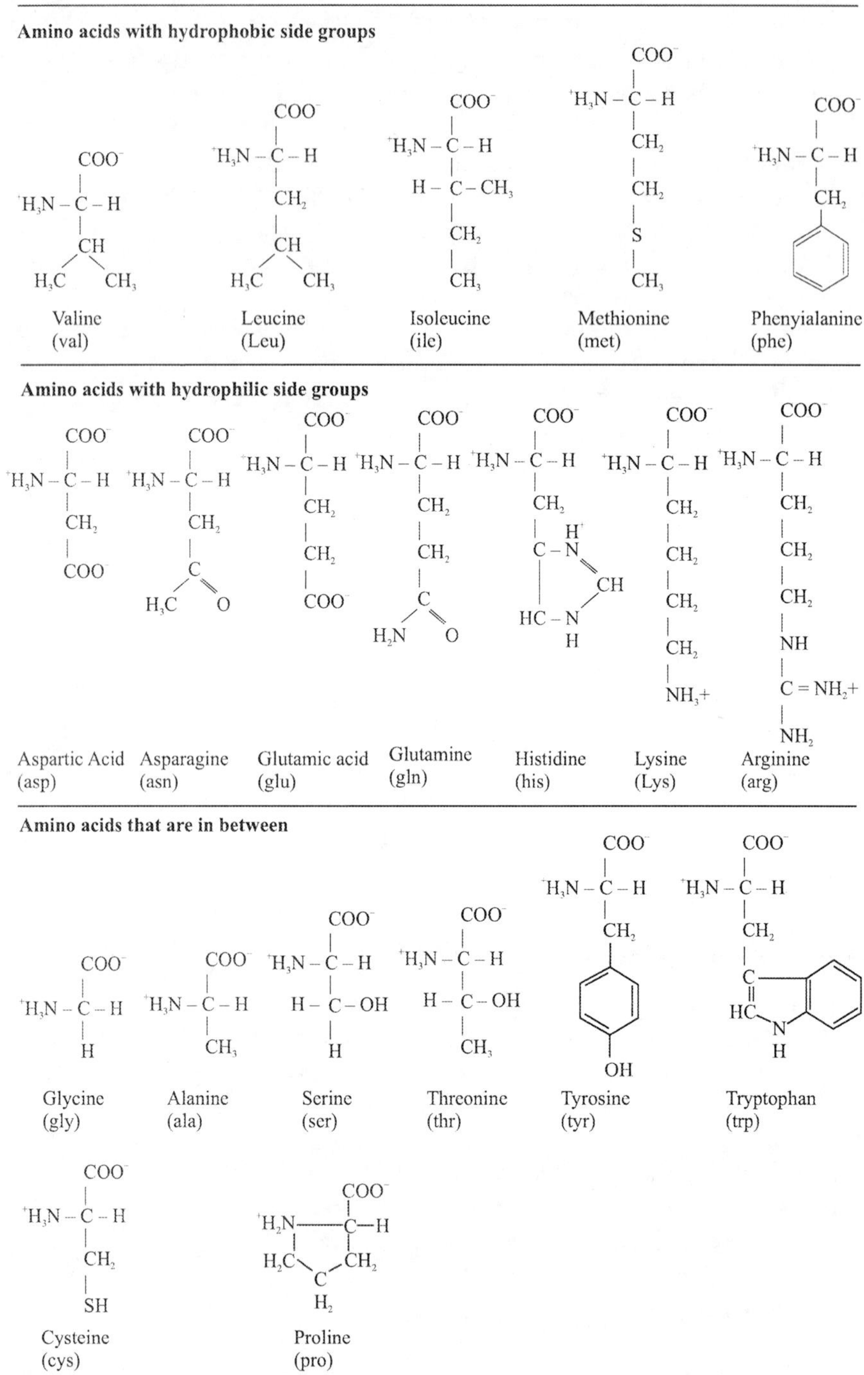

Figure 11.2: Structure of twenty standard amino acids.

hexagonal plates (cystine). They may be either tasteless (tyrosine), sweet (glycine and alanine) or bitter (arginine). They have high melting point (above 200°C) and often result in decomposition. Amino acids are soluble in polar solvents such as water and ethanol but they are insoluble in non-polar solvents such as benzene and ether. Some amino acids like tryptophan, tyrosine, histidine and phenylalanine absorb ultraviolet rays at 260–290 nm. This property enables the identification of not only these amino acids but also the proteins which contain them.

Although over 100 amino acids have been shown to be present in various plants and animals, only 20 of them (L-isomers) are found as constituent of most natural proteins. These 20 amino acids of proteins are often referred to as **standard, primary or normal amino acids** to distinguish them from others. The almost universal use of 20 amino acids for the synthesis of various protein molecules in "one of nature's enigmatic rules." These have, therefore, rightly been called as the **"magic 20"**, a phrase coined by Francis Crick.

Ionization and Electrochemical Properties of Amino Acids

Amino acids are readily soluble in water and ionize as acids or bases. This behaviour of amino acids is important in understanding structural and functional properties of proteins. Amino acids with one carboxyl and one amino group ionize as Zwitterions or hybrid ions, bearing both positive NH_4^+ and negative COO^- charge in aqueous solutions. The net effect of these charges makes the amino acid electrically neutral and it does not move in electric field. Amino acids can be crystallized as zwitterions to produce crystal lattice. Zwitterions are amphoteric (can act as proton acceptor and donor) and their solutions are called ampholytes. Based on the number of amino and carboxylic groups in any given amino acid, it is quite plausible to ascertain its nature upon ionization. For example, lysine, arginine and histidine bearing two amino groups and one carboxyl group ionize as positively charged ions (anions) while aspartic acid and glutamic acid bearing two carboxylic groups and one amino group will ionize as negatively charged species (cations). At low pH, both amino and carboxyl groups are protonated. However, with pH increase, first carboxyl and then amino group lose their proton (H^+). For the standard 20 amino acids, the pK (pH at which the group is half dissociated) is in the range of 1.8–2.9 for the α-carboxyl group and 8.8–10.8 for α –amino group. Amino acids having ionizable groups on side chain 'R' (e.g. aspargine, glutamine, and tyrosine) have a distinct pK. Addition of each molecule of base to the amino acid results in removal of one proton from one amino acid. A graph plotted between pH change and base addition will show two distinct stages representing removal of one proton from glycine similar to behaviour of any diprotic acid. The titration curve of glycine shows that at low pH (or high hydrogen ion concentration) both the amino group and the carboxyl group are fully protonated so that the amino acid is in the cationic form. As the amino acid in solution is titrated with a strong base, such as NaOH, it loses two protons, first from the carboxyl group (pK = 2.3) and then from the amino group having pK values in the alkaline range 8.8–10.8. Acidic amino acids such as aspartic acid and glutamic acid have pK values of 3.9 and 4.1, respectively whereas pK values of basic amino acids lysine and arginine are 10.8 and 12.5, respectively. In peptides and proteins where all C_α carboxyl and amino (except terminal amino and carboxyl group) group are engaged in formation of peptide bond, only R-chain groups are free to ionize.

Amino acids can be classified on the basis of type of R-group or side chain and its behaviour, specially, polarity (i.e. behaviour in aqueous solutions). The behaviour of R-group varies from hydrophilic (polar) to hydrophobic (non polar). Based on polarity of R-groups, amino acids can be classified into five types, viz. (i) non-polar aliphatic (ii) non-polar aromatic (iii) polar uncharged (iv) polar negatively charged (v) polar positively charged (**Table 11.1**).

Several amino acids are also synthesized *de novo* in the body during the various metabolic pathways. These are derived from various carbohydrates and lipid skeletons. These are synthesized within individual cells. Although these are called as non-essential but that does not mean that these are not significant. Rather, these amino acids are of special significance. However, since these can be synthesized in the body itself, their incorporation in diet is not necessary. Alanine, arginine, aspartic acid, asparagine, cystine, glutamic acid, glycine, histidine, proline and serine are non-essential amino acids.

Generally, majority of proteins are made out of combinations of 20 amino acids. Most of them are α-amino acids while proline and hydroxyproline are imino acids. The genetic code consisting of 64 codons specifies for 20 amino acids. Some special derived amino acids made from modifications of other amino acids are given in **Table 11.2**. In addition to routine amino acids which generally make various proteins, more than 200 non-protein amino acids have been observed in nature. Some of the examples of such non-protein amino acids having specific physiological importance are listed in **Table 11.3**:

PROTEINS

Proteins are large molecules made from smaller units called amino acids. There are twenty amino acids commonly found in both plant and animal proteins. There are at least eight amino acids that the body cannot make itself which need to be obtained from the food we eat. These are isoleucine, leucine, lysine, methionine, phenylalanine, threonine, tryptophan and valine. Additionally, infants need supplementation of histidine and taurine. Proteins are necessary for maintaining tissues and for sustaining growth. They are also used to make hormones and other physiologically active substances. The structure levels of proteins were first defined by Linderstrom-Lang and are often referred to as primary, secondary, tertiary and quaternary (**Figure 11.3**). Three of these structural levels (primary, secondary and tertiary) can exist in molecules composed of a single polypeptide chain, whereas,

Table 11.1 General properties of standard amino acids

Amino Acid	Abbreviation and Denotation	Mol. wt.	pI	% Occurrence in proteins
Nonpolar, aliphatic R group				
Glycine	Gly (G)	75	5.97	7.5
Alanine	Ala (A)	89	6.01	9.0
Valine	Val (V)	117	5.97	6.9
Leucine	Leu (L)	131	5.98	7.5
Isoleucine	Ilu (I)	131	6.02	4.6
Proline	Pro (P)	115	6.48	4.6
Aromatic R groups				
Phenylalanine	Phe (F)	165	5.48	3.5
Tyrosine	Tyr (Y)	181	5.66	3.5
Tryptophan	Trp (W)	204	5.89	1.1
Polar, uncharged R groups				
Serine	Ser (S)	105	5.68	7.1
Threonine	Thr (T)	119	5.87	6.0
Cysteine	Cys (C)	121	5.07	2.8
Methionine	Met (M)	149	5.74	1.7
Asparagines	Asn (N)	132	5.41	4.4
Glutamine	Gln (g)	146	5.65	3.9
Negatively charged R groups				
Aspartate	Asp (D)	133	2.77	5.5
Glutamate	Glu (E)	147	3.22	6.2
Positively charged R groups				
Lysine	Lys (K)	146	9.74	7.0
Arginine	Arg (R)	174	10.76	4.7
Histidine	His (H)	155	7.59	2.1

the fourth (quaternary) involves interactions of polypeptides within a multichained protein molecule.

The basic primary structure of a protein is relatively simple and consists of one or more linear chains of a number of amino acid units. This linear, unfolded structure or the polypeptide chain often assumes a helical shape to produce the secondary structure. This, in turn, may fold in certain specific patterns to produce the twisted three-dimensional or the tertiary structure of the protein molecule.

Table 11.2 Characteristics of some derived amino acids

Amino acid	Characteristics
Cystine	Formed by disulphide (—s—s—) linking between two cysteine molecules.
Hydroxyproline	Produced by hydroxylation of proline after translation.
Hydroxylysine	Produced by hydroxylation of lysine residue. Both hydroxyproline and hydroxylysine are found in collagen and gelatin only.
Thyroxine	It is produced from tyrosine in the thyroid gland and is part of thyroglobulin protein.
Diaminopimelic acid	It is a common constituent of bacterial cell wall glycopeptides.
Desmosine and Isodesmosine	These are the modifications of lysine and are found in elastin of muscles.
Mono-, di-, and tri- methyl lysines	These modifications of lysine are found in histones (basic proteins associated with DNA) and myosine of muscles.

Table 11.3 Characteristics of some non-protein amino acids

Amino acid	Characteristics
β-alanine	A product of pyridine breakdown; Component of Coenzyme A and vitamin pantothenic acid.
Hydroxyproline	Produced by hydroxylation of proline after translation found in collagen and gelatin.
Hydroxylysine	Produced by hydroxylation of lysine residue. Both hydroxyproline and hydroxylysine are found in collagen and gelatin only.
Thyroxine	It is produced from tyrosine in the thyroid gland and is part of thyroglobulin protein.
Diaminopimelic acid	It is a common constituent of bacterial cell wall glycopeptides.
Desmosine and Isodesmosine	These are the modifications of lysine and are found in elastin of muscles.
Mono-, di-, and tri-methyl lysines	These modifications of lysine are found in histones (basic proteins associated with DNA) and myosin of muscles.

Finally, certain proteins are made up of subunits of similar or dissimilar type of the polypeptide chains. These subunits interact with each other in a specific manner to give rise to quaternary structure of the protein. This, in fact, defines the degree of polymerization of a protein unit.

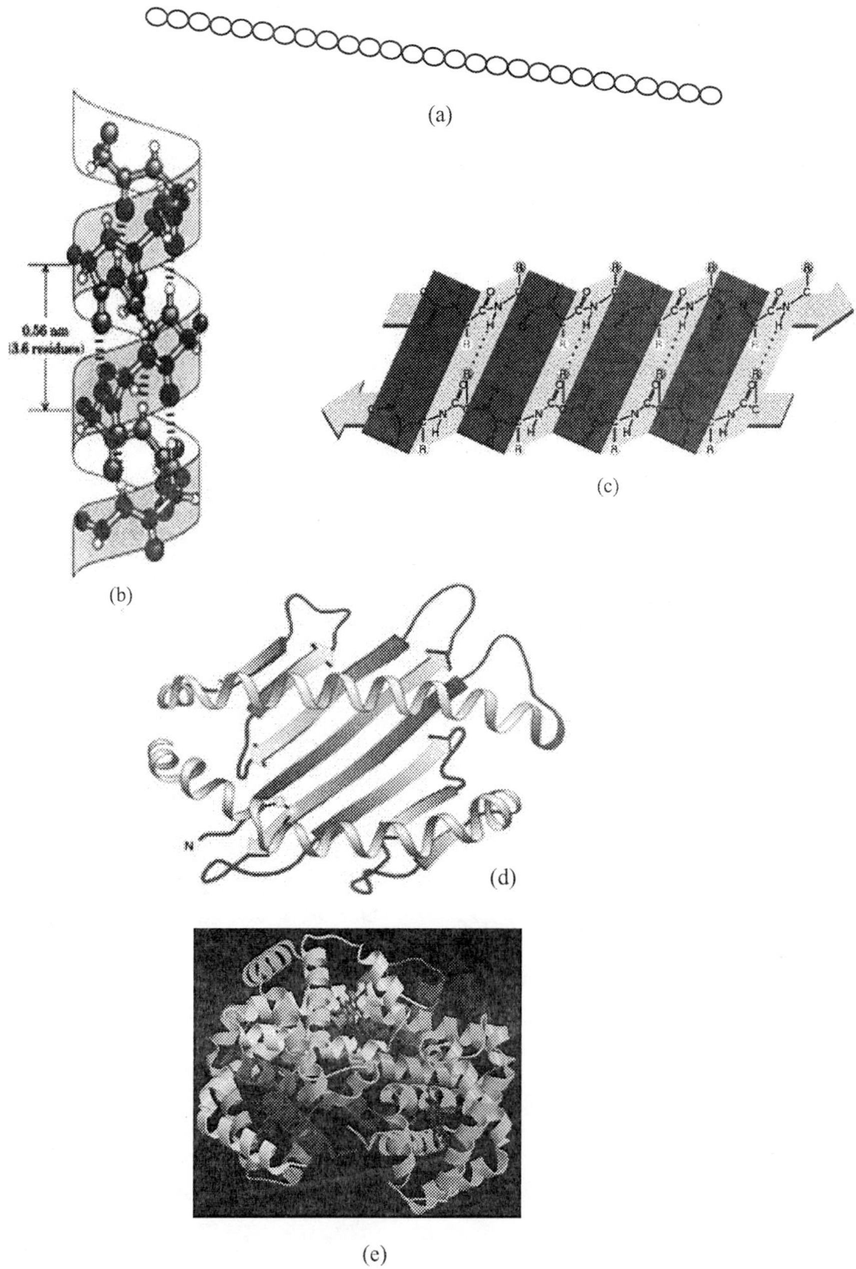

Figure 11.3: Structure of protein molecule (a) linear primary structure (1°), (b) α-helix (2°), (c) β-plated sheet (2°), (d) tertiary structure (3°) and (e) quaternary structure (4°)

PRIMARY STRUCTURE: LINEAR AMINO ACID SEQUENCE

The primary structure of the protein refers to the number and sequence of amino acids, the constituent units of the polypeptide chain. The mode of linkage of the amino acids in the proteins is the peptide bond which links the α-carboxyl group of the one amino acid residue to the α-amino group of the other. The proteins may consist either of one or of more such peptide chains.

Peptides

The amino acid units are linked together through the carboxyl and amino groups to produce the primary structure of the protein chain. The bond between two adjacent amino acids is a special type of amide bond, known as the **peptide bond** and the chain, thus formed is called a **peptide chain (Figure 11.4).** Each peptide chain is of considerable length and may possess from 50 to millions of amino acid units. Depending on the number of amino acid molecule composing a chain, the peptides may be known as **a dipeptide** (containing 2 amino acid units), **a tri-peptide** (containing 3 amino acid units) and so on. If a peptide is made up of not more than 10 amino acids, it is called **an oligo-peptide**; beyond that it is **a polypeptide.** Polypeptide when they are made up of over 100 amino acids are, sometimes, called as macropeptides

Rigid and Planar Peptide Bond: In the late 1930s, Linus pauling and Robert Corey, demonstrated the α-carbons of adjacent amino acids are separated by three covalent bonds, arranged as C_α-C-N-C_α. They also demonstrated that the amide C-N bond in a peptide is somewhat shorter (0.132 nm or 1.32 Å) than the C-N bond in a simple amine (0.149 nm or 1.49 Å) and that the atoms associated with the bond are co-planar. This indicated a resonance or partial sharing of two pairs of electrons between the carboxyl oxygen and the amide nitrogen. The oxygen has a partial negative charge and the nitrogen has partial positive charge, which sets up

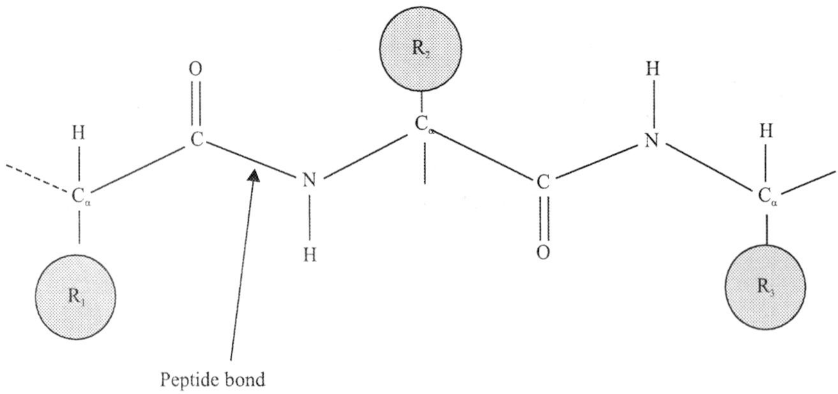

Figure 11.4: Schematic illustrating a peptide bond between two amino acids.

a small electric dipole. The four atoms of the peptide group (C, H, O, and N) lie in a single plane, in such a way that the oxygen atom of the carbonyl group and the hydrogen atom of the amide nitrogen are *trans* to each other. Virtually, all peptide bonds in proteins occur in ***trans*** configuration. They concluded that the amide C-N bonds are unable to rotate freely because of their partial double bond character. The backbone of a polypeptide chain thus is separated by substituted methylene group **–CH(R)–.** The rigid nature of peptide bonds limits the number of conformations that can be acquired by a polypeptide chain.

However, rotation is permitted about the bond between the nitrogen and α carbon atoms of the main chain (N–C_α) and between α carbon and carbonyl carbon atoms (Cα–C). The degree of the rotation at the N-C_α bond is called ***phi*** (ϕ) and that between Cα –C bond is called ***psi*** **(ψ).** Both ϕ and ψ are considered to be at 0° in the conformation in which the two peptide bonds are connected to single α carbon are in the same plane. Theoretically, ϕ and ψ can have any value between –180° and +180°. However, many values of ϕ and ψ are prohibited by steric hindrance between atoms in the polypeptide backbone and the amino acid side chains. The conformation in which ϕ and ψ are both 0° is prohibited for this reason. Some of the plausible conformations of peptides are shown in the **Figure 11.5** and are discussed below:

- **(i) The planar peptide group:** The oxygen and hydrogen atoms are on opposite sites of the C-N bond. This is *trans* configuration.
- **(ii) The three bonds between sequential Cα carbons in a polypeptide chain:** The N-C_α and C_α–C bonds can rotate, with bond angles designated as ϕ and ψ respectively.
- **(iii) Limited rotation around two (N–C_α and Cα–C) of the three types of bond in a polypeptide chain:** C-N bonds are in the planar polypeptide groups, which make up one-third of all the backbone bonds, are not free to rotate.
- **(iv) The two coplanar peptide bonds flanking an α carbon:** By convention, ϕ and ψ are both defined as 0° when the two peptide bonds flanking an α carbon are in the same plane. In a fully stretched out polypeptide chain, ϕ = ψ = 180°.

SECONDARY STRUCTURE

If the peptide bonds were the only type of linkage present in proteins, these molecules would have behaved as irregularly coiled peptide chain of considerable length. Contrastingly, proteins show a great deal of regular structural properties, indicating the presence of a regular coiled structure in these molecules. Rather the functional proteins such as enzymes have highly specified structures. This involves the folding of the chain which is mainly due to the presence of hydrogen bonds. Thus, folding and hydrogen bonding among neighbouring amino acids results in

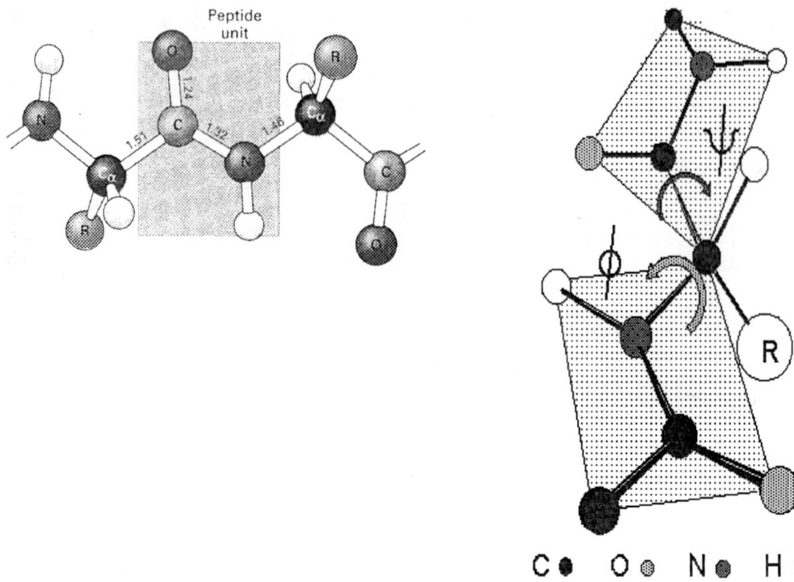

$\phi = \psi = 180°$ is extended conformation
ϕ : Cα to N–H
ψ : C=O to Cα

Figure 11.5: Bond angles and planar arrangements of atoms in peptide.

the formation of rigid and helical structure called as α helix. This corresponds to the secondary structure of proteins, which refers to the hydrogen bonding ability based on spatial relationship of amino acids that are near to each other in the polypeptide.

The conformation of any polypeptide chain can be elucidated if the values of ϕ and ψ for each amino acid residue in the chain are known. In a fully stretched polypeptide chain, $\phi = \psi = 180°$. Ramachandran (1963) demonstrated that an amino acid residue in a polypeptide chain cannot have just any pair of values of ϕ and ψ. By assuming that atoms behave as hard spheres, allowed ranges of ϕ and ψ can be predicted and visualized in steric contour plots referred to as **Ramachandran Plots**. Such a plot for poly-L-alinine (or any amino acid except glycine and proline) shows three separate allowed ranges in diagram. One of them contains ϕ-ψ values that generate the antiparallel β-sheet, the parallel β-sheet and the collagen helix. A second region has ϕ-ψ values that produce right-handed α-helix, and a third region, the left-handed α helix. However, left-handed α-helices are not found in proteins because they are thermodynamically unfavorable (**Figure 11.6**). Based on the nature of hydrogen bonding, Pauling and Corey (1951) identified two regular types of secondary structure in proteins: alpha (α-helix) and beta pleated sheet (β-pleated sheet).

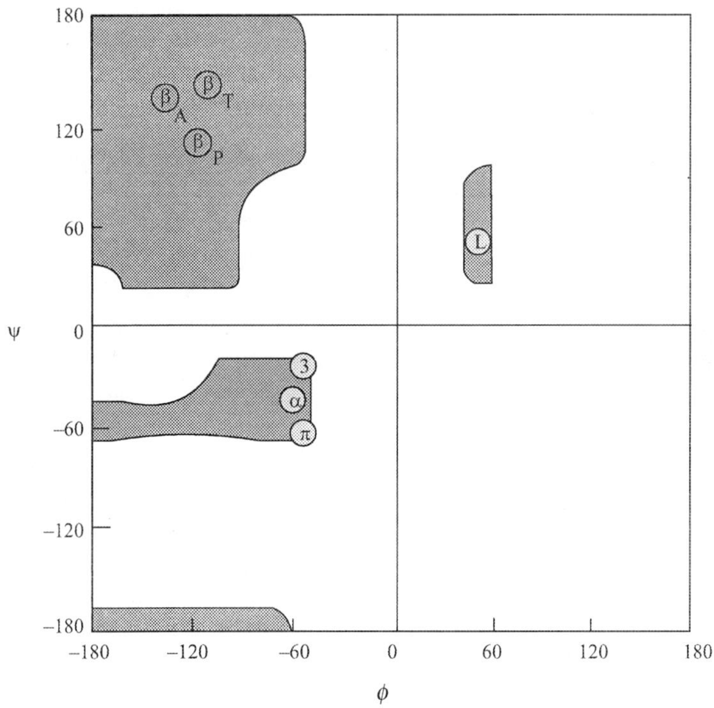

Figure 11.6: Ramchandran plot – Steric contour plot for predicting protein structure based on pair of values of ϕ and ψ.

Alpha (α)-helix: The polypeptide backbone of an alpha helix is twisted by an equal amount having each α-carbon with a phi angle of approximately -57° and a psi angle approximately -47°. A complete turn of the helix contains an average of 3.6 amimoacyl residues, and the distance it rises per turn is 0.54 nm (5.4 Å). Proteins contain only L-amino acids, which make a right-handed stable α-helix and only right-handed α-helices occur in nature (**Figure 11.3b**).

The stability of α-helix arises primarily from hydrogen bonds formed between the oxygen of the peptide bond carbonyl and hydrogen atom of the peptide bond nitrogen of the fourth residue down the polypeptide chain. The ability to form the maximum number of hydrogen bonds, supplemented by van der Waals interactions in the core of this tightly packed structure, provides the thermodynamic driving force for formation of α-helix. Many α-helices have predominantly hydrophobic R groups on one side of the axis of the helix and predominantly hydrophilic ones on the other. These **amphipathic helices** are well adapted to the formation of interfaces between polar and nonpolar regions such as the hydrophobic interior of a protein and its aqueous environment. Clusters of amphipathic helices can create a channel, or pore that permits specific polar molecules to pass through hydrophobic cell membranes. α- keratin is rich in hydrophobic residue including Phenylalanine,

Isoleucine, Valine, Methionine and Alanine. These give water resistant properties to keratin. Example of α- helix protein:- α- keratin (hair, wool, feathers, nails, claws, horns, tortoise shell and much of the outer layer of skin). Oxygen binding proteins (myoglobin and hemoglobin) also contain α- helix chain.

Beta (β) Sheet: The second recognizable regular secondary structure in proteins is β-Sheet. The amino acid residues of a β-Sheet, when viewed edge-on, form a zigzag or pleated pattern in which the R groups of adjacent residues point in opposite directions. Unlike the compact backbone of the α-helix, the peptide backbone of the β-Sheet is highly extended (**Figure 11.3c)**. But like the α-helix, β-sheet derive much of their stability from hydrogen bonds between the carbonyl oxygens and amide hydrogens of peptide bonds. However, in contrast to the α-helix, those bonds are formed with adjacent segments of β-sheet. β-sheet protein structure is highly flexible and soft in nature (**Figure 11.7**). β-Sheet can be arranged either to form a parallel β-Sheet, in which the adjacent segments of the polypeptide chain, proceed in the same direction amino to carboxyl, or an antiparallel sheet, in which they proceed in opposite directions. Either configuration permits the maximum number of hydrogen bonds between segments, or strands, of the sheet. Most β-Sheets are not perfectly flat but have a right handed twist. Clusters of twisted strands of β-sheet form the core of many globular proteins. Example of β-sheets include silk protein fibroin, proteins of spider webs, etc.

Loops and Bends: Roughly half of the residues in a 'typical' globular protein dwell in α-helices and β-sheets and half in loops, turns, bends, and other extended configuration features. Turns and bends refer to short segments of amino acids that join two units of secondary structure such as two adjacent strands of an antiparallel β-sheet. A β turn involves four aminoacyl residues, in which the first residue is hydrogen-bonded to the fourth, resulting in a tight 180° turn. Proline and glycine often are present in β turns. Loops are regions that contain residues beyond the minimum number necessary to connect adjacent regions of secondary structure. Loops are irregular in configuration; and are very important in functioning of proteins. Helix-loop-helix motifs play a role in the oligonucleotide-binding part of DNA-binding proteins such as repressors and transcription factors. Structural motifs such as the helix-loop-helix motif that are intermediate between secondary and tertiary structures are often termed **supersecondary structures.** While loops lack apparent structural regularity, they exist in a specific configuration stabilized through hydrogen bonding, salt bridge, and hydrophobic interaction with other portions of the proteins. This structural flexibility enables such regions to act as switches that affect protein structure and function.

TERTIARY STRUCTURE

The term tertiary structure refers to the unique three-dimensional conformations that globular proteins acquire as they fold into their biologically active structure.

(a)

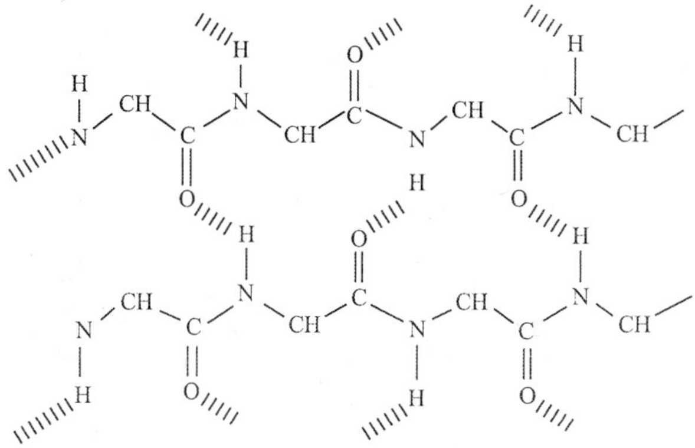

Part of parallel β sheet

(b)

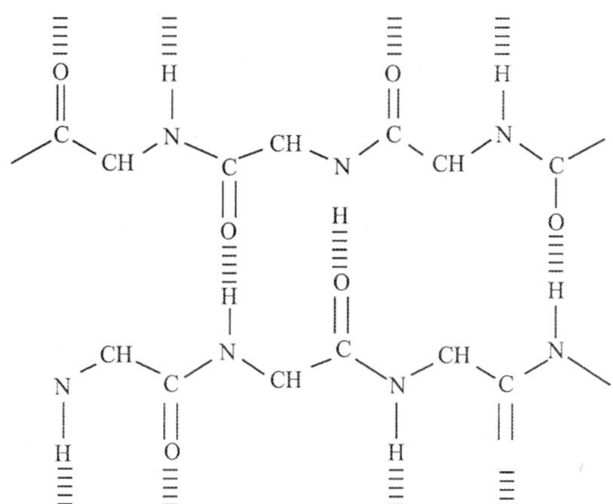

Part of antiparallel β sheet

Figure 11.7: Structure of (a) parallel β-sheet and (b) antiparallel β-sheet.

Protein folding, a process in which an unorganized, nascent linear molecule gains a highly organized structure, occurs as a consequence of the interactions among the side chains in their primary structure. Many polypeptides fold in such a fashion that amino acid residues that are distant from each other in the primary linear structure come into close proximity with each other. Because of efficient packing as the polypeptide chain folds, globular proteins are compact. Large globular proteins often contain more than one independent units called domains. A variety of interactions stabilize tertiary structure including hydrophobic, electrostatic and covalent interaction (**Figure 11.3d**).

Hydrophobic Interactions

As a polypeptide folds, hydrophobic R groups are brought into close proximity because they are excluded from water. Then the highly ordered water molecules in solvation shells are released from the interior, increasing the disorder (entropy). This brings about the favourable entropy change, which is a major driving force in protein folding.

Electrostatic Interactions

These are some weaker but important electrostatic interactions (ion-dipole, dipole-dipole, van der Waals). They are significant in the interior of the folded protein and between subunits or in protein-ligand interactions. In proteins that consist of more than one polypeptide chain, each polypeptide is called a subunit. Ligands are molecules that bind to specific sites on larger molecules such as proteins

Hydrogen Bonds

A significant number of hydrogen bonds form within a proteins interior and on its surface. In addition to forming hydrogen bonds with one another, the polar amino acid side chains may also interact with water or with the polypeptide backbone. When a hydrogen (H) atom is covalently bonded to an electronegative atom, particularly an oxygen or nitrogen (N) atom, it develops a slight positive charge and participates in H-bonding with another electronegative atom.

Covalent Bonds

Covalent linkages are created by chemical bonding in polypeptides structure during or after its synthesis. The most common covalent bonds in tertiary structure are the disulphide bridges found in many extracellular proteins. In extracellular environments these strong linkages partly protect protein structure from denaturation in extreme pH or salt concentrations. Most intracellular proteins lack disulphide bridges because of high cytoplasmic concentrations of reducing agents.

QUATERNARY STRUCTURE

Many proteins, especially those with high molecular weights, are composed of more than one polypeptide chains. Each such polypeptide component is referred as a subunit. Subunits in a protein complex may be identical or different. Multisubunit proteins in which some or all subunits are identical are referred to as oligomers. Oligomers are composed of protomers, which may consist of one or more subunits. A large number of oligomeric proteins contain two or four subunit protomers, referred to as dimers and tetramers, respectively. Such multisubunit proteins have several reasons for their organization and occurrence (**Figure 11.3e**).

The synthesis of separate subunits may be more efficient than substantially increasing the length of a single polypeptide chain. In supramolecular complexes such as collagen fibres, replacement of smaller worn-out or damaged components can be managed more effectively and rapidly. The complex interactions of multiple subunits help regulate a protein's biological function.

In a multisubunit complex protein, polypeptide subunits are held together by noncovalent interactions such as the hydrophobic and electrostatic interactions, hydrogen bonds and covalent bondings. Among these, the hydrophobic effect is clearly the most important because the structures of the complementary interfacing surfaces between subunits are almost similar to those observed in the interior of a globular protein. Covalent cross-links also significantly stabilize some multisubunit proteins. Examples of these include the disulphide bridges in the immunoglobulin (Ig) and the desmosine that connect four polypeptides of elastin.

STRUCTURAL AND FUNCTIONAL ASPECTS OF PROTEINS

Looking at the versatility and presence of proteins in almost all the important aspects (both structural and functional) these can be considered as the most important class of biomolecules. These are found to play significant role in almost all the functions, e.g. all the reactions occurring in living beings are catalyzed by enzymes. An elaborate mechanism and variety of enzymes exists and basically all enzymes are proteins. Locomotion and movement of cells and organisms, ranging from bacteria to humans, depends on proteins such as actin, myosin and flagellin. Transport of gases, minerals and ions requires specific proteins such as haemoglobin, porins, permease, etc. Cytochromes operate in the electron transport chain as electron carrier proteins. Proteins act as receptors for hormones and other signalling molecules (signal transduction). Gene expression and regulation involves a variety of proteins such as transcription factors, repressors, catabolite activator protein (CAP). These proteins act as key components in development and functioning of the cell.

Apart from functional aspects, proteins, as called appropriately, are the basic building blocks of the cells. The very structure of cells and the surrounding extracellular matrix, in which they are embedded, is largely made of protein. Proteins assume a variety of shape, structure and properties to suit the very existence of life or survival of the organism. Keratin is one such protein dispersed over so many animals forming the external layer, i.e. skin and horn, hooves, nails, hair, etc. It is this versatility of the protein which should be appreciated.

Some peptides are also known to play important roles in physiology of organisms. Angiotensin II, an octapeptide, is believed to play a major role in hypertension as it causes constriction of blood vessels leading to rise in blood pressure.

Enkephalines are pentapeptides that regulate conduction of nerve impulses involving pain to the brain and act as natural analgesics. Incorporation of a single incorrect amino acid at the sixth position in the 146 amino acid long β-protein chain of haemoglobin leads to sickle cell anemia.

The quaternary protein structure involves the clustering of several individual peptide or protein chains into a final specific shape. A variety of bonding interactions including hydrogen bonding, salt bridges, and disulfide bonds hold the various chains into a particular geometry. There are two major categories of proteins with quaternary structure – fibrous and globular. Silk is a fibrous protein and hydrogen bonding on amide groups on different chains is the basis of beta-pleated sheet in silk proteins. Keratin in wool and hair is composed of coiled alpha helical protein chains.

Globular Proteins

Globular proteins have a combination of the above types of structures are usually aggregate as clumps or sphericals. Major examples include insulin, haemoglobin, antibodies (also called as immunoglobulins) and most enzymes.

Antibodies are immunoglobulins that save the body from antigens by immobilizing them and bringing about their elimination. Many hormones are proteins that act as messengers and coordinate bodily activities. Examples of protein hormones include insulin, oxytocin, and somatotropin. Structural proteins are fibrous and tough proteins that make structural components of the cells or organs. Collagens and elastin provide support for connective tissues such as tendons and ligaments while keratins provide a tough cover to tissues. Several proteins serve as reservoirs of amino acids. These are called as storage proteins. Milk protein or casein and egg protein or ovalbumin are examples of storage proteins.

SUGGESTED READING

Anfinsen C.B. 1973. *Principles that governs the folding of polypeptide chain.* Science. 181:223–230.

Creighton T.E. 1993. *Proteins: Structure and molecular Properties*. 2nd ed., W.H. Freeman, New York.

Greenstein J.P. and Winitz M. 1961. *Chemistry of the Amino Acids*. Vol. 3. Wiley.

Haurowitz F. 1963. *The Chemistry and Function of Proteins*. 2nd ed., Academic press, Inc., New York.

Ramachandran G.N. 1963. *Aspects of Proteins Structure*. Academic Press, Inc., New York.

Branden, C., and Tooze, J. 1999 *Introduction to Protein Structure*. Garland Science Publishers.

Stryer, L., 1995 *Biochemistry*, (5th edition) W.H. Freeman and Company.

Nelson, D.L. and Cox, M.M. 2000 *Lehninger Principles of Biochemistry*, (3rd Edition), W.H. Freeman and Company.

REVIEW QUESTIONS

1. What are standard amino acids? Write a note on their physical and chemical properties.
2. Write a note on types and chemical nature of amino acids.
3. What is meant by primary, secondary and tertiary structure of a protein? How protein conformation is important in functioning of protein molecules.
4. Write a note on structure and biological importance of proteins.

UNIT XII

Lipids

Lipids are a heterogeneous group of compounds including fats, wax, oil, steroids and other related compounds. These are relatively insoluble in water but are readily soluble in organic solvents such as alcohol, benzene, ether, etc. and yield fatty acids upon hydrolysis. Conventionally, lipids also include many biological compounds that can be dissolved in non-polar solvents such as chloroform and ether, etc. Insolubility of lipids in water shows the hydrophobic nature of their hydrocarbon structure although lipids also contain small polar groups. All the biological systems employ ample amount of water necessitating need of a water insoluble, hydrophobic structural component for compartmentalization of aqueous cytoplasm, i.e. cell membrane. This barrier should also allow transport of molecules and ions and therefore should not be very rigid. Lipids are extensively distributed throughout the living organisms and play an important role in functions such as structure of cell membrane, propagation of nerve impulse, cellular transport and insulation. Lipids provide the quasi-fluid character to the cell membrane which is necessary for transport of molecules.

BIOLOGICAL IMPORTANCE

Lipids perform several important biological functions. They make structure of plasma membrane and are important stored energy resource. Excess of carbohydrates are converted into the fat which is stored in various tissues and utilized at the time of requirement or when sugar is exhausted. Upon oxidation, lipids generate more heat per gram than carbohydrate or protein. Combustion of the fats yields twice as much energy (about 9 kcal/g) as the burning of an equal weight of carbohydrate (about 4 kcal/g). Fat is the most concentrated form in which potential energy can be stored as it is anhydrous. Fatty acids with their flexible backbone can be stored in much more compact form than the highly spatially oriented and rigid glycogen structure. It remains as a stable and fixed reserve of energy until mobilized by enzyme, which hydrolyze it to glycerol and fatty acid. Fat being a bad conductor of heat, provides excellent insulation against low temperature. Many derivatives of lipids are important building blocks of biologically active materials. Lipoproteins, containing phospholipids are constituents of the cell membranes. Incorporation of fat in diet is also necessary for the absorption of fat-soluble vitamins.

Classification of Lipids

Lipids can be classified into three types as simple, compound and derived lipids.

(i) Simple Lipids (Triglycerides and Triacylglycerols)

These are esters of the fatty acids with various alcohols and include two types, viz. fats and waxes.

Fats

These are esters of fatty acids with glycerol in which three molecules of fatty acids combine with one molecule of glycerol to form triglycerides (**Figure 12.1**). For example, three molecules of butyric acid on condensation with one molecule of glycerol yield tributyrin.

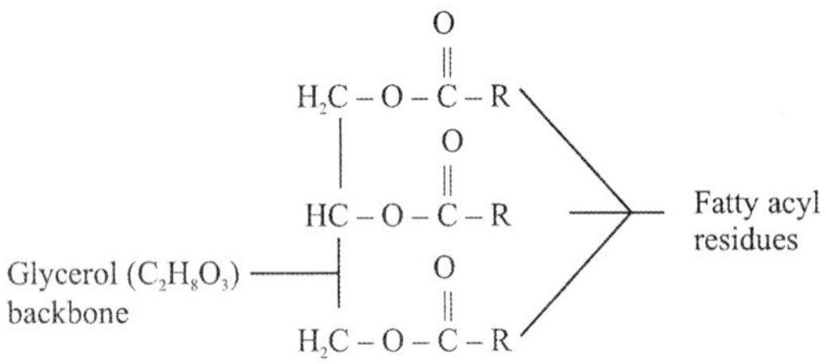

Figure 12.1: Structure of a triglyceride.

A triglyceride containing three similar fatty acids is called as **simple** triglyceride whereas when the fatty acids are different, it is called as a **mixed** triglyceride. Natural fats are mainly composed of mixed glycerides. As these glycerides have no free acid or basic group, they are often termed as neutral fats. Fats are lighter than water and become solids at low temperature but may remain liquid at room temperature. The melting point of any fat depends upon chain length and degree of saturation of constituent fatty acids. Oils are fats in liquid state.

Fatty Acids

Fatty acids are aliphatic carboxylic acids generally having even number of carbon (excepting those found in bacteria) because these are synthesized from acetic acid units with an input of two carbons. Fatty acids occur mainly as esters in natural fats and oils. Fatty acids are named after the hydrocarbon with the same number and arrangement of carbon atoms, with **–oic** being substituted for the last **–e** of alkane. Thus, saturated acids end with suffix **–anoic** while unsaturated fatty acids end with **–enoic**. Carbon atoms are numbered from the carboxyl carbon and the adjacent carbon atoms are referred as α, β, and γ- atoms while the terminal methyl carbon is termed as ω-carbon. The number and position of the double bond in unsaturated fatty acids is denoted by Δ e.g. Δ^9 denotes a double bond between 9th and 10th carbon in oleic acid (octadecanoic acid). Another denotion for this double bond can be ω9 meaning presence of double bond on the 9th carbon from the ω-carbon.

$$\overset{\omega}{CH_3}(CH_2)_7\overset{\Delta^9}{CH=CH}(CH_2)_7COOH$$

Glycerol is common to all fats and, therefore, their properties vary as per the attached chain (s) of fatty acids. The general formula of fatty acids is $CH_3\ (CH_2)_n$ COOH or RCOOH, where ***R*** is a hydrocarbon chain. The value of ***n*** varies from zero in acetic acid to 86 in mycolic acid found in Mycobacteria. The carboxylic acid (–COOH) group of a fatty acid is strongly polar and ionizes in water at cytosolic pH to lose a proton as:

$$RCOOH \rightleftharpoons RCOO^- + H^+$$

The hydrocarbon chain is insoluble in water and is non-polar. Fatty acids therefore possess both, hydrophilic as well as hydrophobic character (amphipathic). Based on the absence or presence of double or triple bonds, fatty acids can be grouped into two types: saturated and unsaturated. Saturated fatty acids have only single bonds in the hydrocarbon chain while unsaturated fatty acids have at least one C = C double bond in the chains. Saturated fatty acids are solids at room temperature because of regular nature of their aliphatic chains. Unsaturated fatty acids are all liquid at room temperature, because the *cis*- double bond interrupts the regular packing of the chains.

Saturated fatty acids

Saturated fatty acids bear maximum possible number of attached hydrogen to satisfy all the valencies of constituent Carbon atoms (**Table 12.1**). These can be envisaged as based on acetic acid in which $-CH_2$ is added progressively.

Unsaturated fatty acids

Unsaturated fatty acids contain at least one or more double bonds having general formula $R-CH=CH\ (CH_2)_nCOOH$. Unsaturated fatty acids may be classified in mono-saturated (monoenoic), polyunsaturated (more than one double bonds) and eicosanoids which are compounds derived from eicosa (20 Carbon) polyenoic fatty acids, e.g. leukotrienes, lipoxins, prostanoids, prostglandins, etc. Most eicosanoids exhibit important physiological and pharmacological properties (**Table 12.2**). Generally the double bonds range between 1 and 6 and occur after 9, 12, 15, 18……carbon atoms.

Table 12.1 Example of some saturated fatty acids

Name	Formula	C-number	Common source
Acetic acid	CH_3 COOH	2	Vinegar
Butyric acid	$CH_3(CH_2)_2COOH$	4	Butter
Caprylic acid	$CH_3(CH_2)$ COOH	8	Coconut oil, palm oil
Myristic acid	$CH_3(CH_2)_{12}COOH$	14	Coconut oil, Nutmeg oil
Palmitic acid	$CH_3(CH_2)_{14}COOH$	16	Common in all animal and plant fats
Stearic acid	$CH_3(CH_2)_{16}$ COOH	18	Common in all animal and plant fats
Arachidic acid	$CH_3(CH_2)_{18}$ COOH	19	Peanut oil

Table 12.2 Some common unsaturated fatty acids

Name	Formula	C-number (Family)	Source
Palmitoleic acid	$CH_3(CH_2)_5CH = CH(CH_2)_7COOH$	16 (ω7)	Sardine oil, milk fat
Oleic acid	$CH_3(CH_2)_7CH = CH(CH_2)_7COOH$	18 (ω9)	Olive oil, pork fat
Linolenic acid	$CH_3CH_2CH = CHCH_2CH = CHCH_2$ $CH = CH(CH_2)_7COOH$	18 (ω6)	Linseed and Hempseed oil
Arachidonic acid	$CH_3(CH_2)_4(CH = CHCH_2)_4(CH_2)_2COOH$	20 (ω6)	Peanut oil and animal fats.

Fatty acids having more than one double bonds are termed as polyunsaturated fatty acids. According the number of double bonds, fatty acids are termed as monoenoic (1), dienoic (2), trienoic (3), and so on. Unsaturated fatty acids may have either odd or an even number of carbon atoms.

Odd-numbered fatty acids are common in plants while even-numbered fatty acids are mostly found in animals. Increase in the number of double bonds results in decrease in the melting point. In general, the melting points of unsaturated fatty acids are lower than saturated fatty acids and are oils (liquids) rather than solid at room temperature. The hydrocarbon chain of a saturated fatty acid possesses zigzag configuration with the bond angle between carbon-carbon being 109^0. Unsaturated fatty acids show geometric isomerism and generally occur in the relatively less stable *Cis* form rather than more stable *Trans* form.

Waxes

Waxes are esters of higher molecular weight fatty acids with alcohols other than glycerol or with sterols such as cholesterol (**Table 12.3**). Fatty acids and alcohols of waxes have 24–36 carbon atoms. Carnauba wax, found in carnauba palm of Brazil, is made up of myricyl alcohol (30 C) and cerotate (26 C) whereas common bee wax comprise of palmitic acid (16C) and myricyl alcohol.

Table 12.3: Some important alcohols and acids found in waxes

S.No.	Alcohols	Acids
1.	Lauryl $C_{12}H_{25}OH$	Myristic $C_{13}H_{27}$
2.	Cetyl $C_{16}H_{33}OH$	Palmitic $C_{15}H_{31}COOH$
3.	Octodecyl $C_{18}H_{37}OH$	Cerotic $C_{25}H_{51}COOH$
4.	Caranaubyl $C_{24}H_{49}OH$	Melissic $C_{29}H_{59}COOH$
5.	Ceryl $C_{26}H_{53}OH$	
6.	Mericyl $C_{30}H_{61}OH$	
7.	Cocceryl $C_{30}H_{60}OH$	

Waxes are insoluble in water due to complete saturation of carbon chains and have high melting points than fats (**Figure 12.2**). These are chemically inert and resistant to digestion by lipases and hence find wide use in polishes while living beings use them to make an impervious coating for avoiding microorganisms and loss or entry of water. The feathers of birds and most leaves are coated with wax whereas human ear is also protected by wax. Carnauba wax, lanolin, bees wax, and spermaceti are used in preparation of cosmetics, polishes, candles and ointments. Paraffin waxes are different in being mixtures of higher-molecular weight alkenes and are not esters.

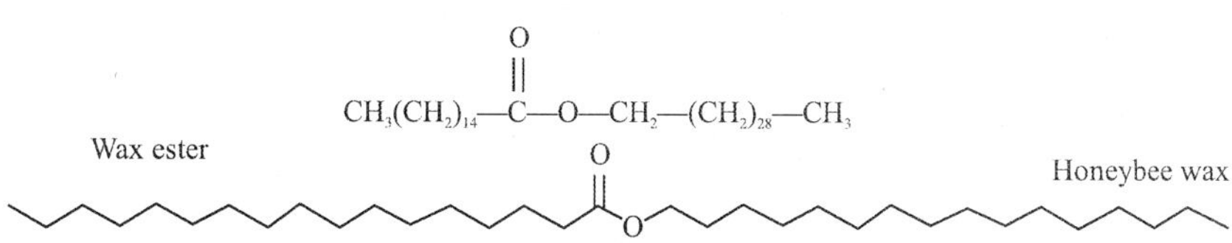

Figure 12.2: Structure of wax.

(ii) Compound Lipids

Compound lipids have some extra groups or elements such as phosphorous, nitrogen, sulphur, or protein apart from regular fatty acids and alcohols (**Figure 12.3**). Some types of such compound lipids are listed below:

(a) Phospholipids

These are the most common compound lipids containing phosphoric acid, glycerol and fatty acids. These are found widely dispersed as structural components of cell membranes (**Figure 12.4**). Phospholipids are classified into phosphoglycerides, phosphoinositides and phosphosphingosides. Among these, **phosphoglycerides** are the major phospholipids found in cell membranes and include two fatty acid molecules or 'tails' esterfied to the first and second hydroxyl groups of glycerol

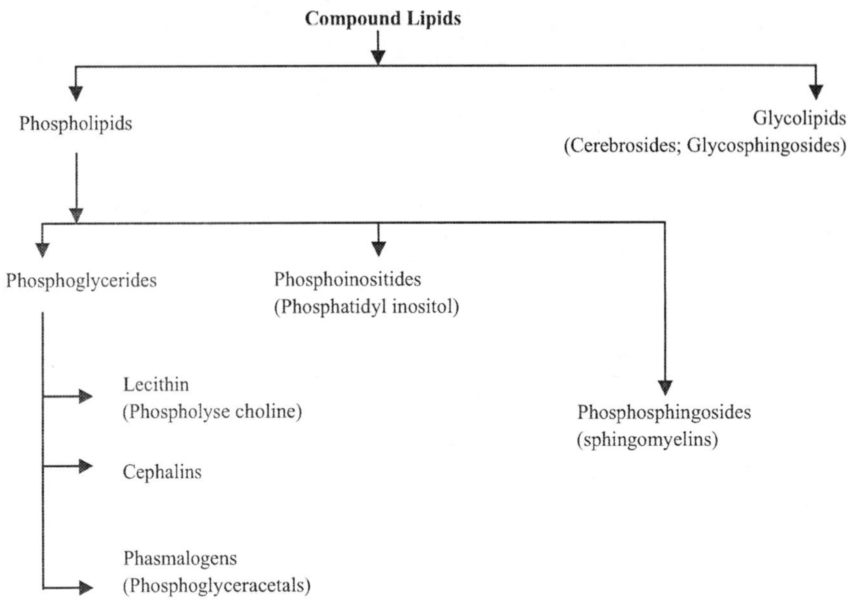

Figure 12.3: Classification of compound lipids.

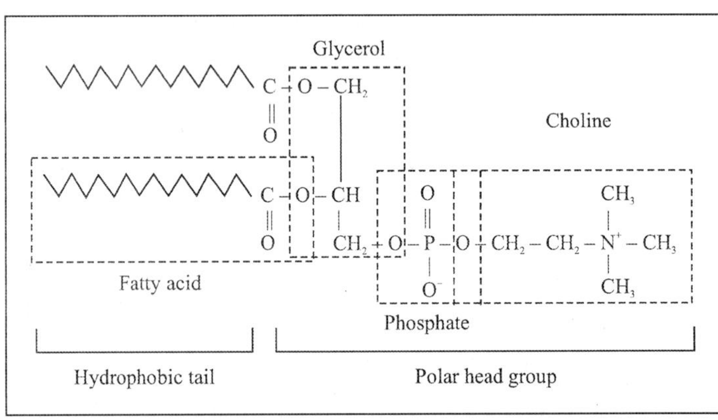

Figure 12.4: Structure and constituents of phospholipids.

while the third hydroxyl group of glycerol forms an ester linkage with phosphoric acid (**Figure 12.5**). **Lecithins** are extensively distributed in nature among various living organisms, e.g. in soybean, yeasts, glandular and nervous tissues of animals. The lecithins are essential for normal transport and utilization of other lipids. In addition to glycerol and two moles of fatty acids, the lecithins also contain phosphoric acid and a nitrogenous base, choline, either at the end or at middle carbon atom of glycerol unit. Therefore, two forms of lecithins, α and β are recognized. Upon complete hydrolysis, lecithin yields choline, phosphoric acid, glycerol and two

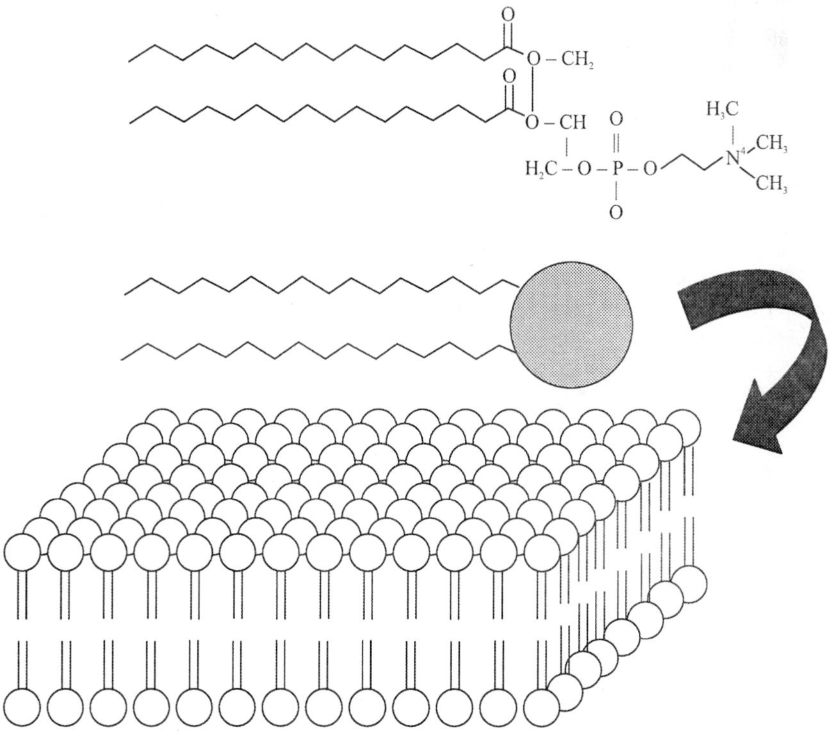

Figure 12.5: The polarity of the constituent phospholipids provides fluidity to the cell membrane.

moles of fatty acids. **Cephalins** are closely related with lecithins in animal tissue. These are similar in structure to the lecithins except that choline is replaced by either ethanolamine or serine. Serine is the biochemical precursor of ethanolamine. Accordingly, two types of cephalins are recognized, phosphatidyl etanolamine and phosphatidyl serine. Like lecithins, the cephalins also exist in 2 forms, α and β, based upon the relative positions of the two substituent fatty acids. Cephalins are relatively less soluble in alcohol than lecithins. **Plasmalogens** represent about 10% of the phospholipids of the brain and muscles. They are similar to lecithins but have one of the fatty acids replaced by an unsaturated ether. Since the nitrogen base can be choline, ethanolamine or serine, these types of plasmalogens are accordingly distinguished as phosphatidyl choline, phosphatidyl ethanolamine and phosphatidyl serine (**Figure 12.6**).

(i) Phosphoinositides

Phosphoinositides are a kind of phospho-glyco-lipids, found in the brain tissues and soybeans and are of considerable importance because of their role in cellular transport. These phospholipids have a cyclic hexa-hydroxy alcohol called inositol as the stereoisomer, myo-inositol. Phosphinositides upon hydrolysis yield 1 mole of glycerol, 2 moles of fatty acid, 1 mole of inositol alongwith 1, 2, or 3 moles of phosphoric acid from mono-,di- or triphosphoinositides, respectively.

(ii) Phosphospingosides

These occur specially in the myelin sheath of nerve cells. These differ from other phospholipids in having a nitrogenous base sphingosine or dihydro-sphingosine in place of glycerol. Sphingomyelins are electrically charged molecules due to phosphocholine as their polar head group (**Figure 12.7**). Hydrolysis of the phosphosphingosides yields equal moles of fatty acid, phosphoric acid, choline and sphingosine or dihydrosphingosine.

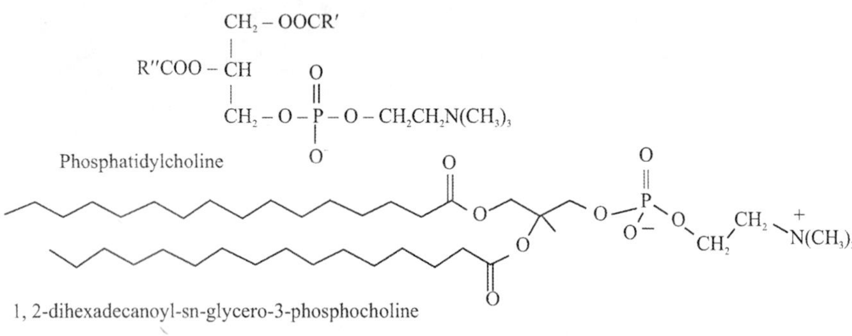

Figure 12.6: Structure and constituents of α phosphatidylcholine.

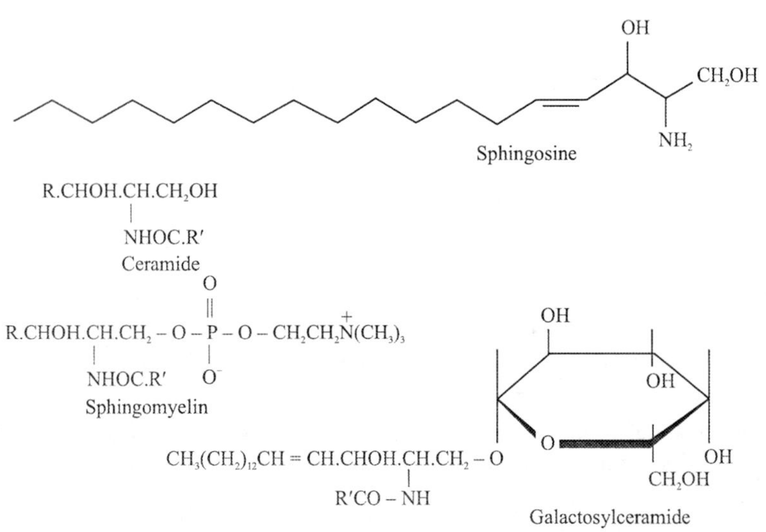

Figure 12.7: Structure of sphingosines: ceramide, sphingomyelin and galactoceramide.

(b) Glycolipids

Glycolipids are made up of 2 long hydrocarbon tails (hydrophobic part), one or more sugar residues (hydrophilic part). These are mainly localized in several animal tissues including brain. Abnormally high deposition of glycolipids in brain causes Niemann-Pick disease. Gycolipids appear in high concentration in the liver and the spleen in Gaucher disease. Structurally, cerebrosides are similar to phosphosphingosides. They are made up of a high molecular weight fatty acid, sphingosine and have galactose or glucose instead of choline. They have no electric charge because their polar head groups are neutral. Following is the list of glycolipids alongwith their particular fatty acid:

(a) Kerasin – contains saturated C 24 lignoceric acid (CH_3 $(CH_2)_{22}COOH$).
(b) Phrenosin – contains a 2-hydroxy derivative of lignoceric acid called cerebronic acid (CH_3 $(CH_2)_{21}COOH$).
(c) Nervon – contains an unsaturated homologue of lignoceric acid called nervonic acid (CH_3 (CH_2) $CH = CH$ $(CH_2)_{13}COOH$).
(d) Oxynervon – contains a 2-hydroxy derivative of nervonic acid called oxynervonic acid (CH_3 $(CH_2)_7CH = (CH_2)_{12}CH$ (OH) $COOH$).

Gangliosides

Gangliosides are a type of glycolipid that makes up about 6% of the membrane lipids in the gray matter of the brain. These are found in high concentration in ganglions of nervous system and also in spleen and red blood cells.

Sulfolipids

Sulphur containing glycolipids are widely found in plant chloroplasts and some chromatophores of photosynthetic bacteria. As the sulfur in this compound is present as a sulphonic group in a hexose, this may also be included under a class of compounds called sulfolipids.

(iii) Derived Lipids

Derived lipids include products derived out of hydrolysis of simple and compound lipids and other structurally related compounds including steroids, terpenes, fatty acids, fatty aldehydes, etc.

Steroids

One of the most important derived lipids, steroids are made up of fused and fully saturated ring system called **cycloperhydrophenanthrene** or **sterane**. Steroids do not contain any fatty acids and hence are non-saponifiable (cannot be hydrolyzed by heating with alkali to yield soaps of their fatty acid). Thus these may be separated from associated fats after the saponification of fats. The hexagonal rings of steroids are different from benzene rings as C atoms are fully saturated with hydrogen and the double bonds, if any, are shown as such. The 'angular' methyl groups occur at C-10 and C-13 while a hydroxyl group is present on C-3. A characteristic side chain at carbon number **17** is used as a basis for classification of steroids. Some most important steroids are described in the following text (**Figure 12.8**).

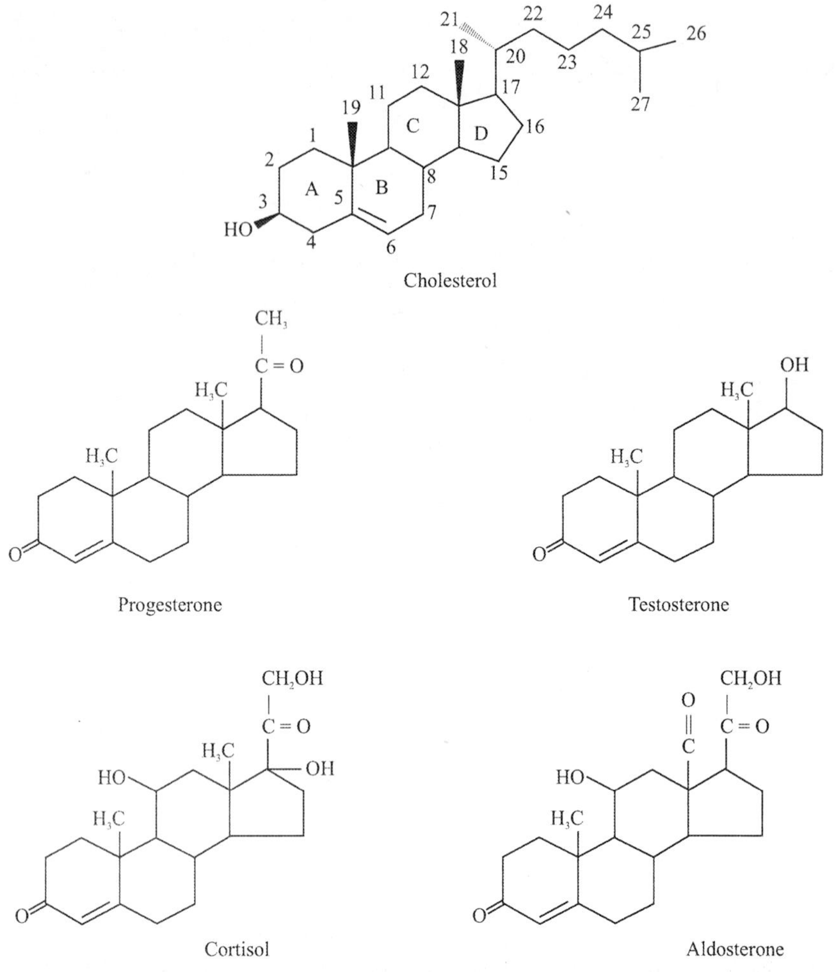

Figure 12.8: Structures of some biologically important sterol derivatives.

Cholesterol

Cholesterol is one of the most infamous lipid due to its direct relevance in cardiovascular diseases in humans. However, it is the main sterol of humans and other higher animals and makes important ingredient of cell membranes and plasma lipoproteins. Among microorganisms, cholesterol is mainly found in the cell membrane of mycoplasma. Sterols are known to have a stabilizing effect on cell membrane because of their planer chemical structure. More importantly, this steroid is the precursor of many other biologically important steroids including hormones and bile acids. Cholesterol is abundant in fish liver oil, brain and spinal cord. It is found associated with nerve cells and nervous tissues. It is markedly absent in plants. It is derived from cholestane or $C_{27}H_{48}$ and has molecular formula $C_{27}H_{46}O$.

In addition to an OH group at C3, there is double bond at C5. Reduction of the double bond gives rise to two products, coprostanol and cholestanol. The hydroxyl group constitutes its polar head while the rest of the molecule is hydrophobic. It has melting point of 149°C. In blood, about 2/3rd of cholesterol occurs as esters of unsaturated fatty acid and remaining 1/3rd as free alcohol. **Some of the important sterols are as follows:**

(i) **Ergosterol:** Ergosterol has a molecular formula $C_{28}H_{44}O$ with one OH group at C_3 and 3 double bonds at C_5, C_7 and C_{22} and the parent compound is ergostane, $C_{28}H_{50}$. It is found in ergot and other fungi including yeasts.

(ii) **Coprostanol:** (*copro - fecal*) It is produced in the intestine as a result of bacterial action on the double bond of cholesterol.

(iii) **Cholestanol:** It occurs as a minor constituent of the sterols of blood and other tissues.

(iv) **Lanosterol:** A major constituent of wool fat, it is a C 30 compound with two methyl groups at C4 and a third angular methyl group on C 14. There are two double bonds at C_8 and C_{24}.

Bile acids (C 24 sterols)

Bile acids are important end products of cholesterol metabolism. About 20 natural bile acids have been identified. All these are derived from a C 24 parent steroid, cholanic acid and resemble coprostanol in having rings A and B in *cis* form. The most abundant bile acids in human bile are cholic acid (20–60% of the total bile acid), chenodeoxycholine acid (30–50%) and deoxycholic acid (5–20%). Bile acids differ from each other in the number and position of OH groups which are all in α configuration. In the bile acids, the number of OH group(s) may be 1, 2, or 3. The side chain of bile acids is usually made up of 5 carbon atoms and bears the carboxyl group. The salts of these conjugated acids have high surface activity and hence promote the intestinal absorption of lipids like cholesterol. They are water-soluble and act as powerful detergent in biological systems.

C 21, C 19 and C 18 steroids

This category includes majority of steroid hormones. Some of important steroid hormones are described below:

Adrenocorticoid hormones

These are C 21 corticoid hormones produced in adrenal glands. These may be divided into two groups according to function. Mineralocorticoids regulate the concentrations of ions, mainly Na^+ and K^+, and glucocorticoids control carbohydrate metabolism. Aldosterone is one of the most important mineralocorticoids. Increased secretion of aldosterone enhances the reaborsption of Na^+ and Cl^- ions in the kidney tubules and loss of K^+. Since Na^+ concentration controls water retention in the tissues, aldosterone also controls tissue swelling.

Cortisol is the major glucocorticoid. Function of cortisol is to increase the glucose and glycogen concentration in the body. This is done at the expense of other nutrients. Fatty acids from fat storage cells and amino acids from body proteins are transported to the liver, which, under the influence of cortisol, produces glucose and glycogen from the sources.

Sex hormones

The most important male sex hormone is testosterone. This hormone, which promotes the normal growth of the male genital organs, is synthesized in the testis from cholesterol. During puberty, increased testosterone production leads to secondary male sexual characteristics as male voice and facial and body hair.

The most important female sex hormone, estradiol, is synthesized from the corresponding male hormone (testosterone) by aromatization of the A ring. Estradiol, together with its precursor progesterone, regulates the cyclic changes occurring in the uterus and ovaries known as the menstrual cycle.

Terpenes

Among the nonsaponifiable lipids found in plants are many hydrocarbons known as *terpenes*. These hydrocarbons and their oxygenated derivatives always have less than 40 carbon atoms. The simple terpenes are called *monoterpenes* and conform to the formula $C_{10}H_{16}$, which those with the formula $C_{15}H_{24}$ are called as *sesquiterpenes*, and with $C_{20}H_{48}$ as *triterpenes*. Terpenes with 40 carbon atoms include compound called carotenoids (**Figure 12.9**). **Monoterpenes and sesquiterpenes** give rise to fragrances of many plants (volatile C_{10} and C_{15} terpenes). These and their oxygenated derivates occur as components of the essential oils, some of which are used as perfumes. Important monoterpenes are geraniol (from rose oil), limonene (from lemon oil) and menthol (from peppermint oil). Among the sesquiterpenes, farnesol is widely distributed but is present in essential oils in small amounts. **Diterpenes** are usually found as constituents of resins and balsams. The two resin acids, abietic and sapietic acids are tricycle diterpenes. Vitamins A_1 and A_2 and their aldehydes, called retinenes, are important monocycle derivates of diterpenes. **Phytol** is an acyclic diterpene and is obtained from hydrolysis of chlorophyll. Mainly two kinds are **triterpenes** are found in nature, ***lanosterol*** a tetra cyclic alcohol, and ***squalanee*** an acyclic hydrocarbon, found in the olive oil and other vegetable oils. Each squalene molecule has six double bonds. **Polyterpene** is found in latex of many tropical plants. Most commonly used and important is the natural rubber. A molecule of rubber is composed of about 500 to 5000 isoprene units joined in a long straight chain.

Carotenoids

This class of compounds includes tetraterpenes widely found in both animals and plants. However, these are exclusively produced by plants. These occur in

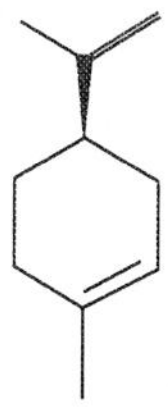

Limonene
(found in the skin
of citrus fruits)

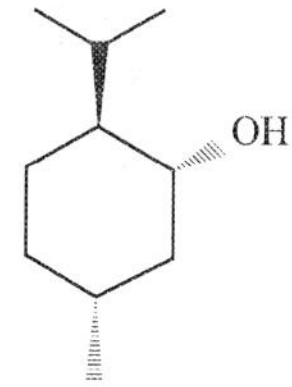

Menthol
(peppermint)

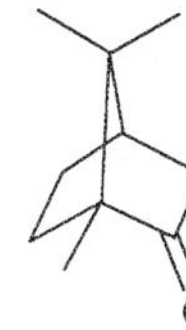

Camphor
(camphor tree)

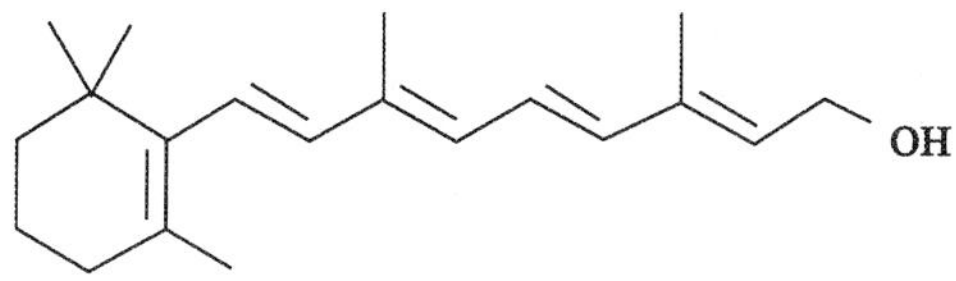

Vitamin A
(involved in chemistry of vision)

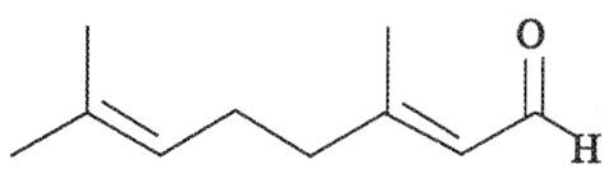

Citral
(lemon grass)

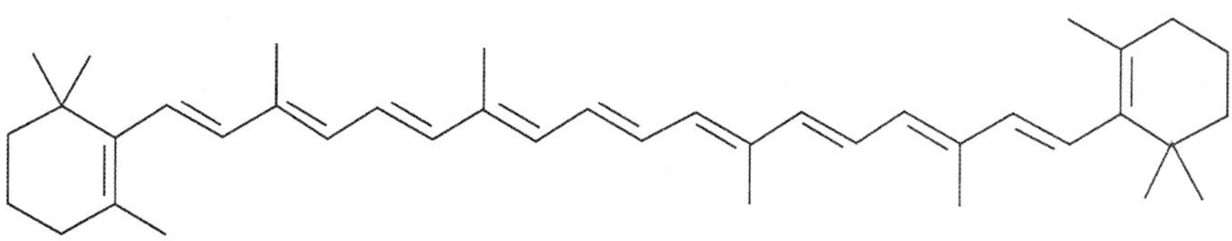

β-carotene
(found in carrots and other vegetables)

Figure 12.9: Structure of some common terpenes.

unsaponifiable residue of the plants and animal lipids. Carotenoids have many conjugated double bonds resulting in their spectral properties and making them appear red or yellow. Because they impart colours, they are called as *lipochromes* or *chromolipids*. **Lycopene** is the carotenoid pigment of tomato, and many other fruits. It is a highly unsaturated, unbranched, long chain hydrocarbon and is composed of two identical chains having formula $C_{20}H_{28}$ joined together by a double bond between C_{15} and C_{15}'. A molecule of lycopene in all contains as many as 13 double bonds, of which 11 are conjugated. **Carotene** has same molecular formula as that of lycopene. These can be separated from plant sources using $CaCO_3$ column chromatograph. Three types of carotenoids are observed in nature **(i) α-carotene-** violet crystals **(ii) β-carotene-** red crystrals (iii) **γ-carotene-** dark red crystals (**Figure 12.10**).

Xanthophylls are identified by the presence of hydroxyl groups in the ionone rings of carotene, in the para position to the long chain. The xanthophylls of the leaf are derived from α-carotene while that of corn is a derivative of β-carotene. Astaxanthine, the red colored pigment of crustaceans is also a xanthophyll.

lycopene ($C_{40}H_{56}$; red; λ_{max}, 476 nm)

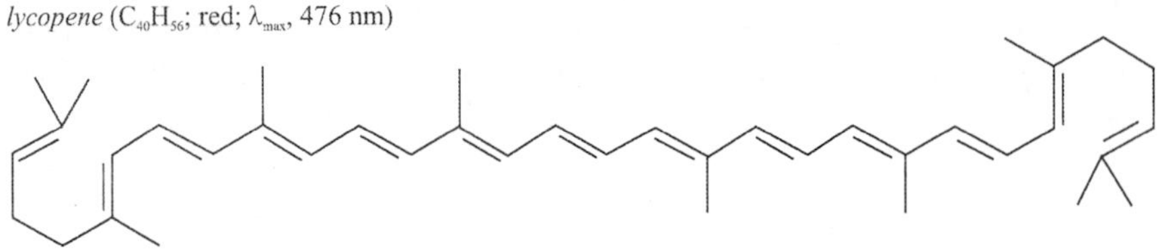

γ-carotene ($C_{40}H_{56}$; orange; λ_{max}, 460 nm)

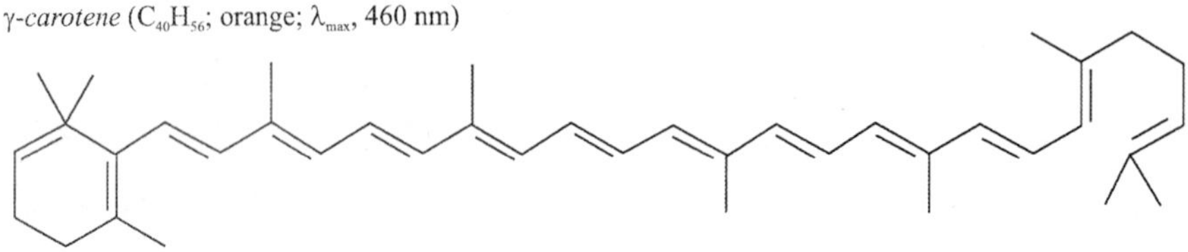

α-carotene ($C_{40}H_{56}$; orange; λ_{max}, 456 nm)

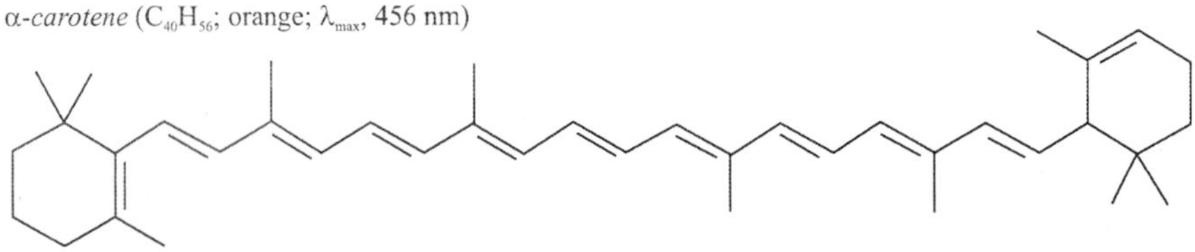

β-carotene ($C_{40}H_{56}$; orange; λ_{max}, 463 nm)

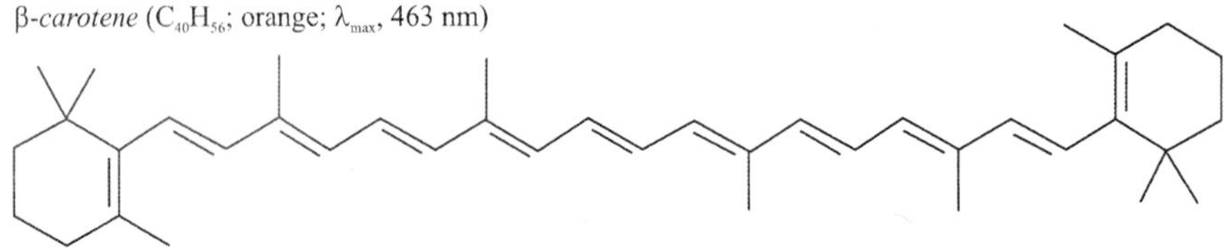

Figure 12.10: Structure and spectral properties of some plant pigments.

PHYSICAL CHARACTERISTICS OF FATS AND OILS

Lipids are sparingly soluble in water, colourless and odourless compounds with specific gravity about 0.86. Unsaturated fatty acids having double bonds show *cis*-trans isomerism. The melting point of fats depends on the chain length of the constituent fatty acid and its degree of unsaturation. Fats containing saturated fatty acids from C4 to C8 are liquid at room temperature but those containing C10 or higher saturated fatty acids are solids and their melting points increase with increasing chain length. With the introduction of double bond in the fat molecule, the melting points lower considerably. Lipid mass is converted into a number of small lipid droplets by the process of emulsification. Fats may be emulsified by shaking either with water or with emulsifying agents such as soaps, gums etc. The process of emulsification is of great metabolic significance in biological system. Fats have to be emulsified before their absorption in the intestine. This process is done with the help of the bile juice secreted from liver. Fats are bad conductor of heat. A layer of fat below the skin provides insulation and acts as a blanket for warm-blooded animals.

CHEMICAL PROPERTIES OF FATS AND OILS

Fats exhibit a great deal of chemical reactivity based upon their unsaturation (double bond) and presence of free carboxyl (–COOH) groups.

Reactions involving –COOH groups

(i) Hydrolysis: Lipases hydrolyse fats to yield fatty acids and glycerol. Lipases are produced by a wide range of organisms. Lipases from bacteria are of particular interest with respect to biotechnological applications. Stored edible fats are subject to rancidity caused by microorganisms.

(ii) Saponification: The hydrolysis of fats by alkali resulting in the formation of glycerol and salts of fatty acids (soaps) is called saponification. Soaps are of common use in our daily life. These are sodium or potassium salts of higher fatty acids which clean objects due to their emulsifying action.

Reactions involving double bond

(i) Hydrogenation: Unsaturated fatty acids, whether free or bound in lipids, can be hydrogenated to yield saturated fatty acids in the presence of metal catalysts such as platinum, palladium or nickel. This reaction is of high commercial significance since it transforms the inexpensive and unsaturated liquid vegetable fats into solid fats.

(ii) Halogenations: Double bonds of unsaturated fatty acids can take up halogens like Bromine (Br_2) and Iodine (I_2) at their double bond(s) at room temperature in acetic acid or methanol solution. The ability of many unsaturated fatty acid to take up Iodine indicate its degree of unsaturation and is represented by Iodine number.

(iii) Oxidation: Unsaturated fatty acids are susceptible to oxidation at their double bonds. Their oxidation by oxidizing agents such as ozone or potassium permanganate leads to formation of aldehydes and glycerol.

Reaction involving –OH groups

Dehydration: Fats, when heated in the presence of the dehydrating agent such as $NaHSO_4$ yield an unsaturated aldehyde called acrolein out of glycerol moiety which can be easily identified by its pungent odor.

Characterization of Fats

With the availability of a wide range of compounds among lipids it is imperative to have some key identification characteristics for ascertaining their type, quality, purity and composition. Following criteria are used in characterization of fats (**Table 12.4**).

Table 12.4 Characteristic analytical values for some fats and oils of plant and animal origin

Fat/Oil	Saponification number	Iodine number	Reichert-Meissl test	Acetyl number	Acid number
Plant fats					
Castor oil	175–187	80–90	1.4	146–150	0.2–0.4
Linseed oil	188–195	170–195	1.0	4.0	1.0–4
Olive oil	190–195	80–90	0.6–1.5	10–11	0.2–3
Coconut oil	254–262	7–10	0.6–7.5	2.0	2.5–6
Animal fats					
Cod liver oil	180–190	140–180	–	–	0.5–5
Human fat	194–198	65–69	0.4	–	–
Butter	196–200	35–42	–	2.7–8.6	–
Cow's milk	216–235	26–25	–	–	–

(i) Acid number: It is the amount of KOH in milligram required to neutralize the free fatty acids present in 1 gram of fat. The acid number indicates the quantity of free fatty acid present in the fat.

(ii) Saponification number: It is the amount of KOH in milligram required to saponify 1 gram of fat completely. The number indicates the average chain length of the fatty acids in the fats and varies inversely with the chain length of the fatty acids. Therefore, shorter the average chain length of the fatty acids, the higher will be the saponification number.

(iii) Iodine number: It is the amount of iodine in grams that combine with 100 grams of oil or fat. It is a measure of the degree of unsaturation of a fat or oil; a high iodine number indicates a high degree of unsaturation of the fatty acids of the fats.

(iv) Reichert-Meissl number: It is the amount of 0.1 N KOH in millilitres required to neutralize the soluble, volatile fatty acids derived from 5 grams of fat. The Reichert-Meissl number measures the quantity of short chain fatty acids in the fat molecule.

(v) Acetyl number: It is the amount of milligrams of KOH required neutralize the acetic acid. It is estimated by saponification of 1 gram fat after it has been acetylated. The acetyl number is the measure of the number of OH groups in the fat.

LIPID METABOLISM

Lipids are of immense metabolic importance because as an energy resource they generate more energy than other organic compounds. Fats are hydrolyzed by lipolytic enzyme (lipases) to release glycerol and fatty acids.

$$\text{Fats} + H_2O \xrightarrow{\text{Lipase}} \text{Glycerol} + \text{Fatty acid}$$

Glycerol further reacts with ATP to form glycerol phosphate which is oxidized to glyceraldehyde -3- phosphate. This may be used in the synthesis of glycogen, glucose or fructose by glycolytic reversal or it can directly participate in glycolysis to yield energy. Remaining fatty acid is oxidized by special sequential reactions of two types, namely, α- and β- oxidation.

The oxidation of long-chain fatty acids to acetyl-CoA is a common energy-yielding pathway in animals and some bacteria (**Figure 12.11**). Complete oxidation of a typical fatty acid, palmitic acid, yields 2,380 kcal/mole.

$$CH_3-(CH_2)_{14}-COOH + 23\ O_2 \rightarrow 16\ CO_2 + 16\ H_2O + 2380\ \text{kcal / mole}$$

The mechanism of even chain saturated fatty acid oxidation was worked out by Franz Knoop (1914). He found that fatty acids are sequentially oxidized at their β-carbon and hence this mechanism was referred as β-oxidation of fatty acids.

The complete oxidation of fatty acids to CO_2 and H_2O occurs in the mitochondria, where the transfer of electrons from the fatty acids to oxygen is used to generate ATP. The combustion occurs in two consecutive steps (i) The fatty acid is sequentially oxidized so as to convert all of its carbons to acetyl-

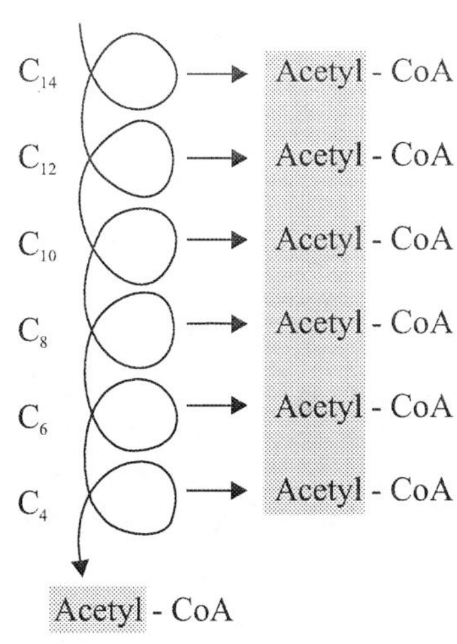

Figure 12.11: Successive reductions in length of the fatty acid by 2-carbons during the β-oxidation of fatty acids.

coenzyme A and (ii) The acetyl-coenzyme A is oxidized by the reactions of the citric acid cycle. Both stages generate ATP by oxidative phosphorylation.

The oxidation of fatty acid in carried out in, liver and other tissue mitochondria. However, these cannot be oxidized unless carnitine (3-hydroxy, 4-trimethyl ammonium butyrate) is present for catalyzing the oxidation. This is possibly because free fatty acid or fatty acyl CoA can not penetrate the inner membranes of mitochondria. The acetyl-CoA is translocated from outside the mitochondrion to the actual internal site of β-oxidation in the mitochondria by enzyme carnitine acetyl-CoA transferase. The acyl carnitine molecule readily passes through the membrane and is converted to acetyl-CoA in the matrix.

β-oxidation is the prominent pathway for oxidation of fatty acids. Other two types of oxidation pathway of fatty acids α and ω also exist. α-oxidation involves removal of one C-atom at a time from the carboxyl end of the molecule and is found in seeds and leaf tissue of plants and some brain cells.

β-oxidation of Fatty Acid

Fats or triacylglycerols are good storage compound but are not readily available as fuels because of their insolubility in water. The stability of the C-C bonds of fatty acid is overcome by activation of the carboxyl group at C-1 by addition of coenzyme A, which allows stepwise oxidation of the fatty acyl group at the C-3 position or the beta (β) carbon, and hence the oxidation of fatty acid is called **β-oxidation.** Free fatty acids cannot enter the mitochondria membrane and must undergo a sequence of three enzymatic reactions in order to enter the mitochondrial matrix.

The pathway involves following steps (**Figure 12.12**):

(i) *Fatty acid activation*

ATP driven Activation of fatty acids by conversion of fatty acid into the corresponding CoA-thioester in presence of coenzyme A and specific acyl coA-synthetase.

$$\underset{\text{Fatty acid}}{RCH_2\,CH_2\,COOH} + \underset{\text{Coenzyme A}}{CoASH} \xrightarrow[\text{Thiokinase , Mg++}]{\text{ATP} \rightarrow \text{AMP + PPi}} RCH_2\,CH_2\,CO - ScoA$$

Alternatively, synthesis of acyl-coA derivatives may involve transfer of CoA to the fatty acid.

$$\text{Propionyl} - CoA + R\,CH_2COOH \longrightarrow \text{Propionic acid} + R\,CH_2\,CO - ScoA$$

(ii) α, β-*dehydrogenation of acyl-CoA*

Dehydrogenation of the product involving removal of two hydrogen from α- and β- carbon atoms of the fatty acids to yield α, β- unsaturated fatty acyl coA

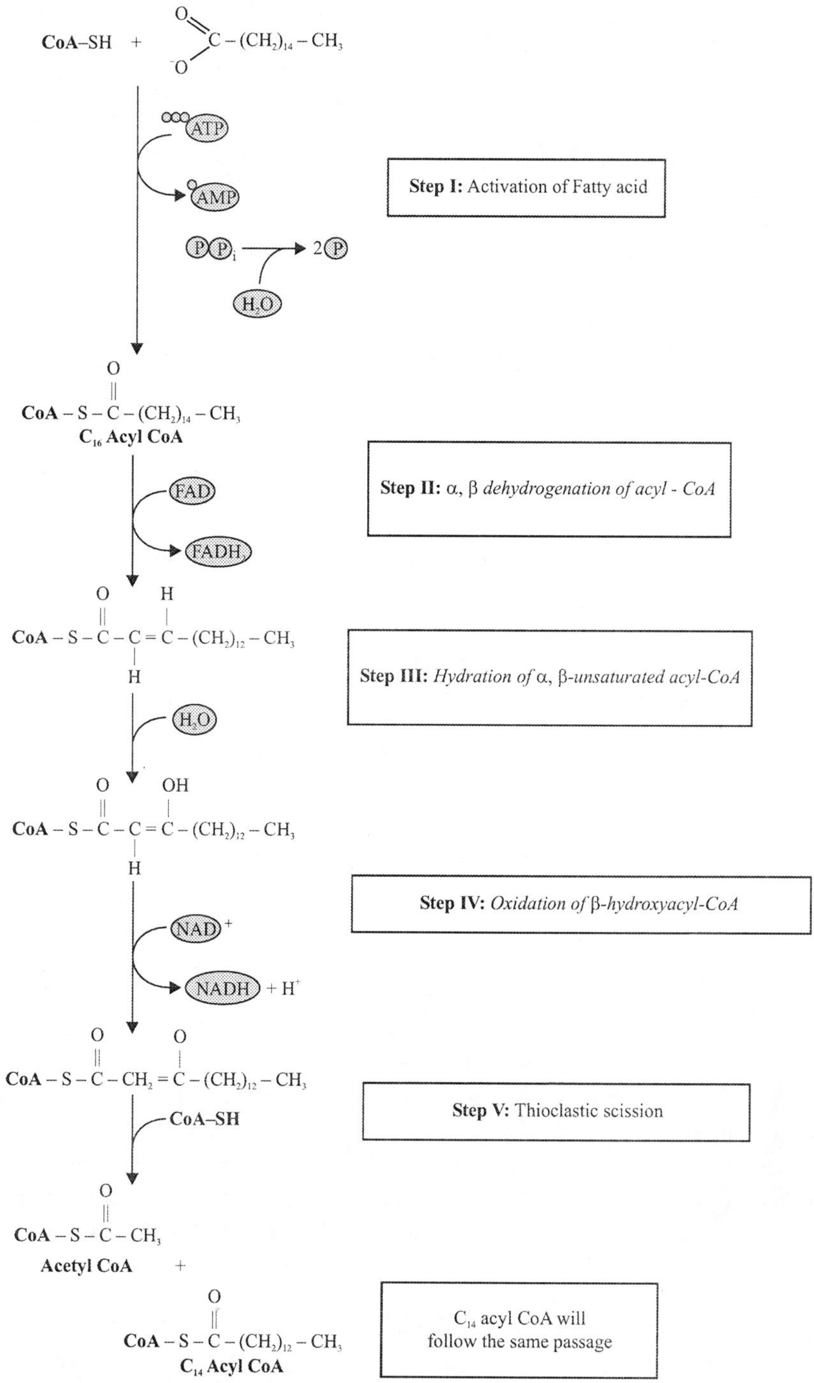

Figure 12.12: Steps involved in the β-oxidation of fatty acid.

derivatives. Various types of dehydrogenases with respect to number of carbon atoms (long or short chain) have been found to catalyze these reactions:

$$\underset{\text{Acyl Coenzyme A}}{RCH_2\,CH_2\,CO-ScoA} \xrightarrow[\text{FAD} \;\rightarrow\; \text{FADH}_2]{\text{dehydrogenase}} \underset{\alpha,\beta \text{ unsaturated acyl coA}}{RCH=CH\,CO-ScoA}$$

(iii) *Hydration of α ,β -unsaturated acyl-CoAs*

Addition of the water molecule to above product in presence of enoyl-CoA hdyrase to yield β-hydroxyacyl –CoA derivative.

$$\underset{\alpha,\beta \text{ unsaturated acyl coA}}{RCH\,CH\,CO-ScoA} \xrightarrow[H_2O]{\text{hydrase}} \underset{\beta \text{ hydroxy acyl coA}}{RCHOHCH_2\,CO-ScoA}$$

(iv) *Oxidation of β- hydroxyacyl- CoA*

The β-hydroxy group of the above product is hydrogenated to the keto group in presence of β-hydroxy–CoA dehydrogenase and NAD^+ to yield the respective β-ketoacyl –CoA ester. NADH + H^+ thus yields 3ATPs when oxidized in ETC (electron transport chain).

$$\underset{\beta\text{-hydroxy acyl coA}}{RCHOHCH_2\,CO-ScoA} \xrightarrow[\text{NAD} \;\rightarrow\; \text{NADH}_2]{\text{dehydrogenase}} \underset{\beta \text{ keto acyl coA}}{RCOHCH_2\,CO-ScoA} + NADH + H^+$$

(v) *Thiolysis or Thioclastic scission*

In the last step β-ketoacyl –CoA ester is oxidized and cleaved by β-ketothiolase to liberate acetyl-CoA from the fatty acid:

$$\underset{\beta \text{ keto acyl coA}}{RCOHCH_2\,CO-SCoA} + HSCoA \xrightarrow{\beta\text{-ketothiolase}} \underset{\text{acyl coA}}{RCO-SCOA} - \underset{\text{acetyl coA}}{CH_3CO\text{–}SCoA}$$

The entire process may be repeated to achieve successive reduction in length of the fatty acid by 2-carbons each time. For example, in the C-16 fatty acid (palmitic acid) undergoes the passage 7 times through the sequence, in each passage losing two carbons as acetyl-CoA. At the end of seven cycles, the last two carbons of palmitate are left as acetyl-CoA. This overall result is the conversion of 16-carbon chain of palmitate to 8 acetyl-CoA molecules. In nutshell, β-oxidation of fatty acids can be represented as follows:

$$\text{Fatty acyl-CoA} + \text{FAD} + \text{NAD}^+ + \text{HS-CoA} \rightarrow \text{Fatty acyl-CoA (2 C shorter)} + \text{FADH}_2 + \text{NADH} + \text{H}^+ + \text{acetyl-CoA.}$$

Thus, in every passage of β-oxidation there is formation of 5 ATP molecules while one ATP gets used up in the first step. In this way, palmitic acid (C-16) will yield 35-1 = 34 ATPs in 7 passages. Also, 8 molecules of CH_3CO-SCoA (acyl coA) produced will yield 12 × 8 = 96 ATPs. Thus a total of **130 ATPs** will be produced by complete β-oxidation of one palmitic acid molecule. The shortened acyl-CoA undergoes another cycle of oxidation, starting with the reaction catalyzed by acyl-CoA dehydrogenase. β-ketothiolase, hydroacyldehydrogenase and enoyl-CoA hydratase all exhibit specificity with respect to the length of the acyl group. Therefore, by repeated turns of the cycle, a fatty acid is degraded to acetyl-CoA molecules with one being produced in every turn until the last cycle. The β oxidative system of fatty acid metabolism is found in all organisms. However, in bacteria grown in the absence of fatty acids, the β oxidative system is practically absent but is readily included by the presence of fatty acids in the growth medium. The bacterial β oxidation system is completely soluble and hence is not membrane-bound.

α-OXIDATION OF FATTY ACIDS

β-oxidation is the most prominent pathway for oxidation of fatty acids. However, other two types of oxidation pathway also occur living systems. α-oxidation involves removal of one carbon at a time from the carboxyl ends of the fatty acid molecule. α-oxidation was first found in seeds and leaf tissues of plants and some animal brain cells. Firstly, the hydroxylation of fatty acid is catalyzed by a mitochondrial enzyme, monoxygenase that requires O_2, Mg^{2+}, NADH and a heat-stable cofactor. Alteration of the α-hydroxyl fatty acids to CO_2 and the next lower unsubstituted acid appears to take place in the endoplasmic reticulum and in presence of O_2, Fe^{2+} and ascorbate. In the α oxidation of fatty acid, only free long-chain fatty acids serve as substrate without any involvement of CoA and without generation of any high energy phosphate compounds. α-oxidation is a specialized mechanism for the oxidation of odd member fatty acids such as phytanic acid. Phytanic acid, a product of phytol, cannot be oxidized by α-oxidation pathway due to presence of 3-methyl group. Fatty acids are oxidized by perooxidases by decarboxylation followed by reduction to an aldehyde having one carbon less than original fatty acid.

$$RCH_2\,CH_2CH_2\,COOH \xrightarrow[\text{Fatty acid peroxidase}]{H_2O_2} RCH_2\,CH_2\,CHO + CO_2 + H_2O + O_2$$

$$RCH_2\,CH_2\,CHO \xrightarrow[\text{NAD} \;\rightarrow\; \text{NADH}_2]{\text{Aldehyde dehydrogenase}} RCH_2\,CH_2\,COOH$$

The resulting fatty acid thus produced may be further oxidized successively to make an α-oxidation spiral. This kind of oxidation may result in formation of fatty acids with odd numbers of carbon. Also, the aldehydes formed may get oxidized to long chain alcohols.

ω-Oxidation or Co-oxidation of Fatty Acids

Omega (ω) oxidation of fatty acids is usually a minor pathway of fatty acid metabolism. However, failure of β-oxidation pathway under unusual conditions (in absence of carnitine) may lead to accumulation of fatty acids in the cell which in turn initiates ω-oxidation of fatty acids.

The ω-oxidation is responsible for the oxidation of alkanes in the animal, plant and bacterial systems. The mechanism proceeds by hydroxylation of the terminal methyl group of the fatty acid to a primarily alcohol. In animals, the cytochrome P_{450} system is the hydroxylase responsible for this alkane hydroxylation; whereas in bacteria, rubridoxin is the intermediate electron carrier which feeds electrons to ω-hydroxylase system. The intermediate product, RCH_2OH is oxidized to an aldehyde by an alcohol dehydrogenase, which in turn is oxidized to a carboxylic acid by an aldehyde dehydrogenase in both systems. Once formed, the dicarboxylic acid formed may be oxidized from either end of the molecule by the β-oxidation pathway to yield energy.

Oxidation of Fatty Acid via Citric acid Cycle

The acetyl residues of acetyl-CoA produced via β-oxidation of fatty acids may be oxidized to CO_2 via the citric acid cycle. TCA also occurs in the mitochondrial matrix. Acetyl-CoA derived from fatty acid oxidation, enters a routine pathway of catabolism along with acetyl-CoA derived from glucose via glycolysis. Under excessive lipid β-oxidation a large amount of acetyl-CoA residues are generated. Their quantity is high enough to be catabolized by TCA.

SUGGESTED READING

Ansell G.B. and Leaderer E. 1961. *Chemistry of Lipids*. Ann. Rev. Boichem. 30: 71–90

Chapman D. 1969. *Introduction to Lipids*. McGraw-Hill, New York.

Gunstone F.D. 1975. *An Introduction to the Chemistry and Biochemistry of Fatty Acids and their Glycerides.* Halstead Press, New York.

Hawthorne J.N. and Ansell G.B. 1982. *Phospholipids*. Elsvier, New York.

Johnson A.R. and Davenport J.B. 1971. *Biochemistry and Methodology of Lipids*. John Wiley and Sons, Inc., New York.

REVIEW QUESTIONS

1. What are lipids? Discuss their biological importance.
2. Give details about the general structure and kinds of lipids.

3. Describe nature and types of common fatty acids.
4. What are compound lipids? Classify and discuss their biological roles.
5. Write a note on nature of derived lipids.
6. Comment upon the following:
   a) β-oxidation of lipids
   b) Cholesterol
   c) Reichert-Meissl number

UNIT XIII

Enzymes

Enzymes are biological catalysts that increase the rate of reaction without getting affected in the process. All reactions are regulated thermodynamically and have an energy barrier that needs to be overcome. The energy of activation is needed to transform the substrate molecules into transition state – an intermediate state between reactants and product. The transition state has the highest free energy of components of a reaction. The Gibbs free energy of activation ($\Delta G^{\ddagger}$) is equal to the difference in free energy (ΔG) between transition state and the substrate. The enzyme stabilizes the transition state of a chemical reaction and thereby decreases free energy of activation ($\Delta G^{\ddagger}$). During this the energy level of substrates and products remains unaltered. Therefore, the rate of reaction is enhanced but there is no net change in energy of the reaction.

In the absence of enzyme the reaction may not happen at all while in its presence the rate can be enhanced upto 10^7 folds. Thus, there is hardly any need of high temperature and the reaction proceeds smoothly under mild conditions. All enzymes are proteins. They are highly specific for the substrate and catalyze specific reaction only. Thermodynamically, the feasibility of any reaction is determined by the equation:

$$\Delta G = \Delta H - T\Delta S$$

where,

ΔG is change in the free energy of reactants and the products

ΔH is change in enthalpy or heat content of reactants and the products

T is the standard temperature

ΔS is change in entropy of the system

An overall decrease in the free energy ($-\Delta G$) indicates that the reaction is energetically favourable and the products are at a lower energy level than reactants or substrates. However, Gibbs free energy of activation ($\Delta G^{\ddagger}$) is independent of ΔG, i.e. change in free energy. The rate of reaction is governed by $\Delta G^{\ddagger}$. It must be noted that negative free energy change ($-\Delta G$) indicates that the reaction is energetically favourable and will occur spontaneously while if the free energy is positive (+ΔG) then it is non-spontaneous and needs an extra input of energy. In biochemical reactions this input of energy is obtained through coupling of energetically favourable reaction. Enzyme does not alter the equilibrium position of any reaction but accelerate the rate of reaction by attaining equilibrium position quickly. A reaction that takes more than one hour to reach equilibrium, in presence of enzyme, may attain equilibrium in 1second.

Every enzyme has an active site for binding with specific substrate. The active site consists of few amino acids that have a unique stereochemical three-dimensional arrangement. For an easier understanding, the active site can be visualized as a cleft or crevice on the enzyme surface that produces a non-polar environment for binding the substrate. The active site binds substrate (s) or reactants under the influence of electrostatic attraction, H-bonds, van der Waals forces, hydrophobic interactions and reversible covalent bonds to produce enzyme-substrate complex. Catalytically active residues within the enzyme bring about a conformational change in the substrate and transform it to the transition state thereby reducing the energy of activation. After this, the reaction takes place and products are released. Enzyme becomes free and can bind with the other substrate molecule.

Emil Fischer (1894) proposed that the shape of the active site of the enzyme and its substrate is such that they fit in each other like a lock and key (**Figure 13.1**). Later, Daniel Koshland Jr. (1958) introduced an induced fit model of an enzyme in which binding of substrate with enzyme induces a conformational change in the active site of the enzyme to achieve the desired transition state (**Figure 13.2**). The enzyme undergoes a conformational change and also enforces the bound substrate to acquire desired conformation. Binding of glucose induces change in conformation of the enzyme hexokinase, so as to complement the structure of glucose. Enzymes show features of both lock and key model (complementarity) and induced fit model to some extent (conformational change).

Substrate specificity is primarily based upon the spatial arrangement of amino acids forming active site and the steric compatibility of substrate for getting fit into it. Serine proteases (trypsin, chymotrypsin and elastase) are protein hydrolyzing enzymes and as the name suggests, these enzymes have serine residue in their

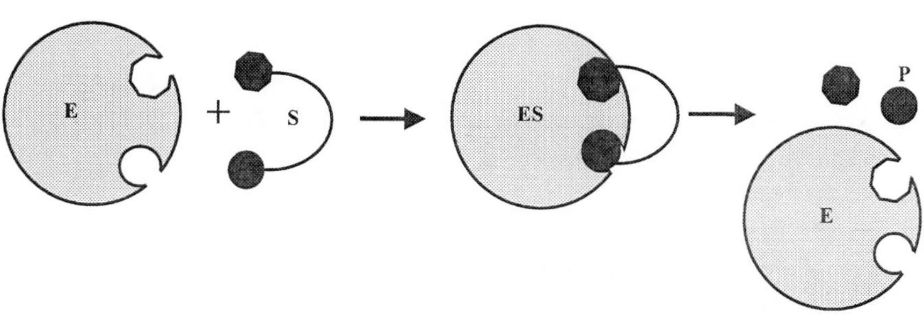

Figure 13.1: Lock and key theory of enzyme action (Fischer, 1894).

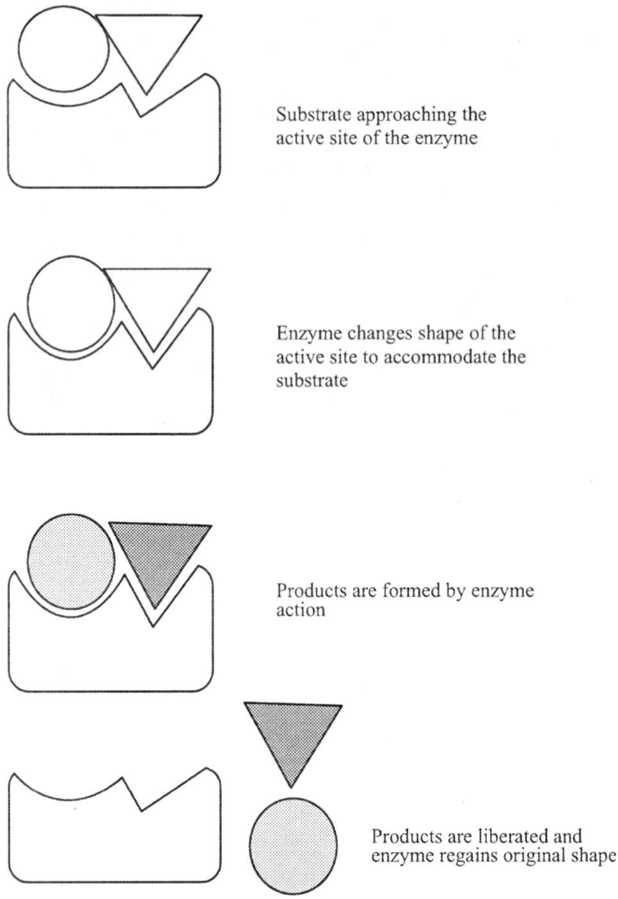

Figure 13.2: Induced fit model of enzyme action (Koshland, 1958).

active site. All these hydrolyze protein molecules from the C-terminus. However, there exists an interesting selectivity of substrates. Trypsin cleaves C-terminus side of proteins having positively charged lysine or arginine residues; Chymotrypsin cleaves on the carboxyl end of proteins containing aromatic and hydrophilic amino acids while elastase specifically cleaves on carboxyl side of proteins having small uncharged side-chains. Such a striking difference in the specificities of these serine proteases is due to the nature of amino acids in their active site which complement their substrates. Trypsin bears a negatively charged Asp residue in its substrate binding site that interacts with the positively charged Lys and Arg residues of substrate. Chymotrypsin bears relatively smaller amino acids (Gly, Ser) in its active site so as to accommodate bulkier amino acids of the substrates. On the other hand, elastase active site bears large neutral amino acids (valine, threonine) so that the enzymes has access to only smaller amino acids of substrate while preventing access to larger ones.

CLASSIFICATION OF ENZYMES

The nomenclature of enzyme is simple to understand. Enzymes are named by adding the suffix – **ase** after the substrate on which they act, e.g. amylase, protease, lipase, phosphatase, inulinase, xylanase, etc.

However, some exceptions to this rule do exist. For example, the names, trypsin and chymotrypsin do not denote their substrates. The uniform code of naming enzymes as provided by International Union of Biochemistry (IUB) provides them a name along with a specific number. The system categorizes all the enzymes into six major classes based on the reaction they catalyze (**Table 13.1**). Each enzyme is categorized by a four-digit classification number for a common universal identity.

Table 13.1 International classification and nomenclature of enzymes

Class	Name	Type of catalyzed reaction	Mechanism	Example
1	Oxidoreductases	Transfer of electrons	$A^- + B \rightarrow A + B^-$	Lactate dehydrogenase
2	Transferases	Transfer of functional groups	A–B + C → A + B–C	Hexokinase
3	Hydrolases	Hydrolysis reactions: addition of water	A–B + H_2O → A–H + B–OH	Amylase, Xylanase
4	Lyases	Cleavage of C–C, C–O, C–N and other bonds resulting in double bonds	A–B → A=B + E–F | | E F	Decarboxylase
5	Isomerases	Group transfer within a molecule (Intramolecular group transfer)	A–B → B–A | | | | C D C D	Maleate Isomerase
6	Ligases	Bond formation coupled to ATP hydrolysis	A + B → A– B	DNA ligase

Examples of international nomenclature and enzyme commission numbers: Trypsin is grouped into hydrolases (class 3) and has been assigned and enzyme commission (EC) number **EC 3.4.21.4** where each number denotes specific characteristic.

(i) Trypsin (EC 3.4.21.4)

- **3.** Hydrolase
- **4.** Hydrolyses peptide bonds
- **21.** Active site contains serine (serine protease)
- **4.** Fourth enzyme in sequence to be recognized as serine protease.

(ii) α- amylase (EC 3.2.1.1)

3. Hydrolase

2. Endohydrolysis of (1→4)-α-D-glucosidic linkages

1. 1 First enzyme in sequence to be recognized as α-amylase.

Coenzymes and Prosthetic Groups

For carrying out reaction, many enzymes require presence of small non-protein units / factors called as co-factors. These may be metal ion(s), e.g. Fe^{2+}, Zn^{2+} or complex organic molecule referred as coenzyme. Metal ion (s) or coenzyme linked covalently to the enzyme is called as a prosthetic group, (e.g. haem in haemoglobin) (**Figure 13.3**).

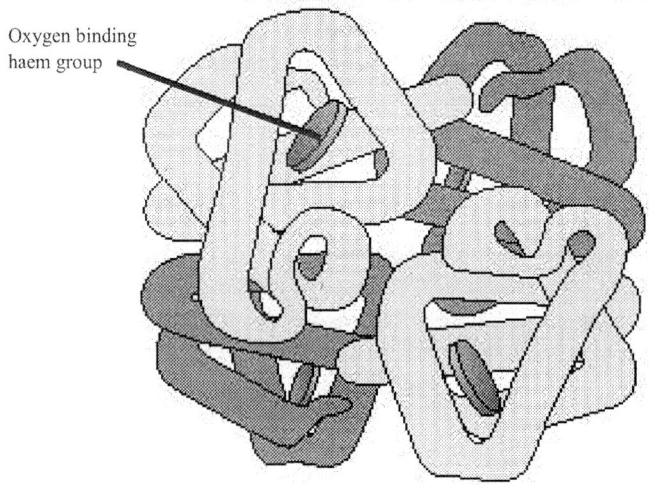

Figure 13.3: Haemoglobin indicating haem group.

A complete unit of catalytically active enzyme with co-factor, metal ion or coenzyme is called as **holoenzyme**. The protein part of the enzyme without co-factor is called as **apo-enzyme**. There are certain coenzymes, (e.g. NAD^+) which are bound by the apoenzyme only during its catalytic cycle and are released as soon as the reaction is over. These co-factors act like co-substrate. Many a co-enzymes are derivatives of vitamin precursors which make an essential component of diet. The deficiency of such components leads to deficiency diseases. Coenzymes NAD^+ and $NADP^+$ are based on a common structure made up of adenine, two ribose sugars linked by a phosphate group and nicotinamide ring (NAD & NADP). $NADP^+$ has an additional phosphate group attached to one of its ribose sugars. These coenzymes have common function of carrying electrons in oxidation-reduction reactions.

However, NAD^+ is used in catabolic reactions (breakdown) while $NADP^+$ is used in anabolic pathways. The reactive entity in both molecules is the nicotinamide ring which can accept or donate electrons during the redox reactions. The transfer of electrons is also associated with the transfer of protons:

$$NAD^+ + H^+ + 2e^- \longrightarrow NADH$$

Another pair of coenzymes, FAD and FMN, are also electron carriers having similar basic chemical structure. These coenzymes have a flavine mononucleotide unit containing the reactive site. FAD has an additional sugar group and an adenine base. FMN and FAD react with two electrons and two protons to complement redox reactions in the following manner:

$$FAD + 2H^+ + 2e^- \longrightarrow FADH_2$$

Isoenzymes (Isozymes)

There are different forms of an enzyme which catalyze the same reaction but exhibit different physical or kinetic properties such as pI (isoelectric point), pH optimum, substrate affinity or inhibition pattern. Such isozymes are products of different genes and occur in different locations. Lactate dehydrogenase (LDH) is a good example of an enzyme having different forms. The enzyme catalyzes reversible conversion of pyruvate into lactate in presence of coenzyme NADH

$$\text{Pyruvate (LDH, NADH)} \longleftrightarrow \text{Lactate}$$

LDH is a tetrameric protein having two different types of subunits, H and M, which differ in amino acid composition. These subunits can combine in varying combinations to make five different isoenzymes: H_4, H_3M, H_2M_2, HM_3 and M_4. These five isoenzymes can be differentiated by polyacrylamide gel electrophoresis. H_4 and H_3M isozymes are found in heart and RBC, H_2M_2 is found in the brain and kidney while HM_3 and M_4 are found predominantly in the liver and skeletal muscles. Thus H units predominate in heart while M units predominate in muscles indicating the relative predominance with respect to tissues. This property of LDH is used to diagnose diseases such as myocardial infarction, hepatitis and muscular disorders based on release and appearance of uncommon type of LDH enzyme in tissues.

Estimation of Enzyme Activity

The activity of an enzyme can be calculated in terms of

(i) Rate of disappearance (e.g. hydrolysis) of substrate, or

(ii) Rate of appearance or formation of product.

A suitable analytical method is therefore required for quantitative estimation of either substrate or product. The rate of disappearance of substrate or rate of formation of product in terms of quantity per unit time (μg / min) is used to represent the unit activity of the enzyme. In addition to above certain crucial factors that may govern the enzyme activity, e.g. pH, temperature, coenzymes, inhibitors, etc. should also be accounted for the resulting rate of enzyme activity. The enzyme activity should be measured in presence of sufficient amount of substrate (concentration) and should not be limited by its insufficiency.

The choice among estimation of substrate or product depends upon the properties of these compounds and availability of a rapid, precise, sensitive and stoichiometric method for their measurement. If the product absorbs light at a specific wavelength then the estimation can be done using spectrophotometer measurement of change in absorbance. Absorbance or optical density is directly proportional to concentration and the rate of change in the absorbance is directly proportional to the rate of enzyme activity in moles of product formed (or substrate consumed) per unit time.

In case a direct measurement is not possible because substrate and products do not absorb light at specific wavelength (for example sugars) then such products can be estimated by measuring absorbance of compounds that are altered or chemically modified (reduced or oxidized) to yield colour in their presence or after reaction (e.g. DNS method for measuring sugars).

Sometimes the enzyme activity is measured by linking or coupling the reaction to another enzyme that brings about a characteristic change in absorbance of the product. These are also being called as linked enzyme assays. For example, activity of glucose oxidase in blood cannot be measured directly because both substrate (glucose) and products (gluconic acid and hydrogen peroxide) do not absorb light at any specific wavelength. However, H_2O_2 can be used to oxidize any compound that yields a coloured compound upon oxidation in presence of peroxidase enzyme. The coloured compound thus produced can be measured spectrophotometrically. In this case the rate of production of H_2O_2 is indicative of glucose oxidase activity which is directly proportional to the rate of formation of chromogen. However, it is important to ensure that the linked enzyme is present in ample amount so that it may not become the rate limiting step.

Examples

(i) DNS method of estimation of glycosidases producing reducing sugars

This method tests for the presence of free carbonyl group (C=O) of the reducing sugars liberated by the action of glycosidases. This involves the oxidation of the aldehyde functional group present in, for example, glucose and the ketone functional group in fructose. Simultaneously, 3, 5-dinitrosalicylic acid (DNS) is reduced to 3-amino, 5-nitrosalicylic acid under alkaline conditions:

$$\text{aldehyde group} \xrightarrow{\text{oxidation}} \text{carboxyl group}$$

$$\text{3, 5-dinitrosalicylic acid} \xrightarrow{\text{reduction}} \text{3-amino, 5-nitrosalicylic acid}$$

The above reaction scheme shows that one mole of sugar will react with one mole of 3, 5-dinitrosalicylic acid. Different reducing sugars generally yield different color intensities and therefore the intensities should be calibrated by preparing respective standard curves.

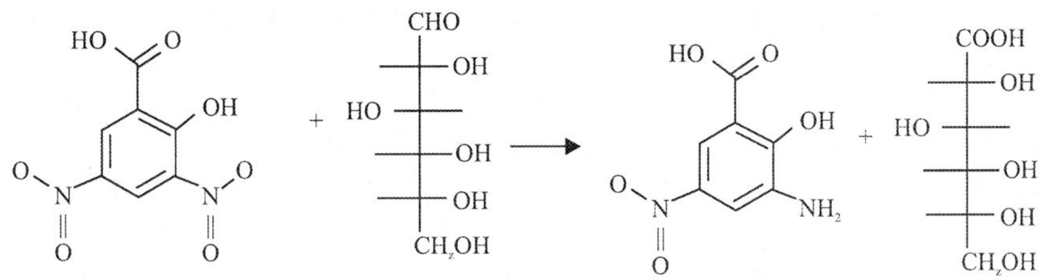

3,5-dinitrosalicylic acid (yellow) Glucose 3-amino-5-nitrosalicylic acid (red-brown) Gluconic acid

(ii) Enzyme linked method for measurement of glucose oxidase:

$$\text{Glucose} + O_2 + H_2O \rightarrow \text{Gluconic acid} + H_2O_2$$

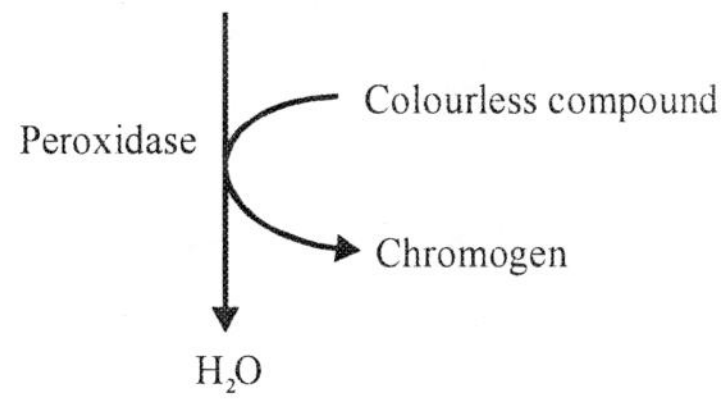

(iii) Rate of formation of NADH or NADPH

These coenzymes absorb ultraviolet (UV) radiations at 340 nm. These are coenzymes whose reduction is coupled with activity of several enzymes. If there is formation of these coenzymes then there will be corresponding increase in absorbance at 340 nm. On the other hand, if an enzyme reaction involves oxidation of NADH or NADPH, then there will be corresponding decrease in the absorption at 340 nm. For example, activity of alcohol dehydrogenase can be estimated by measuring decrease in the absorbance at 340 nm.

$$CH_3CHO + NADH_2 \rightarrow CH_3CH_2OH + NAD$$

Kinetics of an Enzyme Catalyzed Reaction

The study of reaction rates and how they change over the time period in response to experimental conditions is called as kinetics. The rate of any enzyme reaction is referred as velocity. Enzyme velocities are described in terms of rate of reaction at initial time (V_0 at $T = 0$) because the rate of reaction is maximum at time zero due to maximum substrate concentration and absence of product. Enzymes are regulated by feedback inhibition and hence product formed lowers the velocity. In the given figure, the general relationship between velocity and substrate concentration in an enzyme catalyzed reaction is depicted. The relationship is linear initially and the velocity is slowed down as substrate is used up. The enzyme activity is expressed in terms of enzyme units (U). The same unit may be referred as International Unit (IU) of enzyme activity provided that the reaction is conducted under standard conditions. An enzyme unit may be defined as the micromole of substrate transformed or product produced per minute (μmol/min). Katal (kat) is another commonly acceptable unit of enzyme activity which is defined as the catalytic activity which will raise the rate of a reaction by one mole per second under specified conditions. The relation between these expressions is as follows:

$$1\ \mu mol/min = 1U = 16.67\ nkat\ (\text{or nanokat})$$

These expressions of enzyme activity indicate the amount of enzyme activity present in given sample (e.g. culture broth) where there are many unspecified contents or impurities in the broth along with enzyme. Specific activity of the enzyme is defined as the number of enzyme units present per milligram of protein (U/mg). Specific activity indicates the purity of enzyme in the given sample. It increases with the purity of the enzyme and becomes maximum when the enzyme is purified to homogeneity.

Factors Affecting Enzyme Activity

Enzyme activity is dependent upon substrate concentration. Under low concentration of substrate, increment in substrate concentration results in proportional increase of enzyme activity. However, at higher substrate concentration, when the enzyme gets saturated with the substrate any further increase in [S] leads to very small change in V_0. As shown in the **Figure 13.4** the resulting graph between V_0 and [S]

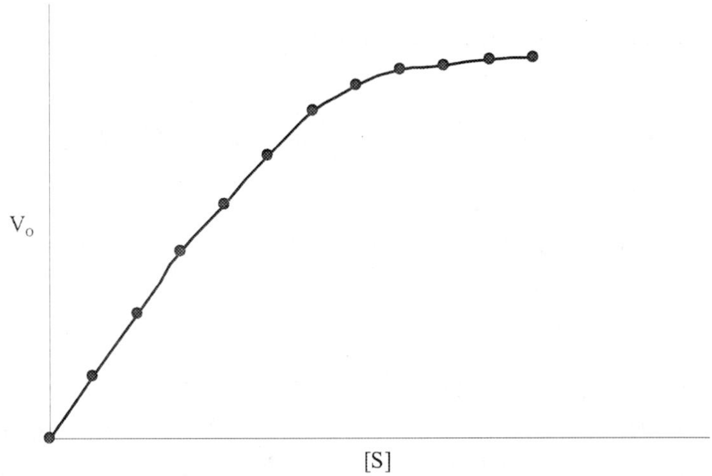

Figure 13.4: Relationship between substrate concentration [S] and initial reaction velocity (V_0).

is a hyperbolic curve. At this juncture, when there is more substrate, any addition of enzyme brings about proportional rise in $V_{o.}$

Temperature plays a significant role in determining the rate of any enzyme catalyzed reaction. It raises the thermal energy of the substrate molecules thereby raising the number of high energy substrate molecules with sufficient energy to overcome energy barrier. However, very high temperature may bring about irreversible denaturation or distortion of the enzyme structure resulting in the loss of enzyme activity. At temperatures lower than the optimum the enzyme activity is reduced but at very low temperatures the enzyme is inactivated completely. However, this inactivation is reversible and the enzyme becomes active once the temperature is raised to normal. A graph between temperature and enzyme activity shows a hyperbola with a distinct maxima (optimum temperature) of activity. Based on temperature relationships the enzymes have been classified as thermostable or thermolabile enzymes. Thermostable enzymes find very useful applications in a variety of industries and scientific experimentation, e.g. Taq DNA polymerase is used in polymerase chain reaction (PCR) and thermostable amylases in starch liquefaction.

Enzyme activity is also affected by pH or the hydrogen ion concentration of surrounding milieu. Smaller changes in pH value bring about reversible changes in the ionization pattern of amino acids of active site while major change in pH denatures the enzyme irreversibly. A graph plotted between enzyme activity and pH will be again a hyperbola or inverted bell with a characteristic maxima indicating optimum pH value for the given enzyme. Based on pH relationship the enzymes can be classified as acidic (acidophilic), neutral or alkalophilic. Enzymes with unusual pH relationship find use in various industries. For example, alkaline proteases are used in detergents; alkaline and thermostable xylanases are used in pulp bleaching in paper industry.

Thermodynamic Consideration of an Enzymatic Reaction

An enzyme provides specific environment in which a given reaction becomes energetically more favourable. The reaction occurs at the active site of the enzyme. Enzyme substrate complex is central to the concept of enzyme action. Catalysts enhance reaction rate by lowering the activation energy. Enzymes affect reaction rate, and do not alter equilibrium.

For any given reaction

$A + B \rightarrow C + D$, the equilibrium constant can be derived as

K_{eq} = [Products]/ [Reactants] = [C] [D] / [A] [B],

where, [A], [B], [C] and [D] indicate the molar concentration of the reactants and substrates.

The equilibrium constant is fixed for any given chemical reaction at a specified temperature. The standard free energy change, ΔG is directly related to $K_{eq.}$ At

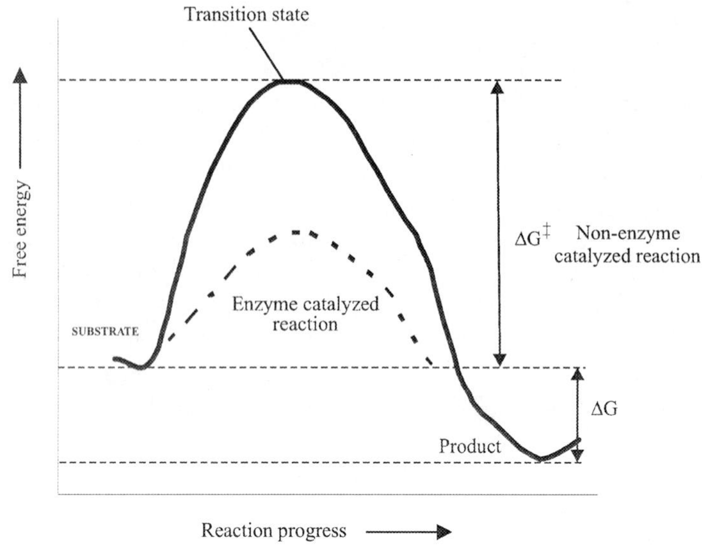

Figure 13.5: Enzyme accelerates the rate of any reaction by lowering down its energy of activation.

equilibrium the concentration of reactants, products and rates of forward and reverse reaction are exactly equal and no further net change occurs. When a reaction is not at equilibrium, the tendency to move towards equilibrium represents the driving force of the reaction, the magnitude of which can be expressed by ΔG.

The free energy change at 298 K, 1 molar concentration of substrate and 1atm pressure is called as standard free energy change, $\Delta G°$.

$$\Delta G° = -RT\ln K'eq$$

where, K'_{eq} is the standard equilibrium constant.

Thus, we see that the standard free energy change of any reaction is just an alternative mathematical expression for equilibrium constant (**Figure 13.5**).

K'_{eq}	$\Delta G°$ (kJ/mol)
0.1	5.7
1.0	00
10	–5.7

Thus, if $K'_{eq} > 1$; **$\Delta G°$ is negative; reaction is spontaneous**

and if $K'_{eq} < 1$; **$\Delta G°$ is positive; reaction is non-spontaneous**

The relationship is exponential and any small change in **$\Delta G°$ brings about a large change in K'_{eq}**. Enzymes do not alter the equilibrium constant but they hasten its attainment. The actual free energy change depends upon concentration of reactants and temperature prevailing at the time of reaction. Thus,

$$\Delta G = \Delta G^\circ + RT \ln [C] [D] / [A] [B]$$

During a reaction moving forward, products increase while reactants decrease, therefore, there is a regular decrease in ΔG which is maximum at the start and reaches zero at equilibrium. Thus, the selective removal of products, if possible can be helpful in maintaining the rate of forward reaction.

Michaelis-Menten Equation of Enzyme Kinetics

It is obvious that addition of enzyme to its substrates under specific conditions initiates conversion of substrate into product (s). The reaction proceeds with time and to ascertain its kinetics it becomes very important to follow it and derive some relationship between rate of reaction and substrate concentration. Michaelis and Menten devised a steady-state model of enzyme kinetics and it is based upon following:

(i) Substrate and enzyme combine reversibly to produce and enzyme-substrate complex

$$E + S \rightarrow ES$$

(ii) Enzyme substrate complex will proceed to form product and enzyme will be released; being reversible, the enzyme substrate complex may disassociate to form E and S

$$E + S \rightarrow ES \rightarrow E + P$$

Rate of any reaction can reveal about the actual catalytic behaviour of the enzyme. Substrate concentration affects the rate of reaction. The effect of substrate concentration on initial velocity is showing in **Figure 13.4**. V_0 increases with the substrate concentration [S] and reaches to a point where it becomes maximum (V_{max}). To understand this phenomenon it is important to understand ES complex.

Enzyme combines reversibly with its substrate:

$$E + S \underset{k_{-1}}{\overset{k_1}{\rightleftharpoons}} ES$$

Then it breaks down into

$$ES \underset{k_{-2}}{\overset{k_2}{\rightleftharpoons}} E + P$$

The breaking down of ES is the slowest or **rate limiting step** in any enzymatic reaction. At any given time of enzymatic reaction, the enzyme is either **free (E) or bound** to the substrate **(ES).** At low concentration of substrate [S], E is free; therefore, reaction rate is directly proportional to [S]. V_{max} is reached when enzyme is saturated with S, so further increase in [S] makes no difference.

Pre-steady state: The concentration of ES complex builds up. The reaction achieves a steady state in which concentration of enzyme substrate complex [ES] remains unaltered or constant.

Steady State Kinetics

The hyperbolic shape of the graph between [S] and V_0 can be expressed by deriving Michaelis Menten equation.

Enzyme combines reversibly with its substrate and forms products as:

$$E + S \underset{k_{-1}}{\overset{k_1}{\rightleftharpoons}} ES \xrightarrow{k_2} E + P$$

k_{-2} is ignored since concentration of product is very little at the start.

Rate of the reaction is determined by (the slowest step, i.e. breakdown of ES into E and P) as:

$$\mathbf{V_0} \ \alpha \ [ES]$$

or

$$\mathbf{V_0} = k_2 [ES] \qquad \textbf{equation (i)}$$

At any given time t, concentration of free enzyme [E],

$$[E] = [E_t] - [ES]$$

where, E_t is total enzyme and ES is the bound enzyme.

Rate of ES formation = k_1 ($[E_t]$ – [ES]) [S]

Rate of ES breakdown = k_{-1} [ES] + k_2 [ES]

At steady state the concentration of ES remains constant and hence

Rate of ES formation = Rate of ES breakdown

$k_1 ([E_t] - [ES]) [S] = k_{-1} [ES] + k_2 [ES]$

Solving this equation for [ES] by multiplying out the left side and simplifying out the right side results in

$k_1 [E_t] [S] - k_1 [ES][S] = (k_{-1} + k_2) [ES]$

Adding the term k_1 [ES][S] both sides gives

$k_1 [E_t] [S] = (k_1[S] + k_{-1} + k_2) [ES]$

Solving this equation for [ES] gives

$[ES] = k_1 [E_t] [S] / (k_1[S] + k_{-1} + k_2)$

$$= \frac{[E_t][S]}{[S] + \frac{(k_{-1} + k_2)}{k_1}}$$

Or

$$[ES] = [E_t] [S] / \{[S] + K_m\} \qquad \textbf{equation (ii)}$$

The ratio $(k_{-1} + k_2) / k_1$ is called K_m or **Michaelis-Menten Constant** and is an important characteristic of the enzyme that denotes the affinity of the enzyme towards the substrate. Michaelis Menten equation describes kinetic behaviour of a number of enzymes that show hyperbolic dependence of initial velocity (V_0) and substrate concentration [S]. V_{max} and K_m are two important and specific characteristics of any enzyme. From equation (i) and (ii), we get

$$\mathbf{V_o = K_2\ [E_t][S] / Km + [S]} \qquad \textbf{equation (iii)}$$

At maximum velocity (V_{max}), the enzyme is fully saturated with the substrate, i.e. [ES] = [E_t]

From equation (i) $\mathbf{V_o}$ = K_2 [ES]

Thus, V_{max} can be defined as K_2[ET] ;

$$V_{max} = K_2[ET]$$

Substituting this value in eq. (iii), we get

$$\mathbf{V_o = V_{max}[S] / Km + [S]} \qquad \textbf{equation (iv)}$$

This is Michaelis Menten equation stating the quantitative relationship between initial velocity V_o, the maximum initial velocity V_{max}, the substrate concentration [S] and the Michaelis Menten constant, K_m.

When V_o reaches half of the maximum velocity (V_{max}):

$$\mathbf{\frac{V_{max}}{2} = V_{max}\ [S] / K_m + [S]} \qquad \textbf{equation (v)}$$

on dividing both sides by V_{max}, we get

$$\mathbf{1/2 = [S] / K_m + [S]} \qquad \textbf{equation (vi)}$$

or $$K_m + [S] = 2\ [S]$$

Thus, $\mathbf{K_m = [S]}$ when V_o reaches half of the V_{max}

The Michaelis Menten equation describes kinetic behaviour of a number of enzymes that show hyperbolic relationship between initial velocity and substrate concentration [S]. V_{max} and K_m are two important and specific characters of any enzyme. K_m shows the affinity of the enzyme towards substrate, larger value of K_m indicates low affinity of the substrate because it signifies that more of substrate is required for reaching half of the maximum initial velocity.

Lineweaver – Burk Plots

As per **Figure 13.6**, V_{max} is achieved as an asymptotic line at inifinite substrate concentration indicating an indeterminate value. However, the Michaelis Menten equation can be transformed by taking reciprocal of both the sides. By measuring V_o at different substrate concentrations and then plotting the double reciprocal plot of 1/ V_o against 1/[S]

$$1/ V_o = \{1/ V_{max} + K_m / V_{max}\}\ 1/[S]$$

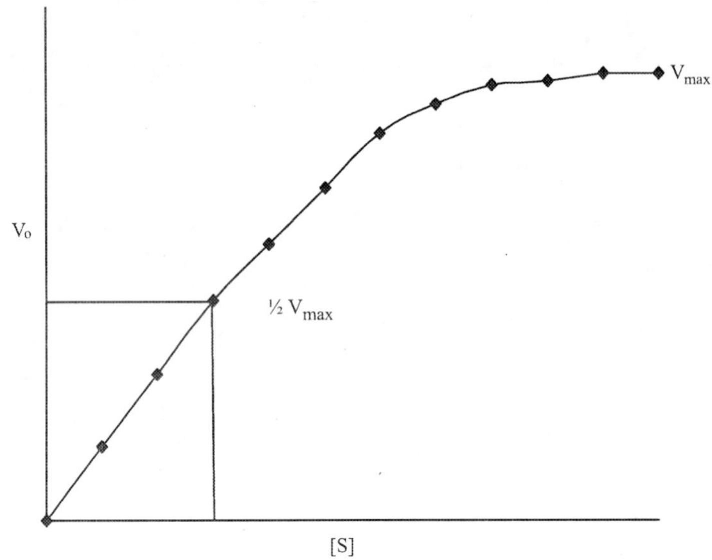

Figure 13.6: The relationship between initial reaction velocity (V_o) and substrate concentration [S].

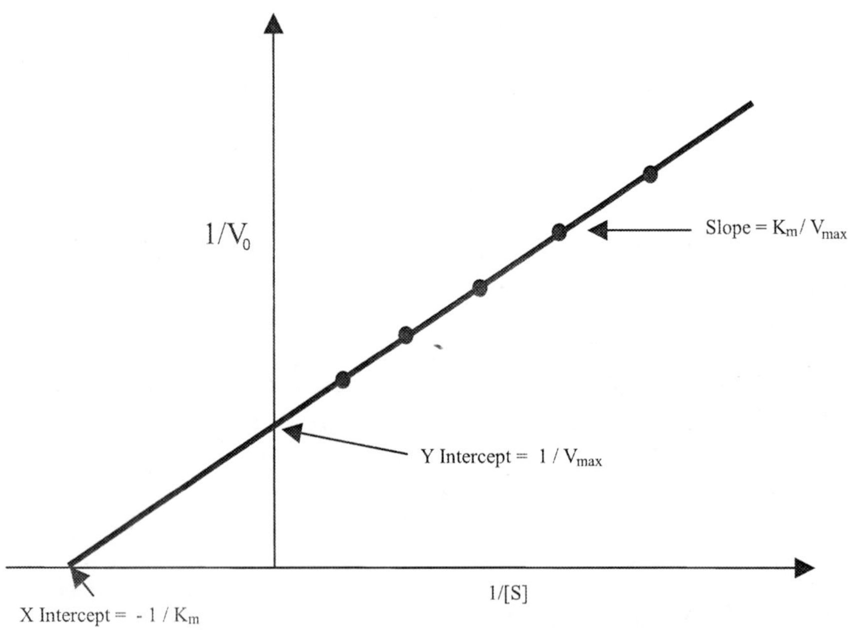

Figure 13.7: The double reciprocal plot or lineweaver – burk plot showing relationship between effect of concentration [S] on initial velocity (V_o) of an enzyme catalyzed reaction.

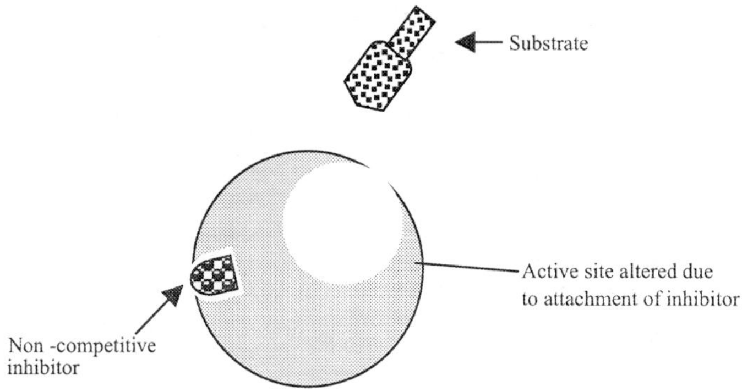

Mechanism of non-competitive Inhibition

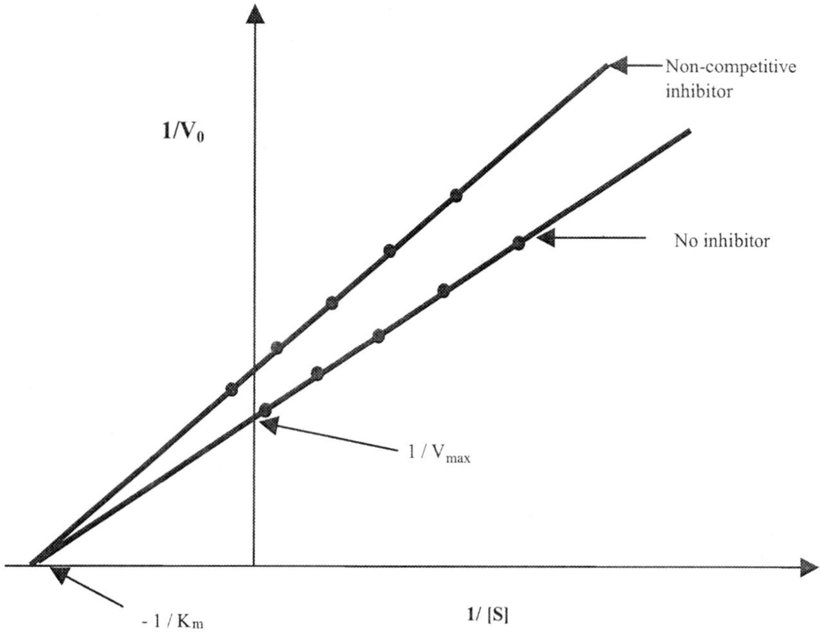

Lineweaver-burk plot showing effect of non-competitive inhibition.

Figure 13.8: Non-competitive inhibition.

$$1/ V_o = \mathbf{K_m / V_{max}} \cdot 1/[S] + \mathbf{1/ V_{max}}$$

The equation can be seen as a equation of straight line, i.e. y = mx + c where m is the slope (K_m/ V_{max}) of the straight line and c is the y-intercept ($1/ V_{max}$). Plot between $1 / V_0$ and 1/[S] yields a straight line with intercept on y-axis at $1/ V_{max}$ and slope is equal to $K_m/V_{max.}$ Using Lineweaver – Burk plot, exact value of V_{max} can be determined (**Figure 13.7**). It is also useful in determining mechanism of enzyme inhibition. Whilst most enzymes follow steady state kinetics, many enzymes, particularly allosteric enzymes do not follow Michaelis–Menten kinetics.

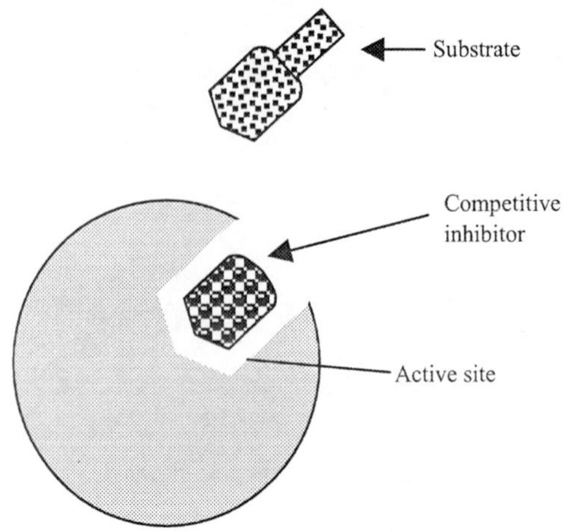

Mechanism of competitive inhibition

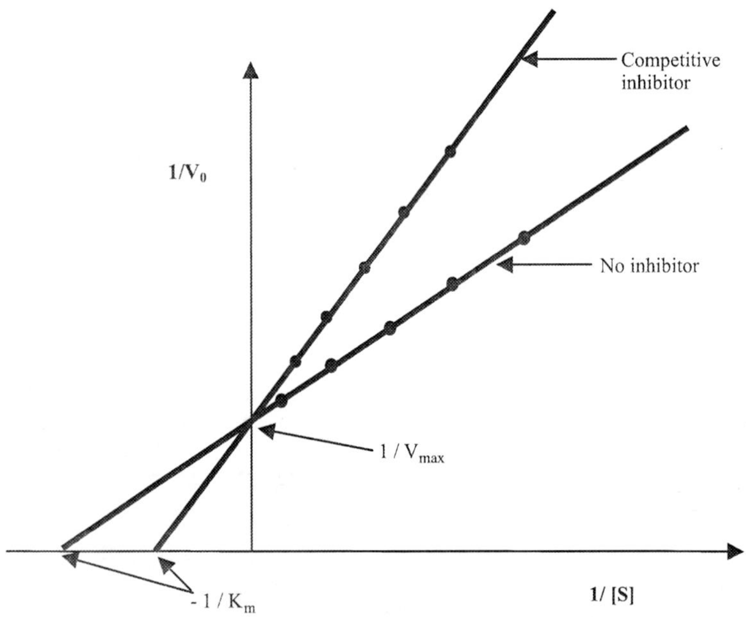

Lineweaver-burk plot showing effect of competitive inhibition.

Figure 13.9: Competitive inhibition.

ENZYME INHIBITION

Enzyme activity may be impeded or interfered by several other molecules present during the reaction. Such molecules that act directly on the enzyme to lower its catalytic efficiency are called as inhibitors. The inhibitor can be any foreign molecule (antibiotic, drug, and toxin) or a part of normal metabolic pathway of the cell. Enzyme inhibition may be of two types: irreversible or reversible. Reversible inhibition can be overcome by removing inhibitor from the enzyme. This is not possible in case of irreversible inhibition, since the inhibitor binds to the enzyme irreversibly (covalently) near the active site thus inactivating the enzyme permanently. Amino acid residues such as serine and cysteine have free reactive groups, –OH and –SH, respectively and compounds such as di-isopropylphosphofluoride (DIPF) react with serine residues in the active site of the enzyme, e.g. acetylcholine-esterase and inactivate the enzyme irreversibly thus preventing the transmission of nerve impulse. Penicillin, a well-known antibiotic, inhibits glycopeptide transpeptidase which is responsible for cross-linking the cell wall during cell wall formation in bacteria. Penicillin binds irreversibly to the serine residues of the active site of the enzyme. Iodoacetamide attaches irreversibly with cysteine residues. The reversible inhibition can again be of two types: competitive or non-competitive. **Figures 13.8 and 13.9** show mechanism and lineweaver burk plots for competitive and non-competitive inhibition. In case of non-competitive inhibition, K_m value remains unchanged while the value of V_{max} is increased.

Competitive Inhibition (Reversible)

A competitive inhibitor is analogous to the usual substrate of the enzyme and competes with the substrate molecules to bind at the active site. Since the binding site is common, either substrate or its competitor (analogue) can bind at the site. The competitive inhibitor binds reversibly to the active site. However, if the substrate concentration is very high the action of competitive inhibitor is overcome by the high number of substrate molecules competing for the active site. During reversible competitive inhibition, the V_{max} remains unaltered but the K_m is increased as the apparent affinity of the enzyme for its substrate decreases in the presence of inhibitor molecule (**Figure 13.9**).

SUGGESTED READING

Atkinson D.E. 1966. *Regulation of Enzyme Action.* Ann Rev. Biochem. 35: 85–124.

Fersht A. 1985. *Enzyme Structure and Mechanism.* 2nd ed. W.H. Freeman and Co., New York.

Gutfruend H. 1972. *Enzyme: Physical Principles.* Wiley-Intescience, New York.

Price N.C. and Stevens L. 1982. *Fundamentals of Enzymology.* Oxford University Press, Oxford.

Sumner J.B. and Somers G.F. 1953. *Chemistry and Methods of Enzymes.* Academic Press, Inc., New York

Bailey, J.E. and Ollis, D. F. 1977. *Biochemical Engineering Fundamentals*, McGraw-Hill Book Company, London.

REVIEW QUESTIONS

1. Define enzymes? Explain how reactions are carried out under mild conditions in presence of enzymes.
2. Explain the mechanism of enzyme action in detail.
3. Write a note on nomenclature and classification of enzymes giving suitable examples.
4. What do you mean by kinetics of an enzyme catalyzed reaction? What is Michaelis-Menten constant?
5. Explain mechanism of competitive and non-competitive enzyme inhibition.

UNIT XIV

Bioenergetics, ATP Generation and Transport of Biomolecules

Metabolism in living organisms is a highly coordinated and concerted activity in which following major functions are accomplished: a. Fixation of solar energy (autotrophy) in chemical form or release of energy rich nutrients from surroundings by oxidation (heterotrophy), b. Conversion of nutrients into characteristic cells, c. Biosynthesis of essential macromolecules of the cell, e.g. Nucleic acid, proteins, polysaccharide, d. Removal of specialized molecules by degradation.

Autotrophs can use atmospheric carbon dioxide as their sole source of C to construct all carbon containing biomolecules. Heterotrophic organisms obtain their carbon from environment in the form of reduced carbohydrates made by autotrophs. There exists a vast interdependent network between autotrophs and heterotrophs where autotrophs fix atmospheric carbon and heterotrophs return it to atmosphere. While the solar energy traverses in a flow, it drives the great cycles of carbon, oxygen, nitrogen and H_2O.

Nitrogen is an essential component of building blocks of organisms. Plants and some bacteria can take up or fix atmospheric nitrogen from surroundings but most animals need it in some organic form (e.g. amino acid). Nitrifying bacteria and denitrifying bacteria also play significant role in the global cycle of nitrogen. The matter is recycled while the energy, while it traverses through different trophic levels, gets dissipated as heat and entropy. That is why, the elements are said to be in cycle while the energy is said to be in flow.

Metabolism is the sum total of all chemical transformations and energy exchanges that occur in a cell. Each small and specific chemical change governed by enzymes is interlinked and leads to formation of a pathway or cycle. These pathway and cycles involve several compounds or metabolic intermediates, called as metabolites. Catabolism denotes the degradative part of metabolism in which organic compounds are degraded into smaller and simpler end products and simultaneously energy entrapped in their bonds is released as free energy. This energy drives the formation of readily available energy sources of the cell, ATP and reduced electron carriers (NADH and NADPH).

Thermodynamically, the feasibility of any reaction is determined by the equation:

$$\Delta G = \Delta H - T\Delta S$$

where,

ΔG is change in the free energy of reactants and the products

ΔH is change in enthalpy or heat content of reactants and the products

T is the standard temperature

ΔS is change in entropy of the system

An overall decrease in the free energy (–ΔG) indicates that the reaction is energetically favourable and the products are at a lower energy level than reactants substrates. It must be noted that negative free energy change (–ΔG) indicates that the reaction is energetically favourable and will occur spontaneously while if the free energy is positive (+ΔG) then it is non-spontaneous and needs an extra input of energy. In biochemical reactions this input of energy is obtained through coupling of energetically favourable reaction. Biological reactions are catalyzed by enzymes. A reaction that takes more than one hour to reach equilibrium, in presence of enzyme, may attain equilibrium in one second.

ATP GENERATION BY ETS

In prokaryotes, the components of electron transport and oxidative phosphorylation are located in the plasma membrane, but in eukaryotes, electron transport and oxidative phosphorylation takes place in the inner membrane of mitochondria. These processes re-oxidize the NADH and $FADH_2$ that occur from the citric acid cycle, glycolysis and fatty acid oxidation and trap the energy released as ATP. Oxidative phosphorylation is by far the major source of ATP in the cell.

Redox Potential

The oxidation of a molecule involves the loss of electrons. The reduction of a molecule involves the gain of electrons. Since electrons are not created or destroyed in the chemical reaction, if one molecule is oxidized, simultaneously another molecule must be reduced; this process is called as oxidation-reduction reaction. Thus, by definition oxidation-reduction reactions involve the transfer of electrons. In the oxidation –reduction reaction:

$$NADH + H^+ + 1/2\ O_2 \rightleftharpoons NAD^+ + H_2O$$

When NADH is oxidized to NAD^+ it loses electrons, while a molecule oxygen is reduced to water, and serves as electron acceptor.

The **oxidation reduction potential**, E' (redox potential) indicates the affinity of a substance for electrons and it is measured relative to that of hydrogen. In biological system, the **standard redox potential** for a substance (E'_0) is measured under standard conditions, at pH 7, and is expressed in **volts**. The standard free energy change of a reaction, ΔGº', can be readily calculated from the change in redox potential E'_0 of the substance and products:

$$\Delta G^{0\prime} = - nF\, E'_0$$

Where *n* is the number of electrons transferred, E'_0 is redox potential in volts (V), $\Delta G^{o\prime}$ is in kilocalories per mole (kcal mol^{-1}) and F is a constant called as constant the Faraday (23.06 kcal $V^{-1}mol^{-1}$). It is obvious from the above formula that reaction with a positive E'_0 will have a negative free energy change or $-\Delta G^{o\prime}$ i.e. the reaction will be exergonic and spontaneous.

Conversely, if a reaction has negative E'_0, the free energy change will be positive and the reaction will be endergonic and non-spontaneous.

In biological systems, the process of electron transport is completed in three steps:

(1) Electron transport from NADH
(2) Formation of an H^+ gradient
(3) Electron transport from $FADH_2$

(1) Electron transport from NADH

Comparing the energetics of the oxidation of NADH:

$$NADH + H^+ + 1/2\, O_2 \rightarrow NAD^+ + H_2O \qquad \Delta G = 52.6 \text{ kcal mol}^{-1}$$

and the synthesis of ATP:

$$iP + ADP + H^+ \rightarrow ATP + H_2O \qquad \Delta G^{01} = + 7.3 \text{ kcal mol}^{-1}$$

It is clear that the oxidation of NADH releases sufficient energy to drive the synthesis of several molecules of ATP. However, NADH oxidation and ATP synthesis do not occur in a single step. The electrons are transferred from NADH to oxygen through a chain of electron carriers collectively called as **electron transport chain** or **respiratory chain (Figure 14.1)**.

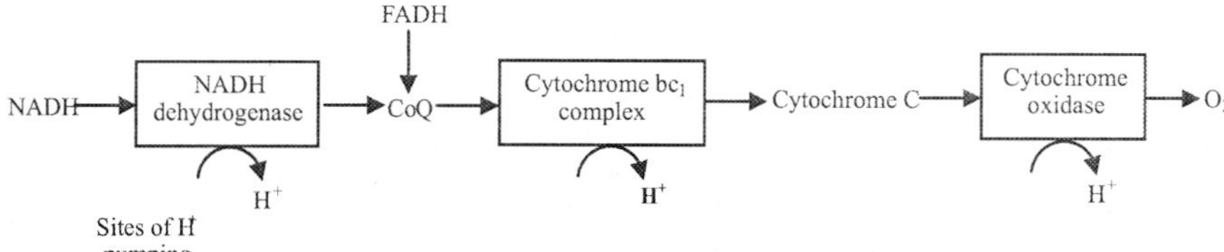

Figure 14.1: Overview of the electron transport chain.

The main part of electron transport chain consists of **three** large proteins complexes placed in the inner membrane of mitochondria, called **NADH dehydrogenase**, the **cytochrome bc$_1$ complex** and **cytochrome oxidase**. Electrons flow from NADH to oxygen through these three complexes. Each complex contains several electron carriers that work one after another to carry electrons down the chain. Two small electron carriers are also needed to link these large complexes; ubiquinone, which is called **coenzyme (CoQ)**, and **cytochrome c**. Electron transport, is completed in five steps:

Step I: NADH to NADH dehydrogenase

NADH dehydrogenase (also called **complex I**) binds the NADH and re-oxidizes it to NAD^+, passing the two electrons from NADH to a prosthetic group called FMN to produce $FMNH_2$. Each electron is accepted together with a hydrogen ion, H^+, such that two electrons and two H^+ are accepted in total. The electrons are then transferred, within the NADH dehygrogenase complex, to **iron-sulphur clusters** (FeS) in iron-sulphur proteins. Within an FeS cluster, an electron is carried by the iron atom which, on accepting the electron, changes from the Fe^{3+}(ferric) state to Fe^{2+} (ferrous) state (**Figure 14.2**). As the electron is passed to another electron carrier, the iron atom of the FeS cluster changes back again to the Fe^{3+} state.

Step II: NADH dehydrogenase to ubiquinone (CoQ)

Electrons from the FeS clusters of NADH dehydrogenase are passed on to ubiquinone, a small lipid-soluble molecule in the inner membrane of mitochondria. This molecule can act as an electron carrier by accepting up to two electrons and two H^+ ions. This leads to conversion of ubiquinone (CoQ) is to ubiquinol ($CoQH_2$).

Step III: Ubiquinol ($CoQH_2$) to cytochrome bc_1 complex

Ubiquinol transfers its two electrons to the next carrier in the chain, the **cytochrome bc_1** complex (also called **complex III**). Cytochrome is a protein which is bound with the haem group that contains an iron atom. As the electron is accepted, the iron atom of haem group changes from ferric state to ferrous state (**Figure 14.2**). The electron from here passes out to the next electron carrier, cytochrome c.

Step IV: Cytochrome bc_1 complex to cytochrome c to cytochrome oxidase

Cytochrome c binds to the cytochrome bc_1 complex and accepts an electron via Fe^{3+} to Fe_2^+. Then it binds to cytochrome oxidase and donates the electron, with the iron atom of the haem of cytochrome c then reverting to the Fe^{3+} state (**Figure 14.2**).

Step V: Cytochrome oxidase to oxygen

Cytochrome oxidase (also called **complex IV**) contains two cytochromes (chytochrome *a* and a_3). Both cytochromes are paired through copper atom. The cytochrome oxidase reaction is a complex reaction; it transfers four electrons from four cytochrome *c* molecules and four H^+ ions to molecular oxygen to form two molecules of water (**Figure 14.2**).

(2) Formation of an H^+ gradient

All of the electron carriers in the electron transport chain cooperate according to their redox potentials. The accepting carrier molecules have a higher affinity for electrons from NADH (more negative redox potential, least affinity of electrons) to oxygen (most positive redox potential, higher affinity for electron). This ensures a unidirectional flow of electrons in E.T.S.

The change in redox potential in the chain is a measure of the energy change occurring. The potential falls during the chain but mainly in three large steps that correspond to the three main protein complexes; complex NADH dehydrogenase, ***cytochrome bc$_1$*** and the cytochrome oxidase complex. The large free energy change at each of these three steps is sufficient enough to pump H^+ ions from the mitochondrial matrix across the inner mitochondrial membrane and into the intermembrane space. Thus, to understand it in simple terms, each of these three complexes is an **H^+ pump** driven by electron transport (**Figures 14.1** and **14.2**). The movement of electron transport along the chain from NADH releases energy that is used to create an H^+ gradient.

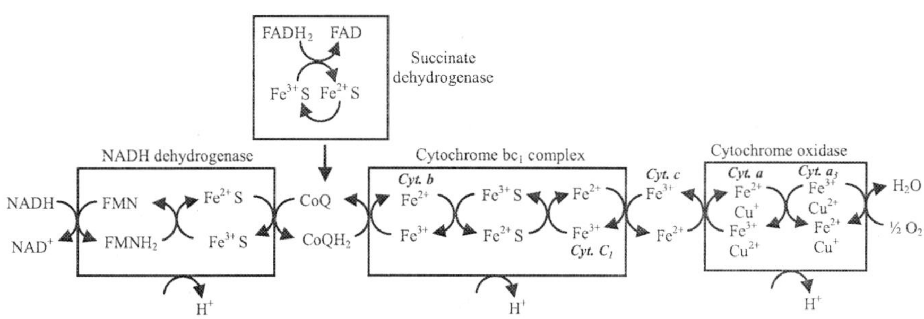

Figure 14.2: Details of electron transport.

(3) Electron transport from $FADH_2$

Succinate dehydrogenase catalyzes the oxidation of succinate to fumarate in the citric acid cycle. The succinate dehydrogenase contains FAD (via **succinate-coenzyme reductase,** also called as **complex II**), an integral protein of the inner membrane of mitochondria. Succinate dehydrogenase is part of this complex but it also contains FeS clusters. During re-oxidation of $FADH_2$, the two electrons pass from the $FADH_2$ to the FeS clusters and are then passed on to ubiquinone (CoQ) (**Figure 14.2**). These electrons then enter the main electron transport chain and cause H^+ ions to be pumped out of the mitochondrion as for the oxidation of NADH, which pumps out three protons.

OXIDATIVE PHOSPHORYLATION

Oxidative phosphorylation is the name given to the synthesis of ATP (phosphorylation) that occurs when NADH and $FADH_2$ are oxidized by electron transport through respiratory chain. The mechanism of phosphorylation was proposed by Peter Mitchell in 1961 and is referred to as **chemiosmostic hypothesis**. This proposes that energy liberated by electron transport is used to create a proton gradient across the mitochondrial inner membrane and that it is this gradient that is used to drive ATP synthesis. Thus, the proton gradient couples electron transport and ATP synthesis, rather than a chemical intermediate. The actual synthesis of ATP is carried out by an enzyme called as **ATP synthase** located in the inner membrane of mitochondria (**Figure 14.3**).

The process of oxidative phosphorylation can be described as follows. Electron transport down the respiratory chain from NADH oxidation cause H^+ ions to be pumped out of the mitochondrial matrix across the inner mitochondrial membrane into the intermembrane space by the three H^+ pumps; NADH dehydrogenase, cytochrome bc_1 complex and cytochrome oxidase. The free energy change in transporting an electrically charged ion across a membrane is related both to its electrical charge and the concentration of the species. The pumping out of the H^+ ions generates a higher concentration of H^+ ions in the intermembrane space and an electrical potential, with the side of the inner mitochondrial membrane facing the intermembrane space being positive (**Figure 14.3**). Overall, an **electrochemical proton gradient** is formed. The protons flow back into the mitochondrial matrix through the ATP synthesis. The ATP synthesis is driven by **proton-motive force**, which is sum of the pH gradient (that is chemical gradient of H^+ ions) and the membrane potential. Three ATP molecules are synthesized per molecule of NADH + H^+ oxidized whereas two ATP molecules are produced per $FADH_2$ oxidized.

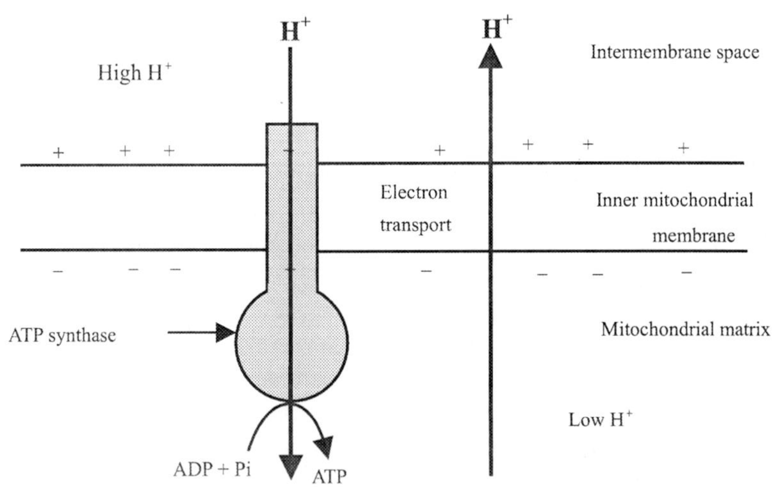

Figure 14.3: The mechanism of oxidative phosphorylation.

Coupling and Respiratory Control

Electron transport is usually coupled to ATP synthesis. This means that electrons flow during the electron transport chain to oxygen and ADP is simultaneously phosphorylated to ATP. It also follows that ATP is not synthesized unless electron transport is occurring to provide the proton gradient. Therefore, oxidative phosphorylation needs NADH or $NADH_2$, oxygen, ADP and inorganic phosphate. The definite rate of oxidative phosphrylation is set by the availability of ADP. If ADP is added to mitochondria, the rate of oxygen utilization rises as electrons flow down the chain and then the rate of oxygen utilization falls when all the ADP has been phosphorylated to ATP. This is referred to as **respiratory control**.

Uncouplers

Some chemicals, such as 2, 4-dinitrophenol (DNP), act as uncoupling agents. DNP and other uncoupling agents are lipid-soluble small molecules that can bind H^+ ions and prevent the formation of a proton gradient. When added to cells, they stop ATP synthesis but electron transport still continues and so oxygen is still consumed.

TRANSPORT OF BIOMOLECULES

A membrane which allows the passage of a solvent but not of a solute is called as *semipermeable membrane*. A membrane which permits one substance to pass through more easily than other is called selectively permeable. A selectively permeable membrane allows the passage of both solvent and solute. Biological cell membranes are selectively permeable rather than semipermeable.

Only small, uncharged molecules such as gases, water, ammonia, glycerol, urea, etc. are able to pass through biological membranes by *free diffusion* but weighty or heavier molecules are not capable to pass through biological membranes. For example, glucose and other sugars cannot traverse through biological membranes. The **polarity** of a molecule also plays a significant role with transport. Biological membranes are impermeable to highly polar compounds, mainly those that are electrically charged, but these are permeable for apolar substances, such as benzene, ethanol, diethyl ether, and many narcotic agents. To be able to take up or release molecules of this type, cells have particular *channels* and *transporters* in their membranes (**Figure 14.4**).

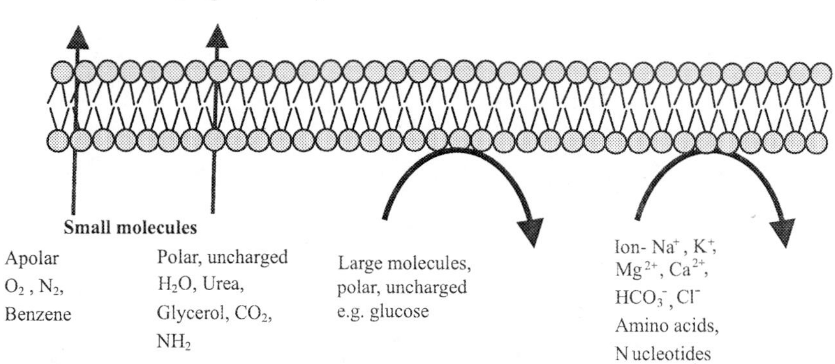

Figure 14.4: Permeability of biological membrane.

Molecules diffuse from an area of their own greater concentration to an area of lesser concentration. The difference among the level of the two concentrations is referred as the *concentration gradient.* Osmosis is the diffusion of water across a semi-permeable membrane in response to the concentration gradient. Biological transport in cells is one of the most important phenomenons.

Transport of molecules across the membrane may be active or passive. Some molecules cross the membrane easily than others while many times it is mediated by carriers. Carrier molecules may also be involved in both active and passive systems.

PASSIVE TRANSPORT OF BIOMOLECULES

Passive transport involves spontaneous transport of molecules through a semi-permeable membrane from their own high concentration to the area of their lower concentration. Transfer of molecule follows rules of simple diffusion and takes place along the concentration gradient, thus there is no expenditure of energy. Passive transport can also be categorized into following two types:

(i) Simple diffusion
(ii) Facilitated diffusion

Simple Diffusion

This type of passive spontaneous transport involves simple diffusion along with the concentration gradient and occurs without any mediation by carrier molecule. Movement of solute takes place from its high concentration to a low concentration area till the concentration gradient disappears. It does not involve any stereo-specificity and is a spontaneous and slow process (**Figure 14.5**). The passage of

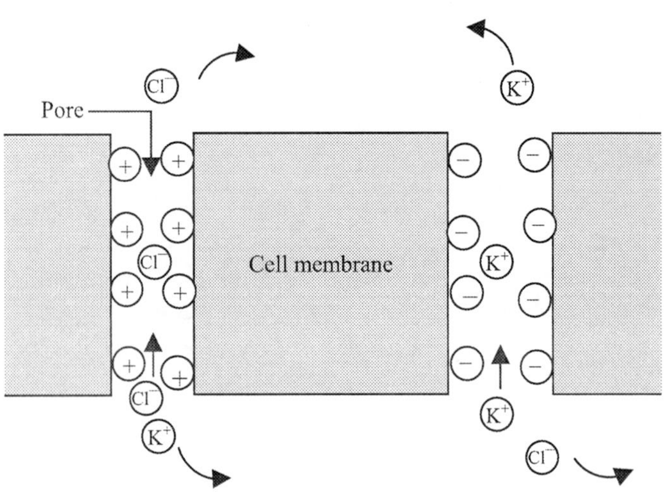

Figure 14.5: Simple diffusion of solutes through pores in a membrane. See that ions are transported based on charge.

the substances through membrane takes place through the porosity formed as a result of transitory movement of proteins in the lipid bi-layer.

Facilitated Diffusion

Very similar to simple diffusion, the facilitated diffusion does not require any energy expenditure and follows the concentration gradient (**Figure 14.6**). However, it differs in being *stereospecific*, meaning that among the two stereo isomers, L and D, only one type is transported selectively. Facilitated diffusion is mediated by special protein molecules or carrier proteins. The process also shows *saturation kinetics*. Increase in the concentration of substance to be transported results in an increase in the rate of transfer up to an asymptotic value.

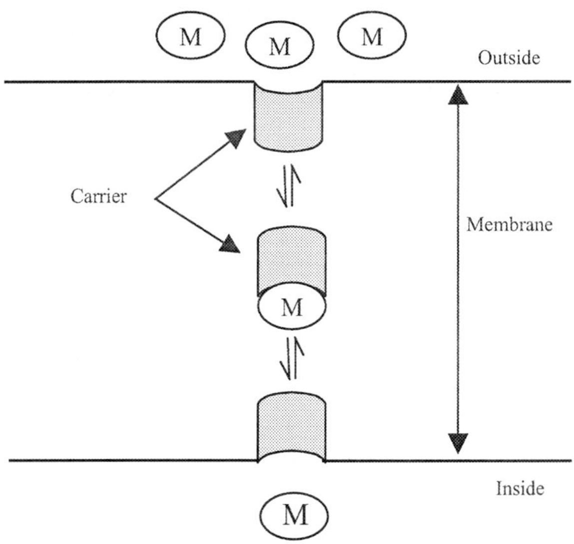

Figure 14.6: Facilitated diffusion of a metabolite (M) by formation of a protein-metabolite complex.

Such carrier proteins that transport specific sugars and amino acids, phosphate, Ca^{2+}, Na^{+} and K^{+} have been isolated. Carrier protein molecules move across the biological cell membranes by thermal diffusion. Carrier proteins of the lipid bilayer bind metabolites at the outer surface of the membrane resulting in a carrier-metabolite complex. The concentration of such complex is high towards the outer side of the membrane. The protein-metabolite complex moves across the membrane towards inside and releases the metabolite inside the cell.

ACTIVE TRANSPORT

In active transport, molecules move from an area of low concentration to an area of high concentration that is against the concentration gradient. Active transport requires energy and is dependent upon ATP supply. If the ATP synthesis stops, transport of Na^+ ions across the membrane is affected. Active transport can be classified into two types. (A) *simple active transport* and (B) *group translocation.*

(A) Simple Active Transport

Simple active transport can again be classified in two types, (i) *Primary active transport* and (ii) *Secondary active transport.*

(i) Primary active transport: It is directly related with chemical energy (ATP) or electrical energy. Examples of primary active transport are Na^+, K^+ transportation in mammals and proton transport by bacterial *ATPase*. Na^+ is pumped out of the cell by sodium pump and K^+ is pumped simultaneously into the cell by a coupled pump. The two pumps work as simultaneously, and in case of failure of any one movement of both Na^+ and K^+ stops. ATPase is an integral plasma protein and works as a monovalent ion pump. The enzyme consists of two polypeptide chains, a large subunit which penetrates the membrane and a small subunit. Na^+, K^+ - ATPase spans entire width of the membrane and bears functional sites exposed on the surface of the membrane.

This ion transport process involves cycle of phosphorylation-dephosphorylation of ATP. Na^+ and K^+ bind to specific site of the ATPase complex (**Figure 14.7**). Bound Na^+ is required for phosphorylation of the larger subunit of ATPase. Phosphorylation of ATPase leads to a conformational change in proteins resulting in a reorientation of the enzyme within the membrane, allowing Na^+ to pass through the membrane. Dephosphorylation of this protein is brought about by bound K^+, resulting in reversion to original conformation protein and K^+.

According to another view such large conformation changes in ATPase may not be taking place. Ion translocation is probably carried out by minor conformational change in ATPase resulting its phosphorylation and dephosphorylation. It has been advised that ATPase has a channel through its centre through which ion transport takes place.

(ii) Secondary active transport: This process can be defined as the movement of molecules accross cell membrane at the expense of primary established gradient of another molecular species. The lactose permease system in *E. coli* and glucose transport of the intestine epithelium of mammals are examples of secondary active transport (**Figure 14.8**). Primary active transport results in the Na^+ being pumped out of the cell and K^+ being pumped into the cell. The electrochemical sodium ion gradient can be then employed for secondary active transport of glucose into the cell against the concentration gradient. Thus, there is glucose-Na^+ co-transport catalyzed by glucose carrier. Such sodium dependent transport has been observed for various amino acids and sugars in different vertebrates and for amino acids in bacteria.

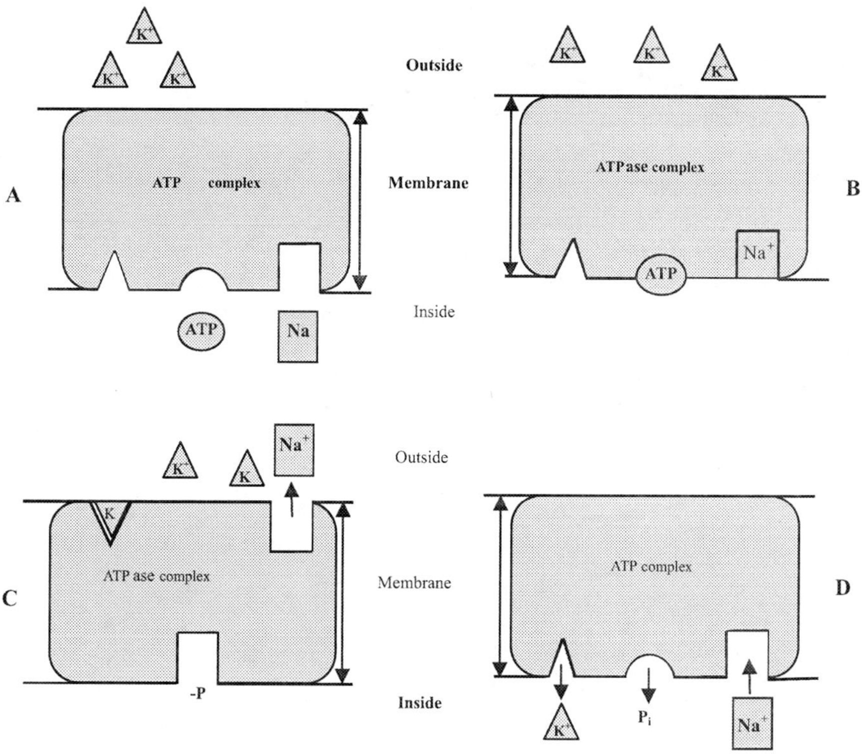

Figure 14.7: Diagram of the Na^+, K^+ ATPase pump. **A.** Na^+ and ATP approach the inner surface of the ATPase complex. **B.** Na^+ and the ATP bind to specific sites on ATPase. **C.** Phosphorylation of ATPase results in conformational change in ATPase and in the binding sites for Na^+, K^+ **D.** Dephosphorylation of ATPase results in reversal of conformation.

The sodium pump maintains a higher concentration of Na^+ outside the cell than on the inner side making the membrane polarized. This results in a tendency for Na^+ to enter the cell. This it does in the form of a carrier-sugar or carrier-amino acid complex.

A 'revolving door' model has been proposed to explain the passage of lactose through the cell membrane for secondary active transport. The carrier protein, acts as a revolving door in the cell membrane (**Figure 14.9**). The carrier protein has a slot, which normally faces outside, into which the substance to be transported get affixed. The carrier protein undergoes a conformational change as soon as the solute enters the site and then it rotates so that the slot comes on the inside. This movement is driven by chemiosmotic potential of the cell. The substance is released into the cell, and the protein now returns to its original immobile form. The secondary active transport is mediated by carrier protein molecule under the influence of chemiosmotic pressure and does not involve direct participation of ATP (**Figure 14.10**).

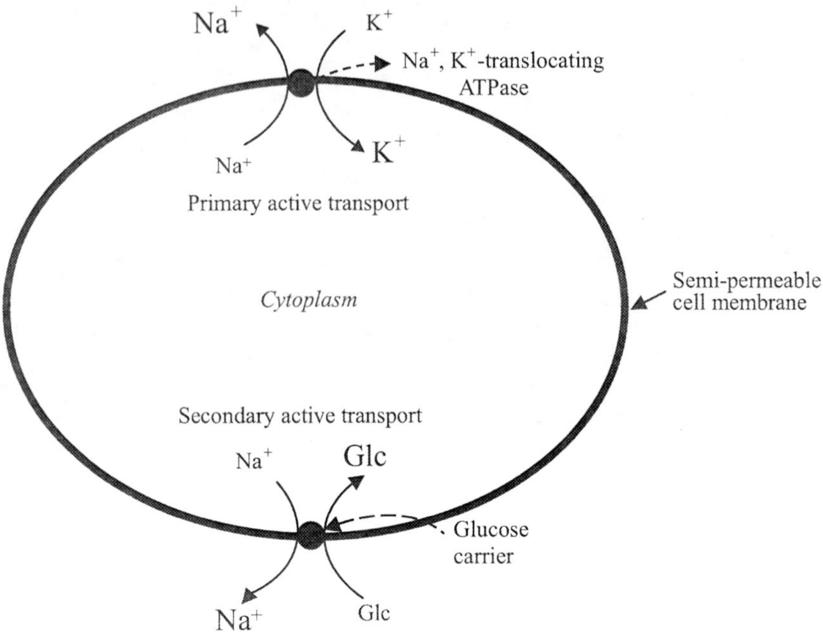

Figure 14.8: Schematic showing active transport in a cell.
Na^+, K^+ active transport at the top, and glucose-Na^+ co-transport in the bottom.

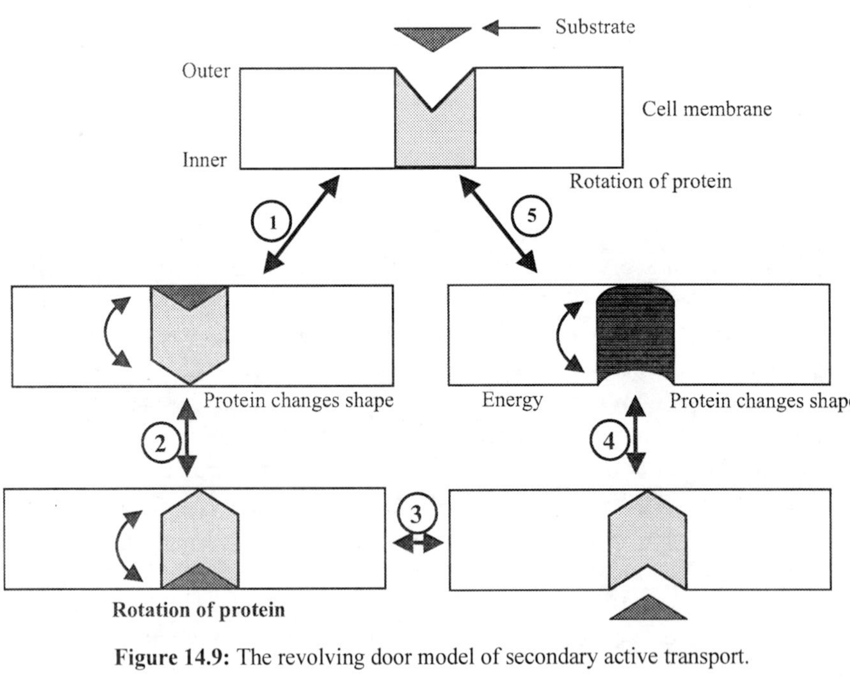

Figure 14.9: The revolving door model of secondary active transport.

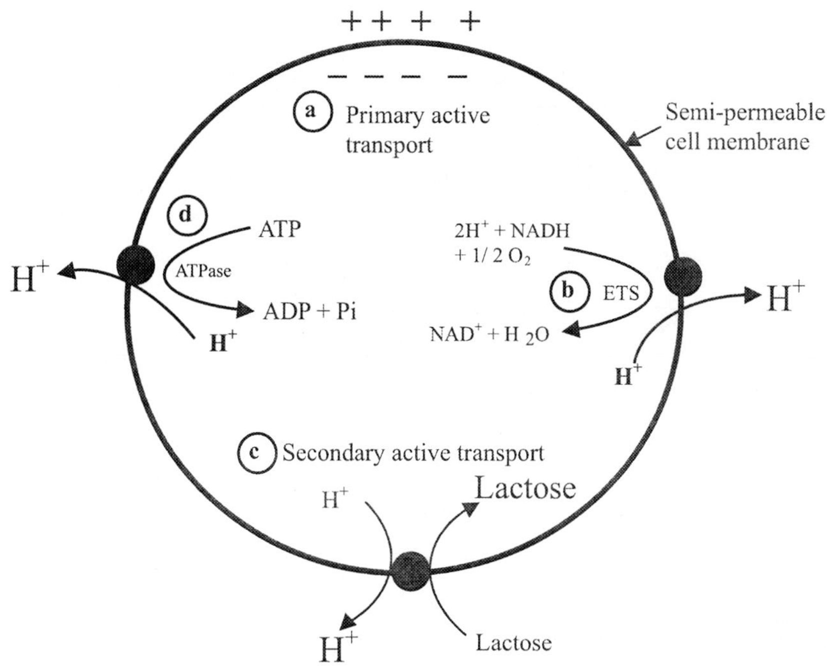

Figure 14.10: Schematic representation of active transport in *E. coli.* **a.** Primary active transport, **b**. Electron transport system **c**. Secondary active transport **d**. ATP primary active transport.

(B) Group Translocation

In this process, the substance to be transported is altered by the enzyme that catalyzes its transport. Transport of a variety of sugars like glucose, fructose and mannitol occurs across cell membranes through the *phosphotransferase system* (PTS). The sugar is released inside the membrane as a phosphorylated derivate which cannot go back through the membrane. The phosphorylated glucose then can directly participate in the metabolic activity, e.g. glycolysis.

The PTS utilizes *phosphoenol pyruvate* (PEP) for energy generation (**Figure 14.11**). PTS comprises two enzymes (I and II) and a heat stable protein (HPr), a small histidine-containing protein also found in the cytosol.

PEP is first dephosphorylated to pyruvate and the phosphate group released is taken up by **Enzyme I,** in the cytosol.

$$\text{PEP + Enzyme I} \xrightarrow{Mg^{2+}} \text{Enzyme I} \sim \text{P + Pyruvate}$$

Enzyme I then transfers the phosphate group to the heat stable protein (HPr). Enzyme II is a complex that contains two enzymes namely, IIa and IIb, Mg^{2+} and phosphatidyl glycerol. Among these, enzyme IIa is sugar-specific and it carries

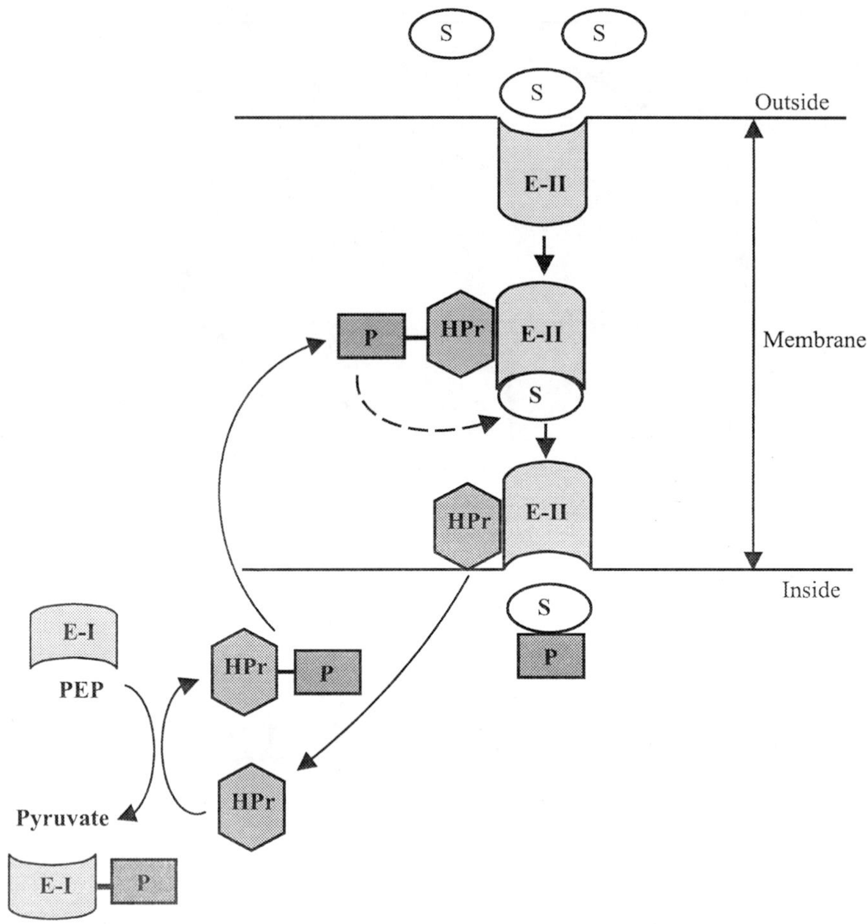

Figure 14.11: The phosphotransferase system of group translocation of sugars.
Abbreviations: PTP- Phosphotransferase system, PEP- Phosphoenol pyruvate
HPr- Heat stable protein, S- Sugar, P-phosphate, E-I- Enzyme I and E-II- Enzyme II

only a specific type of sugar. The cell synthesizes specific enzyme for the type of sugar in which it is growing.

$$\text{Enzyme I} \sim \text{P} + \text{HPr} \longrightarrow \text{HPr} \sim \text{P} + \text{Enzyme I}$$

Enzyme II is located in the outer side of the cell membrane. Enzyme II undergoes a conformational change. The HPr – P complex then joins E-II-substrate complex. The phosphate group is transferred on the sugar molecule and finally sugar is released inside the cell as phosphorylated derivative. These events are repeated for transfer of another incoming sugar molecule.

SUGGESTED READING

Baltscheffsky H., and Baltscheffsky M. 1974. *Electron Transport Phosphorylation.* Ann. Rev. Biochem. 43:871.

Ernster L. 1992. *Molecular Mechanisms in Bioenergetics.* Elsevier.

Kalckar H.M. 1969. *Biological Phosphorylations: Development of Concepts.* Prentice-Hall.

Krebs H.A. and Kornberg H.L.1957. *Energy Transformation in Living Matter*, A Survey. Springer, Berlin.

Bailey, J. E. and Ollis, D. F. 1977. *Biochemical Engineering Fundamentals*, McGraw-Hill Book Company, London.

REVIEW QUESTIONS

1. Explain how ATPs are generated in electron transport system.
2. What is oxidative phosphorylation?
3. Write a note on various types of biological transport mechanisms.
4. What is active transport? Discuss various mechanisms of active transport.
5. What is group translocation? Explain giving suitable example.

UNIT XV

Structure and Replication of DNA

Living organisms are complex systems and need to perform work to sustain. Hundreds of thousands of structural and functional proteins exist inside each cell to perform routine metabolic functions. Proteins also contribute significantly to the structure of the cell (cell membrane, organelles and keratin). All these proteins are produced in the cytoplasm of the cell and then get distributed to perform their functions. An enormous amount of information is required to manage this complex system in a foolproof manner. This information, featuring the detailed and specific structure of the proteins inside cells, is stored in molecules called nucleic acids. Reproduction as well as growth is an essential character of any living organism. During reproduction or cell division the entire information has to be passed on to progeny or daughter cells via a hereditary material.

DNA is present in the cells of all plants, animals, prokaryotes and many a viruses. The chromosomal DNA is also referred as genome signifying that it is hereditary or genetic material. DNA is also present in cytoplasmic organelles such as mitochondria and chloroplasts. In bacteria, often some small, extra chromosomal, self-replicating, circular DNA molecules carrying special genes for antibiotic resistance or conjugation are present. These are called as plasmids. The DNA of all living beings including plants, animals and viruses (smallpox virus, T-even bacteriophages, and polyoma virus) is double stranded. Rare exceptions of single stranded DNA as the genetic material such as Ø × 174 also exist. Most RNA viruses have single stranded RNA molecule as their genetic material (TMV, Influenza virus, polio virus, Bacteriophages F2 and R17) while a few example of double stranded RNA viruses, e.g. Reovirus, wound tumour virus also exist.

The nucleic acids are very large polymeric molecules made up of nucleotides. The backbone of a nucleic acid is made of alternating sugar and phosphate molecules linked together. Each of the sugar groups in the backbone is attached to a heterocyclic nitrogenous base. Together these three components are said to make a nucleotide. Four different types of nitrogenous bases occur in a nucleic acid (adenine, guanine, cytosine, thiamine in DNA; *uracil replaces thiamine in RNA*) and each polynucleotide contains thousands of such bases. The most important and distinguishing feature of a polynucleotide is the order or sequence of nitrogenous bases which is the coded information. This coded information in DNA is transcribed (RNA synthesis) and translated (protein synthesis) for producing structural and functional proteins, while it is passed on to progeny or daughter cells via replication (DNA synthesis). Therefore, nucleic acids serve as a template which determines structure of each protein and acts as physical basis of heredity.

STRUCTURE OF NUCLEIC ACIDS

Sugar Molecule

Deoxyribose is a pentose sugar forming a five membered ring and differs from ribose sugar in having H atom in place of OH at 2′ Carbon (**Figure 15.1**). The "deoxy" prefix means that the ribose molecule has lost oxygen. The missing oxygen happens to be from the second carbon, so the more correct name for deoxyribose is **2-deoxyribose**. Because the last asymmetric carbon atom has a hydroxyl group to the right, the more appropriate name will be D-2-deoxyribose. In RNA backbone D-ribose is present. The sugar molecule is linked to phosphate (at 5′Cor 3′C) and nitrogenous base (at 1′C).

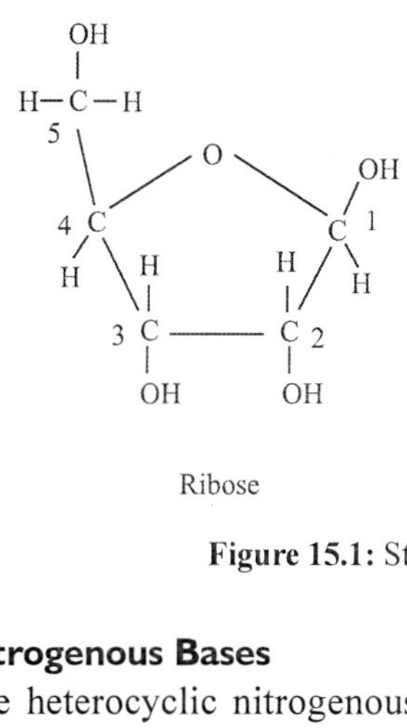

Ribose

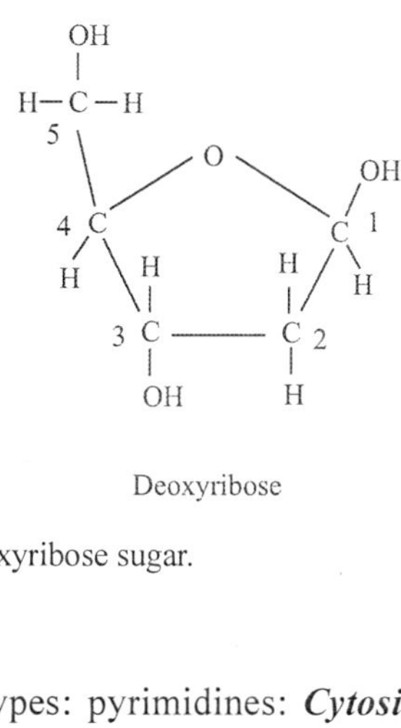

Deoxyribose

Figure 15.1: Structure of tibose and deoxyribose sugar.

Nitrogenous Bases

The heterocyclic nitrogenous bases are of two types: pyrimidines: ***Cytosine, Thymine, Uracil*** (single ringed with nitrogen at 1′ and 3′) and purines: ***Adenine, Guanine*** (double ringed consisting of 5 member imidazole ring attached to a pyrimidine ring) (**Figure 15.2**). Those having one **ring** are similar in structure to the compound **pyrimidine** and hence they are called the pyrimidines or the pyrimidine bases. **Adenine, Cytosine and Guanine are** found in both DNA and RNA; **thymine**, is found only in DNA, and **uracil**, is found only in RNA.

Nucleotides not only make the language of genetics but also play several other significant roles in metabolism. They store and transfer chemical energy ATP, the energy currency of the cell. They make part of the cofactors NAD, FAD,

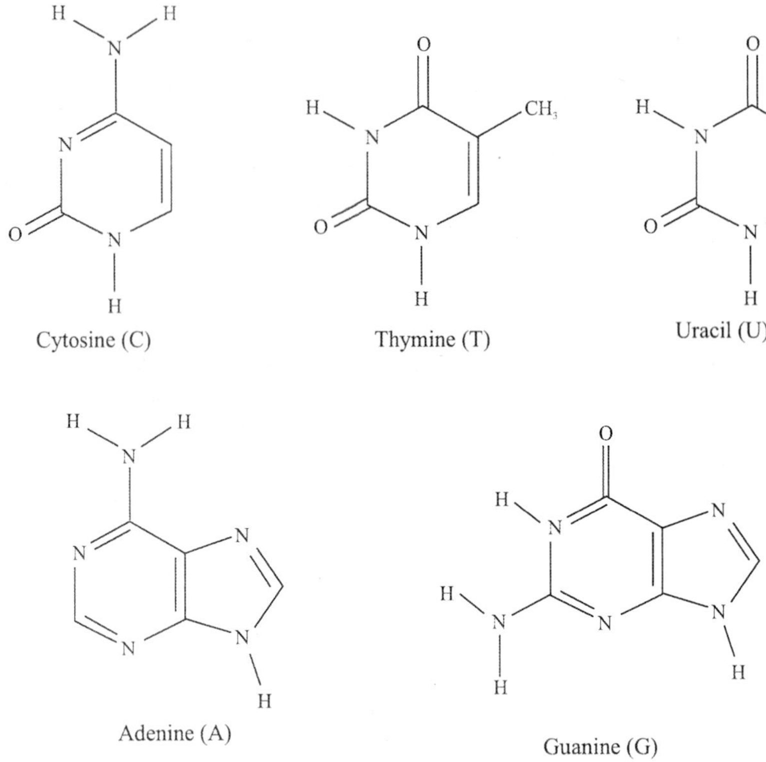

Figure 15.2: Five nitrogenous bases found in nucleic acids.

S-adenosynlmethionine, and coenzyme A. In addition to these, they also make activated biosynthetic intermediates for example cAMP and cGMP, which act as messengers.

Two kinds of pathways are reported for the biosynthesis of nucleotides in the cells: **(i) *de novo* pathways** and **(ii) salvage pathways**. *De novo* synthesis of nucleotides starts with their metabolic precursors: amino acids, ribose 5-phosphate, CO_2 and NH_3. While, salvage pathways recycle the free bases and nucleosides resulting from nucleic acid breakdown.

DE NOVO PATHWAYS OF PURINE & PYRIMIDINE BIOSYNTHESIS

De novo pathways for purine and pyrimidine biosynthesis appear to be identical through out the living beings. The process does not have any of the free nitrogenous bases (ATCGU) as intermediates. Nitrogenous bases are not synthesized and then attached to ribose but the purine ring structure is synthesized one or the few atoms at a time and then attached to ribose sugar. The pyrimidine ring is synthesized as **orotate**, attached to ribose phosphate, and then converted to the common pyrimidine nucleotides. On the other hand, nitrogenous bases are intermediates in the salvage pathways.

Several important precursors are common in the *de novo* biosynthetic pathways for pyrimidines and purines. These precursors are phosphorylated ribose and amino acids. Phosphoribosyl pyrophosphate (PRPP) is the most important precursor in their biosynthesis. Two amino acids act as precursor in each type of pathways: glycine for purines and aspartate for pyrimidines. Glutamine serves as a source of amino groups in 5 different steps in the *de novo* synthesis. Aspartate is also used twice as the source of an amino group in the purine biosynthesis.

The enzyme for '*de'novo*' biosynthesis of purine and pyrimidines are present as multienzyme complexes in the cell. At any given time, the amount of nucleotides is much smaller than the amount required for DNA / RNA synthesis. Hence, it is believed that nucleotide synthesis must occur along with DNA replication and transcription.

Biosynthesis of purine nucleotide: The biosynthetic pathway of purines was deciphered by John Buchnan and G. R. Greenberg in 1950s using isotopic tracer technique. Purine nucleotides found in nucleic acids are adenosine 5' monophosphate (AMP; adenylate) and guanosine 5' monophosphate (GMP; guanylate), containing the purine bases adenine and guanine. In the first step, of the pathway, an amino group donated by glutamine is attached at C-1 of PRPP (step I). The resulting **5-phosphoribosylamine** is highly unstable. The purine ring is gradually built upon this structure.

Next to this, three atoms out of one glycine molecule are added to it and one ATP is consumed in this condensation reaction (step II). In the third step, the amino group of glycine is formylated by N^{10}–formylterahydrofolate (step III), and a nitrogen is contributed by glutamine (step IV). Dehydration and ring closure yields a 5-membered imidazole ring, 5-aminoimidazole ribonucleotide, of the purine nucleus (step V).

Further, a carboxyl group is first added to this ring (step VI). This carboxylation uses the bicarbonate ions generally present in aqueous solutions of the cell. The molecule is rearranged and the carboxylate from the exocyclic amino group is moved to position 4 of the imidazole ring (step VII). Aspartate then donates its amino group in two consecutive steps (VIII & IX); formation of an amide bond is followed by elimination of the carbon skeleton of aspartate. The final carbon is contributed by N^{10} –formylterahydrofolate (step X), and closure of second ring takes place to yield fused ring of the purine nucleus (step XI) (**Figure 15.3 A**). The first intermediate to have a complete purine ring is **inosinate monophosphate** (IMP).

Inosinate is converted to adenylate by the insertion of an amino group derived out of aspartate. On production of adenylate, the reaction derives energy from GTP resulting in synthesis of adenylosuccinate. Adenosine monophosphate (AMP) is produced upon liberating one molecule of fumarate from adenylosuccinate.

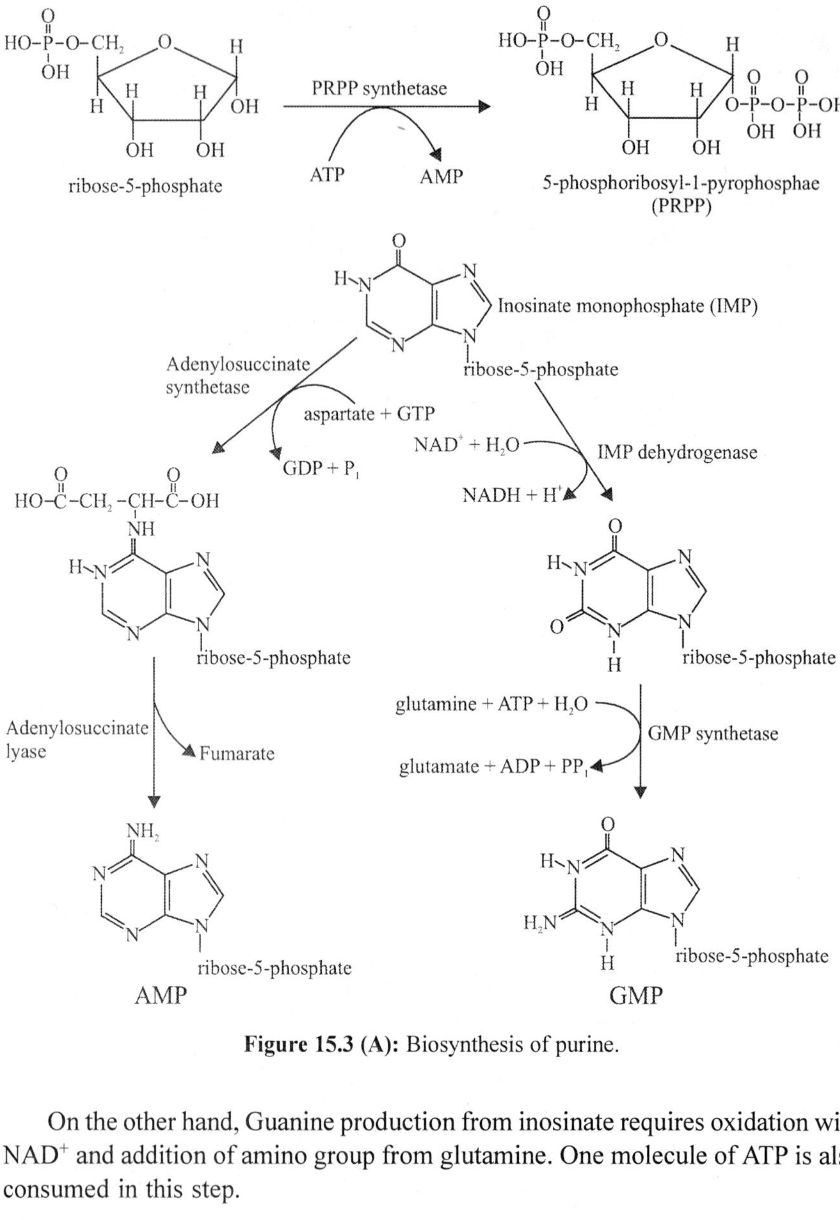

Figure 15.3 (A): Biosynthesis of purine.

On the other hand, Guanine production from inosinate requires oxidation with NAD^+ and addition of amino group from glutamine. One molecule of ATP is also consumed in this step.

Biosynthesis of Pyrimidine Nucleotide

The six membered common pyrimidine nucleotides include cytosine-5′ monophosphate containing cytosine and uridine-5′ monophosphate containing uridine. Their biosynthesis is different from that of purine because ribose 5-phosphate is attached only after pyrimidine ring is made. This biosynthesis requires carbamoyl phosphate (an urea cycle intermediate), aspartate and PRPP. Carbamoyl

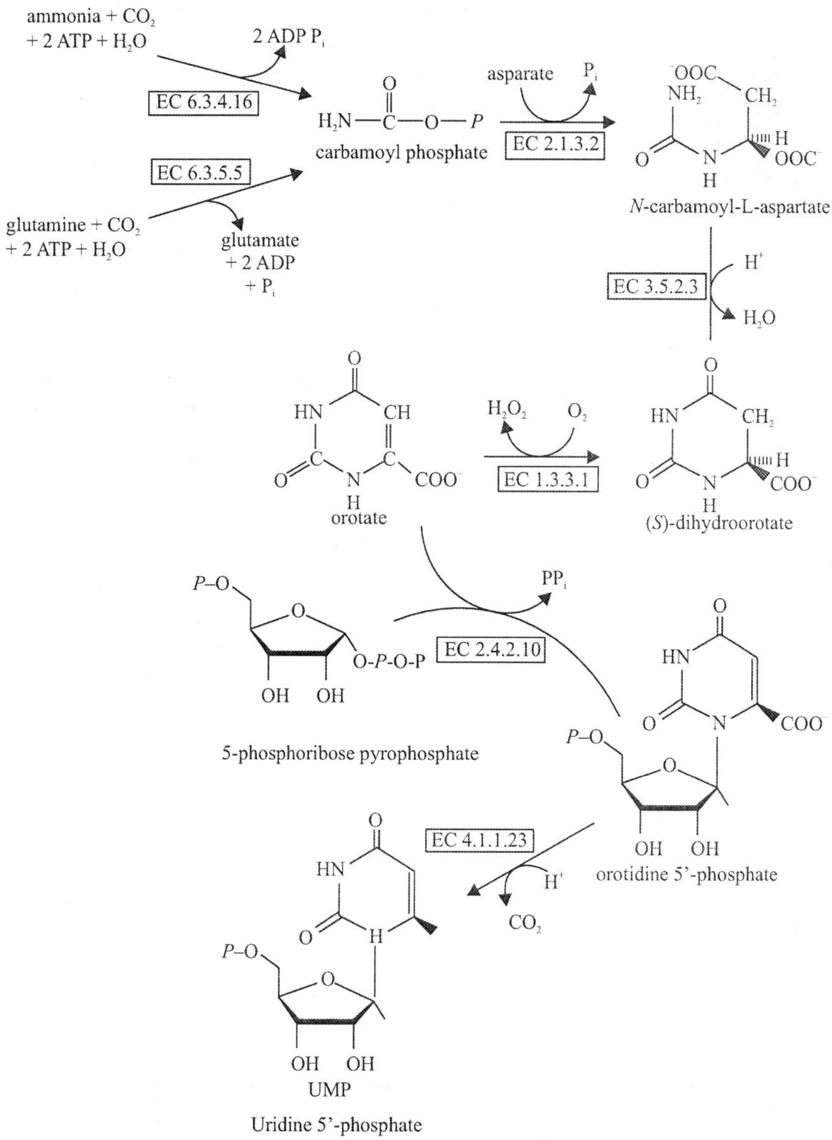

Figure 15.3(B): Biosynthesis of pyrimidine. Enzymes involved in biosynthesis are shown by enzyme commission numbers: EC 1.3.3.1 - dihydroorotate oxidase, EC 2.1.3.2 - aspartate carbamoyltransferase, EC 2.4.2.10 - orotate phosphoribosyltransferase, EC 3.5.2.3 - dihydroorotase, EC 4.1.1.23 - orotidine-5'-phosphate decarboxylase, EC 6.3.4.16-carbamoyl-phosphate synthase, EC 6.3.5.5-carbamoyl-phosphate synthase.

phosphate reacts with aspartate to yield N-carbamoyl aspartate in presence of aspartate transcarbamoylase, an allosteric enzyme. N-Carbamoyl aspartate loses one molecule of water to produce a closed ring, L-dihydrooratate. This compound is then oxidized to pyrimidine derivative orotate in presence of NAD^+. These three steps are catalyzed by carbamoyl phosphate synthetase II, aspartate transcarbamoylase and dihydrooratase which make to be part of a trifunctional single protein (CAD).

Ribose-5-phosphate from PRPP is added to orotate to form orotidylate. This compound is decarboxylated to produce uridylate or UMP (**Figure 15.3 B**). The UMP is further phosphorylated by 2 ATPs to form UTP. Cytidine is prepared by the action of cytidylate synthetase. This reaction utilizes nitrogen of glutamine and consumes one ATP.

Phosphate group

Phosphate groups alternate with deoxyriobose moieties all along the backbone of polynucleotide. The phosphate group can be denoted either as a **phosphate ion** or as **phosphoric acid** molecule. Each phosphate group is linked to 3′carbon atom of one deoxyribose and 5′ of another. The orientation of the sugar in the two strands is opposite to each other making the strands antiparallel to each other and the 3′ end of one strand corresponds with 5′ end of the other.

In case of RNA, since ribose has three free hydroxyl groups at 2′, 3′ and 5′ C, the phosphate can be found on at any of these positions. However, most commonly it is found at 3′ and 5′ positions. Most nucleotides involved in making of DNA and other metabolic reactions are 5′ phosphate esters of nucleosides. The nitrogenous base can exist in two different orientations with respect to the N-glycosidic bond, namely, *syn* and *anti*. However, most natural nucleotides have N-bases in anti conformation.

Nucleotides

A sugar molecule and nitrogenous base combine to make a nucleoside (Deoxyribonucleoside in DNA and Ribonucleoside in RNA). Nucleosides with a phosphate group make nucleotide and depending upon the number of phosphate groups these can exist in the mono-, di-, or tri-phosphorylated forms. The monophosphorylated form of adenosine (adenosine-5′-monophosphate) is abbreviated as AMP while the di- and tri-phosphorylated forms are written as, ADP and ATP, respectively. By convention, the nucleotide is understood to be in the 5′-phosphorylated form. The di- and tri-phosphates of nucleotides are linked by acid anhydride bonds. These bonds have a high $\Delta G°$ for hydrolysis which provides them a high potential to transfer the phosphates to other molecules (phosphorylation). It is this property of the nucleotides that results in their involvement in phosphorylation and transfer reactions in the cell. The nucleotides found in DNA are different from those of RNA in having 2′-deoxy ribose and hence their abbreviations have **"d"** as prefix (**Table 15.1**). The monophosphorylated form of adenosine found in DNA (deoxyadenosine-5′-monophosphate) is written as dAMP. Uridine is not found in DNA and thymine is almost exclusively found in

Table 15.1 Components of nucleotides

Purine/Pyrimindine + Sugar = Nucleoside			Nucleoside + PO_4^{3-} = Nucleotide	
N-base	**Deoxyribonucleosides**	**Ribonucleosides**	**Deoxyribonucleotides**	**Ribonucleotides**
Thymine	Deoxythymidine	Thymidine	Deoxythymidine-5'-monophosphate (5'dTMP)	Ribosylthymine-5'-monophosphate (5'rTMP)
Uracil	Deoxyuridine	Uridine	Deoxyuridine-5'-monophosphate (5'dUMP)	Uridine-5'-monophosphate (5'UMP)
Cytosine	Deoxycytidine	Cytidine	Deoxycytidine-5'-monophosphate (5'dCMP)	Cytidine-5'-monophosphate (5'CMP)
Adenine	Deoxyadenosine	Adenosine	Deoxyadenosine-3'-monophosphate (3'dAMP) Deoxyadenosine-5'-monophosphate (5'dAMP)	Adenosine-3'-monophosphate (3' AMP) Adenosine-5'-monophosphate (5' AMP) Adenosine-5'-triphosphate (ATP) Adenosine-2'-monophosphate (2' AMP) Adenosine-3',5'- cyclic monophosphate (Cyclic AMP or cAMP)
Guanine	Deoxyguanosine	Guanosine	Deoxyguanosine-5'-monophosphate (5'dGMP)	Deoxyguanosine-5'-monophosphate (5' GMP) Deoxyguanosine-5'-triphosphate (GTP)

DNA. Some unusual rare common bases or modified bases are also found to occur in DNA and RNA. The primary modified base in DNA is 5-methylcytosine. Exceptionally, in DNA of PBS viruses, uracil occurs in place of thymine. A variety of modified bases are also seen in the tRNAs.

A variety of nucleotides are also known to play other vital roles in metabolism. **NAD** plays a crucial role in transporting or transferring hydrogens and electrons in respiration and contains an adenosine diphosphate. Another derivative of adenosine diphosphate, **FAD** is also a hydrogen and electron transporter in the respiration process. **Coenzyme A** is also an adenosine diphosphate derivative which is very important in several metabolic pathways and the citric acid cycle. A unique adenosine derivative is the cyclic form, 3'–5'-cyclic adenosine monophosphate, **cAMP**. It is formed in response to activation of receptor coupled adenylate cyclase. It acts as a powerful messenger in signal transduction events from the cell surface to internal proteins, e.g. cAMP-dependent protein kinase (PKA) which is responsible for phosphorylation of number of proteins alterning their activity either positively or negatively. Cyclic-AMP is also involved in the regulation of ion channels by direct interaction with the channel proteins. Nucleotides combine to make a polymeric molecule –Polynucleotide. The DNA of $\phi \times 174$ is a single-stranded circular DNA molecule and the proportion of A = T and G = C is uniquely 1:1.33:0.98:0.75. The replication of this DNA proceeds by a double-stranded replicative form.

NUCLEIC ACIDS

Both DNA and RNA are polymers made up of repeating nucleotide units. The nucleotides bind in a sequence of enzyme (DNA polymerases and RNA polymerases) catalyzed intermolecular dehydration reactions that form phosphate ester bonds. One OH group on the phosphate and the other OH group on the third carbon of the sugar condense and result in polymerization of the nucleotides into a polynucleotide or nucleic acid with a sugar-phosphate backbone from which the nitrogenous bases project perpendicularly.

Nucleic acids have two distinguishing ends, namely, 5′ and 3′ end. These numbers refer to the 5 prime and 3 prime carbons on the terminal sugar where the 5' end bears a phosphate, and the 3′ end a hydroxyl group. DNA and RNA polymerases add new nucleotides to the 3' end of the sugar molecule of previously incorporated nucleotide and hence the direction of nucleic acid synthesis 5' to 3′.

Deoxyribonucleic Acid (DNA)

In most living organisms genetic information is stored in the molecule deoxyribonucleic acid, or DNA. Viruses have both DNA and RNA as genetic material. DNA is produced and contained in the nucleus of living cells. DNA is named so after sugar molecule contained in its backbone (deoxyribose); however, it gets its significance from its unique structure and functions. Four different nucleotide bases in DNA are adenine (A), cytosine (C), guanine (G), and thymine (T).

The structure of DNA molecule was deciphered in the early 1950s by four scientists, James Watson and Francis Crick at Cambridge University and Maurice Wilkins and Rosalind Franklin at King's College, using experimental data and X-ray diffraction patterns. Watson and Crick published a paper in the scientific journal *Nature* describing this research in 1953. They postulated that the DNA molecule is double-stranded and the two strands enfold around each other to form a double helix (**Figure 15.5**).

The Double Helix

The long nucleotides enfold upon each other and through the complementary base pairing among purine and pyrimidine make a double helix. At every step of DNA ladder, a purine and a pryimidine from each strand pair by hydrogen bonding, viz. A = T and G ≡ C. Because of this well-defined nature of pairing the total number of purines in any given DNA is always equal to that of pyrimidines. Adenine joins thymine by two hydrogen bonds at positions 1′ and 6′ while cytosine and guanine are joined by three hydrogen bonds at positions 1′, 2′ and 6′ in the heterocyclic ring (**Figure 15.4**). The double-stranded molecule of DNA thus attains a helical structure in which two complementary strands are coiled together through

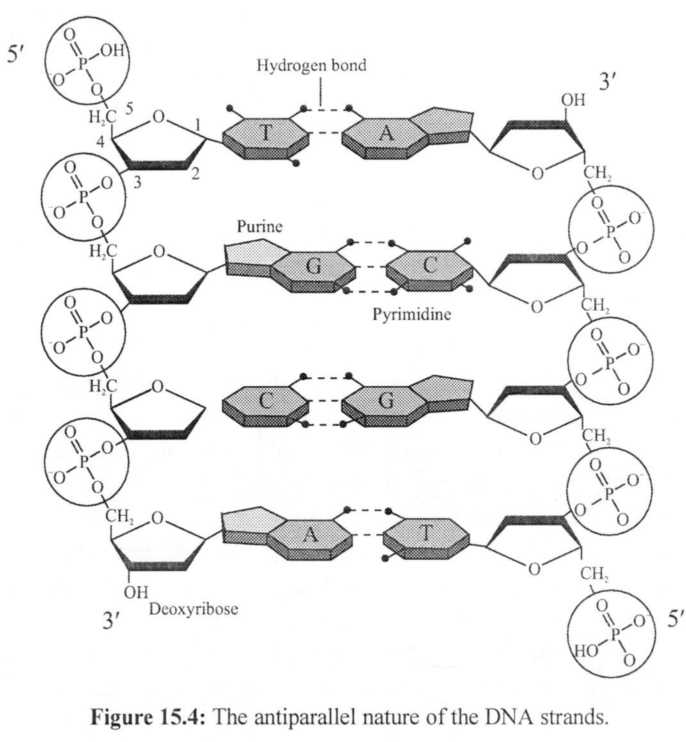

Figure 15.4: The antiparallel nature of the DNA strands.

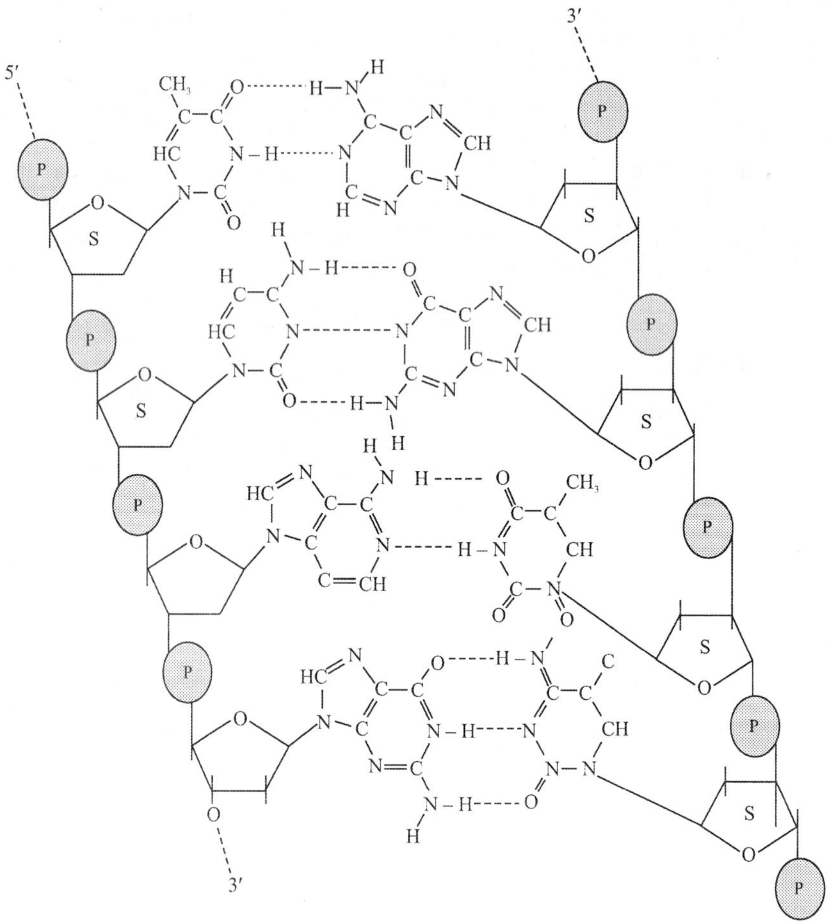

Figure 15.5: H-bonding between nitrogenous bases.

thousands of H-bonds. However, bonding is weak enough to allow separation of the two strands during replication. The nitrogenous bases bond to sugar molecule through one NH group while another NH group is involved in H-bonding. The pyrimidine is linked with its 3′ N-atom to 1′(C) of deoxyribose while the purine is linked with its 9′N to 1′(C) of deoxyribose.

The usefulness and unique properties of DNA can be attributed to its double-helix structure. The nucleotide bases of the DNA molecule form complementary pairs. The nucleotides bind to another nucleotide base in a strand of DNA opposite to the original. This bonding is specific and the bases necessarily complement each other. Thus, adenine always bonds to thymine (and vice versa) and guanine always bonds to cytosine (and vice versa). This bonding occurs across the molecule, leading to a double-stranded system. The two strands are antiparallel with regards to orientation of deoxyribose sugar (3′–5′ and 5′–3′).

The double-stranded DNA molecule has the unique ability that it can make exact copies of itself, or self-replicate. When more DNA is required by an organism during cell division or reproduction, the hydrogen bonds between the nucleotide bases break and the two single strands of DNA separate. New complementary bases are brought in by the cell and paired up with each of the two separate strands, thus forming two new, identical, double-stranded DNA molecules.

STRUCTURE OF DNA

The structure of DNA is an important discovery that revolutionized life science research as a whole. Beginning with Meischer's isolation of nuclein in 1869 it took almost eight decades for deciphering the physical basis of heredity. Experiments of Griffith (1928) and Avery, MacLeod and McCarty (1944) regarding transformation of bacterial cells by DNA established that DNA is the genetic material. In 1947, E. Chargaff showed that the proportion of pyrimidines and purines in any given DNA molecule is similar. Utilizing data and photographs obtained from X-ray diffraction patterns of crystallized DNA, Wilkins and coworkers showed that the purines and pyrimidines are placed at regular intervals along the DNA molecule at a distance of 3.4 Å and the molecule is twisted at every 34 Å, thus there are ten nucleotides per turn as described in their communication published in *Nature* in 1953. Watson and Crick model predicted that DNA would exist as a helix of two complementary antiparallel strands, wound around each other in a rightward or clockwise direction and stabilized by H-bonding between bases in adjacent strands. The bases are in the interior of the helix aligned at almost 90° relative to the axis of the helix. Purine bases form hydrogen bonds with pyrimidines and the experimental data conforms, in any given molecule of DNA, the concentration of adenine (A) is equal to thymine (T) and the concentration of cytosine (C) is equal to guanine (G). Three H bonds make G–C base-pairs more stable than A-T base-pairs. The width or girth of the helix of the DNA is 20Å and the twisting of the strands results in the deep and shallow spiral grooves. These grooves are of unequal size and called as major and minor grooves. The difference in their size is due to the asymmetry and spatial arrangement of the deoxyribose rings and the structurally distinct nature of the base-pairs.

T_M value of DNA

In the process of thermal denaturation, a point is reached at which 50% of the DNA molecule exists as single strands. This point is referred as melting temperature (T_M), and is characteristic of the base composition of DNA molecule. The T_M depends upon several factors in addition to the base composition. These include the chemical nature of the solvent and the type and concentrations of ions in the solution. When thermally denatured DNA is cooled, the complementary strands will again reform the correct base pairs, in a process termed as reassociation or hybridization. The rate of renaturation or reassociation is dependent upon the nucleotide sequence of the two strands of DNA. The antiparallel nature of the helix stems from the orientation of the individual strands. From any fixed position in the helix, one strand is oriented in the $5' \rightarrow 3'$ direction and the other in the $3' \rightarrow 5'$ direction.

Forms of DNA

The double helix of DNA has been shown to exist in several different forms, depending upon sequence content and ionic conditions of crystal preparation. The B-form of DNA is most common and predominates under physiological conditions of low ionic strength and a high degree of hydration. Regions of the helix that are rich in G and C dinucleotides can exist in a novel left-handed helical conformation termed Z-DNA. This conformation results from a 180 degree change in the orientation of the bases relative to that of the more common A- and B-DNA (**Table 15.2**).

Table 15.2 Various conformations or forms acquired by DNA helix

Parameters	A Form	B Form	Z-Form
Direction of helical rotation	Right	Right	Left
Residues per turn of helix	11	10	12 base pairs
Rotation of helix per residue (in degrees)	33	36	–30
Base tilt relative to helix axis (in degrees)	20	6	7
Major groove	Narrow and deep	wide and deep	flat
Minor groove	wide and shallow	narrow and deep	narrow and deep
Orientation of N-glycosidic bond	Anti	Anti	Anti for Py, Syn for Pu
Remarks		most prevalent within cells	occurs in stretches of alternating purine-pyrimidine base pairs

Packaging of DNA in Chromosomes: The Nucleosome Model

DNA molecules have several levels of structural organization. A linear sequence of nucleotides forms primary structure. Two such polynucleotides bond together in a double-strand which twists into a double-helix resulting in its secondary structure. The double-helix then wraps around some basic proteins (histones) making its tertiary structure. These coils wrap and fold around each other and this wrapping and folding continues until the chromosomes are formed.

Organization of DNA in Chromosomes

The DNA of bacteria, e.g. *E. coli* is a circular, double-stranded negatively supercoiled molecule and contains about 4.6 million base pairs (**Figure 15.6**). The bacterial chromosome is packaged into a specific region of the cell called as nucleoid where it is organized as loops bound to a protein scaffold attached to the bacterial cell membrane. The DNA molecule is complexed with several histone –like DNA-binding proteins (HU, HLP-1 and H-NS).

The eukaryotic genome is contained within a specialized cell organelle, nucleus. The eukaryotic genome is almost thousand times bigger than that of *E. coli*. This long DNA is packaged tightly within the chromosomes. Diploid organisms have two sets of each chromosome, excepting sex chromosomes. Chromosomes contain majority of basic histone proteins along with some non-histone proteins. Together the DNA and protein complex is called as chromatin. Each chromosome contains single linear double-stranded DNA molecule. Primarily DNA is bound with histones in almost equal ratio. There are five main types of histones called as H1, H2A, H2B, H3 and H4. These proteins are highly basic in nature and contain a large number of lysine or arginine and positively charged amino acid side chains. Because of positively charged groups, histones bind the negatively charged phosphate groups of DNA molecule. Histones H3 and H4 have highly conserved sequences and differ only in few amino acids throughout the eukaryotes. Perhaps this indicates the importance of histones in securing the genetic material of eukaryotes. The DNA of sperms is too condensed and the histones are replaced with other smaller basic proteins, protamines.

When viewed in an electorn microscope the chromosomes appear as a 'beads on a string structure'. The beads are called nucleosomes and are made up of DNA and histones. The string represents linear double-stranded DNA between two

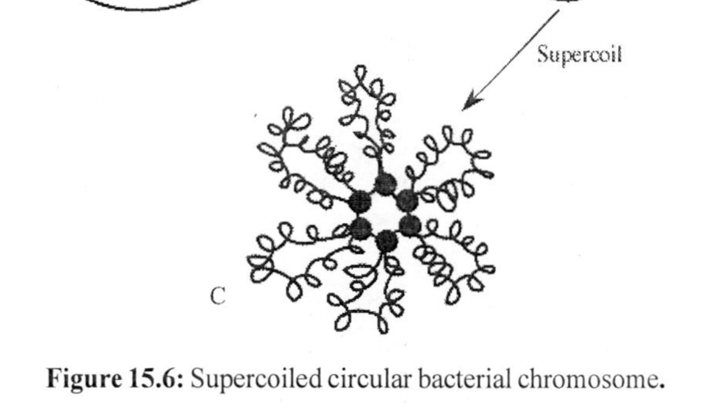

Figure 15.6: Supercoiled circular bacterial chromosome.

nucleosomes (linker DNA) (**Figure 15.7**). The distance between two nucleosomes is 55 base pairs but varies over different organisms in the range of 8–114 bp. When subjected to digestion with nucleases, the linker DNA is degraded while nucleosomes remain intact due to protection offered by histones. Each nucleosome contains 146 bp long DNA twined around a complex of eight histones (two molecules of each of H2A, H2B, H3 and H). The DNA is wound in about 1.8 turns in a left handed fashion. DNA and histone are in contact along the inside face of the DNA super helix. The chromatin as seen through electron microscope seems to be a fiber having 30 nm diameters suggesting that smaller nucleosomes again organize in a higher structure. This fibre is formed with the help of H1 histones that bind linker DNA. These H1 histones interact with each other and pull the nucleosomes together to form a solenoid having six nucleosomes per turn. Further, the 30 nm fiber, *in vivo*, is seen linked randomly to the protein scaffold as loops. The eukaryotic extra chromosomal DNA as seen in mitochondria and chloroplast is much different than eukaryotic chromosome and is similar to bacterial chromosome in being circular.

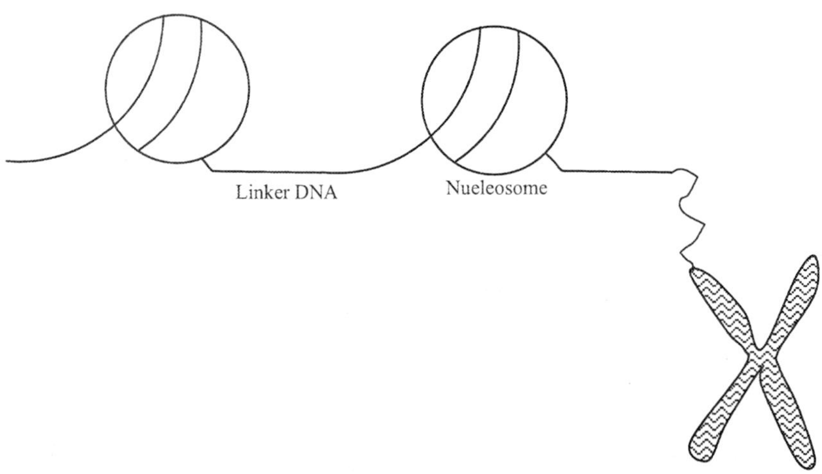

Figure 15.7: Nucleosome model of packaging of eukaryotic DNA in chromosome.

DNA SYNTHESIS / REPLICATION

DNA is the most unique molecule as it can make its own copies. The synthesis of DNA is basically its replication using the pre-existing DNA as template. DNA replication is therefore semi-conservative because each of the two double helices emerging from the parent DNA has one old strand of the parent (**Figure 15.8**).

Resulting two helices are exactly identical with the original helix in terms of base sequence, etc. The two stands of DNA double helix are united by hydrogen bonds between purine and pyrimidine base pairs. The strands wound around each other in a helical fashion. Beginning of the replication is marked by an incision

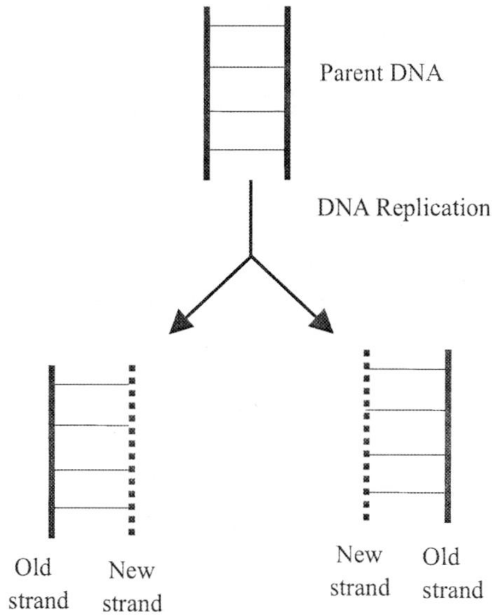

Figure 15.8: Semi-conservative mode of DNA replication.

made by endonuclease on one of the strands of DNA. For replication, DNA binding proteins and DNA gyrase open up the DNA helix. Gyrase binds on a single strand and initiates uncoiling of twined strands. The unwinding of strands results in development of strain in the remaining helix. This strain is relieved by action of a superhelix relaxing protein. A specific RNA polymerase catalyzes formation of a small strand of RNA, called as primer, on one of the DNA template. DNA polymerase III catalyzes the addition of nucleotides to the 3'ends of the newly synthesized RNA primer and thus begins the synthesis of DNA. Later, this RNA primer is degraded by DNA polymerase I which also synthesizes a small DNA segment to replace RNA primer. The newly formed DNA segment is ligated by an enzyme, DNA ligase. The DNA helix is antiparallel, i.e. one strand runs 5′–3′ while the other strand has direction 3′–5′. All DNA polymerases make daughter strands only in the 5′–3′ direction. At least one strand of the DNA is synthesized discontinuously, i.e. in form of **Okazaki fragments** which are later ligated by the enzyme DNA ligase. This is because of the fact that the direction of DNA synthesis is always 5′→3′ (5 prime to 3 prime) while the two parent strands are always antiparallel. This discontinuous strand is called lagging strand.

REPLICATION OF DNA IN PROKARYOTES

The replication of DNA in prokaryotes can best be illustrated with the example of molecular biology workhorse *E. coli.* It contains a single long circular DNA as its genetic material. Replication of *E. coli* DNA always starts at a unique site called origin (*oriC*). The region specifying oriC is characterized by three repeats of a 13 base pairs (bp) sequence (GATCTNTTNTTTT) and four repeats of 9 bp sequence (TTATNCANA). The origin of replication has also been characterized in many plasmids and phages and these are generally rich in A::T base pairs (**Figure 15.9**).

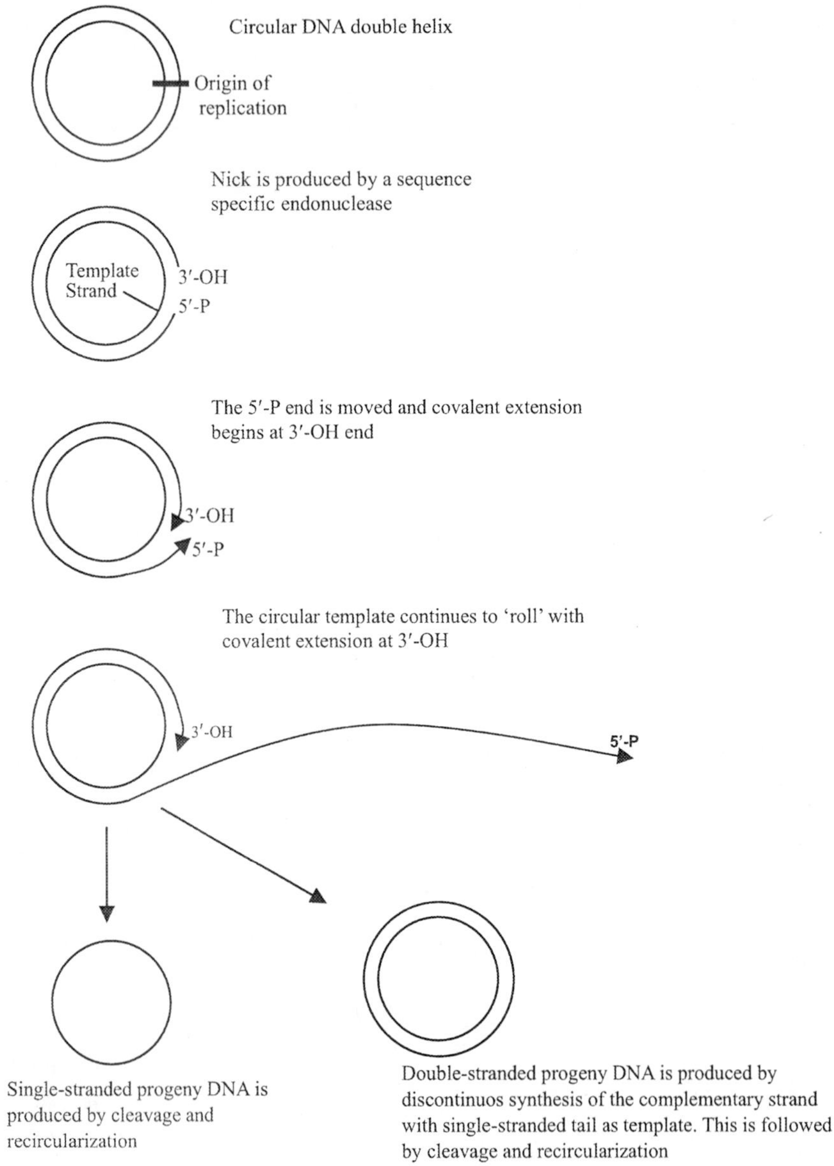

Figure 15.9: Replication of bacterial chromosome (Rolling Circle Model).

This makes melting or uncoiling of this region easy since A::T base pairs contain two H- bonds as compared to triple H-bonds in case of G: C.

The replication of this circular polynucleotide is bidirectional, i.e. replication bubble has two replication forks. DNA synthesis then proceeds in both directions form the single origin. Thus, the region of parent DNA under replication is called as a replication bubble or replication eye and consists of two replication forks moving in opposite directions on the circular bacterial DNA (**Figure 15.10**) The two forks ultimately meet and fuse and two circular daughter DNA molecules, each having one parental and one newly formed strand, are produced.

DNA Polymerases

DNA replication is carried out by specialized enzymes, which carry out condensation of dNTPs to form a polynucleotide chain, DNA polymerases; DNA polymerase is also called as Kornberg's enzyme after A. Kornberg who first isolated it from *E. coli*. It is a single polypeptide (109 kD) and its digestion with trypsin cleaves it into two subunits- (i) Klenow fragment (75 kD) having **polymerase** and 3′–5′exonuclease activity and (ii) a small fragment (38 kD) which has only 5′–3′ exonuclease activity.

Later, in *E. coli* three types of DNA polymerases, i.e. DNA polymerase I, II and III were found. Now it is known that the DNA polymerase I (Kornberg's enzyme) of *E. coli* is not the primary enzyme in DNA replication but has only very specialized function of DNA repair. The chief enzyme in DNA replication is DNA polymerase III which is a hetero-multimeric enzyme with molecular weight more than 250 kD. The 160 kD catalytic core of this DNA polymerase III is made up of an α-subunit (Mr 130 kD) having polymerase activity, one ε-subunit (25 kD) and one θ-subunit (10 kD). ε and θ-subunit are responsible for assembly and proof-reading activities. Two such core enzymes condense to form a dimer Pol III*, with the help of two τ-subunits (71 kD each). One of the α-subunits gets associated with two γ-subunits (52 kD each) and two δ-subunits (32 kD each) resulting in

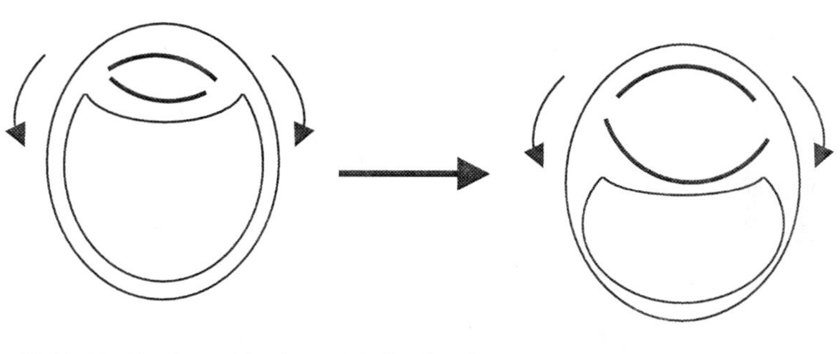

Figure 15.10: Replication of the bacterial circular chromosome. Replication starts at a single origin and proceeds bidirectionally. Eventually, two circular daughter DNA molecules are produced, each having one original and one new strand.

the multimeric polymerase III' having molecular weight 750 kD. Further, this multimeric protein is joined by τ-subunits to complete the holoenzyme (900 kD) which is the complete and fully functional DNA polymerase III. The structure and assembly of Pol III is explained in the **Figure 15.11**.

The DNA polymerase enzyme requires all the four dNTPs, a DNA template and a primer with a free 3′-OH in presence of Mg^{2+} ions. These enzymes are directed by the original parent template from which the polymerase recognizes the

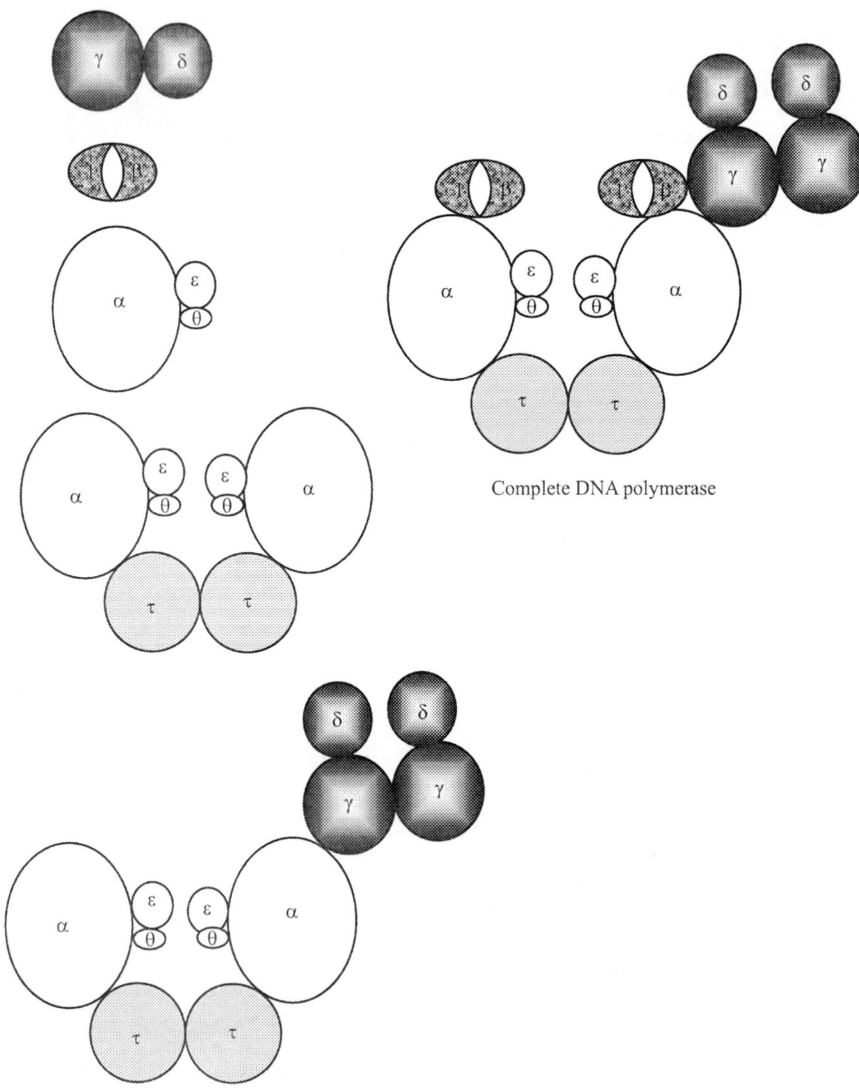

Figure 15.11: The assembly of DNA polymerase.

next nucleotide to be added. Polymerase then adds a complementary nucleotide to the 3′-OH of the primer by creating a 3′–5′ phosphodiester bond and releasing inorganic phosphate. The 3'-OH group makes a nucleophilic attack on α-phosphate group of the incoming nucleotide.

The replication is initiated by binding of a factor DnaA (82 kD protein) that recognizes origin and binds to it. This is followed by binding of a pre-priming factor, DnaT (66 kD) and helicase which causes uncoiling of a small stretch of DNA. The uncoiled region of DNA is joined by SSB (single stand binding proteins), a tetramer of 74 kD DnaB which prevents renaturation of uncoiled DNA. Some ancillary proteins cause the activation of primase which synthesizes RNA primer at the site of origin. The binding of DnaB is also facilitated by another hexameric protein, DnaC. DNA polymerase III adds single nucleotides to the free 3′-OH end of RNA primer and continues. The complete complex of DNA polymerase, primase and ancillary proteins is referred as primosome.

DNA polymerase I corrects or proofreads the wrongly incorporated or mismatched nucleotides by its proofreading activity. A wrong base is removed using 3′–5′ exonuclease activity and hence the error rate in DNA synthesis is very less (below 10^{-8} per base pair). DNA polymerase I also has 5′–3′ exonuclease activity and can also remove nucleotide from the 5′ end of a polynucleotide. This activity is used in removing the RNA primer earlier used for initiating DNA synthesis. DNA polymerase II and III also have all the activities similar to DNA polymerase I but lack 5′–3′ exonuclease activity.

RNA Primer

Presence of RNA primer is essential for DNA replication as DNA polymerase can add nucleotide only to a template at 3′-OH end. The RNA primer is a short segment (5 nucleotide long) synthesized by primase (RNA polymerase). RNA polymerase can make RNA directly on the single strand DNA template since it does not require any primer to begin synthesis. This RNA primer is extended by DNA polymerase III in a bidirectional fashion. After DNA is synthesized by DNA polymerase III, DNA polymerase I removes RNA primer by its 5′–3′ exonuclease activity and replaces it with DNA. The newly synthesized fragment is ligated by DNA ligase. DNA polymerase III cannot remove RNA primer as it lacks 5′–3′ exonuclease activity.

Enzyme Involved in DNA Replication in Prokaryotes

Several enzymes and accessory proteins particitpate in replication of bacterial chromosome. DNA helicase is used for unwinding the double helix. It uses ATP energy in unwinding along with SSBP (single strand binding protein) that prevents the renaturation between base pairs of the unwound DNA. If you visualize a replication fork moving along a DNA helix then the unwinding of DNA would make the DNA rotate rapidly (just like unwinding two strand of a rope). The bacterial chromosome is circular and thus has no ends to rotate. In this situation, the rotation of DNA will result in coiling of the circular chromosome and will halt the process. However, an enzyme, **topoisomerase or gyrase**, breaks a phosphodiester bond in one strand allowing the DNA to rotate freely around the other intact strand. Once new DNA fragment is synthesized, the phosphodiester bond is reformed by the same topoisomerase. The resulting two daughter DNA molecules are intertwined. Theses are separated by another enzyme **topoisomerase** II. This enzyme causes a double strand break in one of the DNA molecule. Topoisomers are the different physical forms of the same DNA molecule. Topoisomerase help intercovert DNA into its different topoisomers. These enzymes help DNA get unwound at a rapid speed. It creates a small nick in one strand, allowing the molecule to uncoil and thus reduces the strain in the helix. As described earlier, Type I topoisomerase cleave only one strand of the DNA molecule and help in relaxation of negative supercoiling of the DNA while topoisomerase II cleave both the strands of DNA molecule and consume ATPs during cleavage.

DNA ligase

The enzyme joins two nucleotide or polynucleotide ends (3'OH and 5' PO_4 groups).

Primase

It is a RNA polymerase that synthesizes primer – a RNA molecule that serves as a template for DNA polymerase to start condensation of nucleotides.

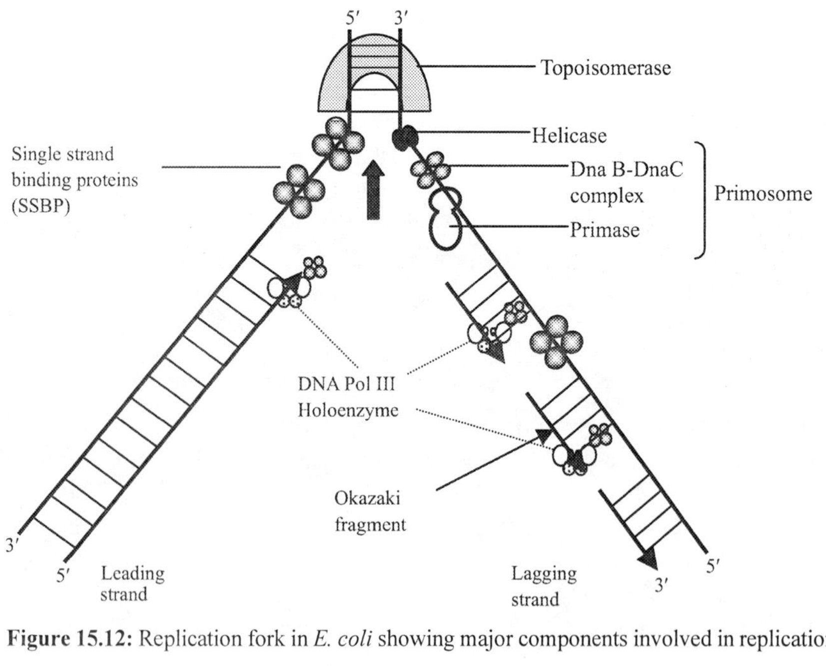

Figure 15.12: Replication fork in *E. coli* showing major components involved in replication.

Helicase

The enzyme helps in formation of single strands in the region undergoing replication by unwinding the two intertwined strands with the expense of energy (ATP).

Factors

Many other factors such as SSBP (single strand binding proteins) and ancillary proteins are also associated with DNA replication.

DNA REPLICATION IN EUKARYOTES

DNA replication is mandatory with the replication of the cell. A eukaryotic cell divides by mitosis to form two identical daughter cells with identical genetic information. A typical eukaryotic cell cycle consists of G1 (gap) phase, S (synthesis) phase where DNA is replicated and G2 phase in which the cell prepares for mitosis. Eukaryotic cells in culture (*in vitro*) have a cycle time of 16–24 hours but in body (*in vivo*) the cell cycle time can be upto 100 days or more. Some cells do not divide after a while and become quiescent are said to go into G_0 phase.

The eukaryote DNA replication gets started and reaches completion with in S phase of cell cycle. The replication starts at multiple loci or origins that are located at every 3–300 kb of DNA and proceeds bidirectionally at every origin. This makes the replication quicker and chromosomal replication is completed within stipulated time period. At each origin there is a replicon consisting of two replication forks moving in opposite directions. The replication continues till all such replication bubbles merge. The origin of replication in some yeasts has been characterized and is referred to as **autonomous replicating sequences (ARS)**. However, it is difficult to identify origins in eukaryotes due to their multiplicity.

Replication origins are activated in a group or clusters of 20–80 replicons that lie close to each other on one part of the chromosome while the other part of the chromosome may not have any of these. Such sets of replication units are activated one by one over the complete length of the chromosome until the complete chromosome is replicated. Regions of chromosome that are transcriptionally inactive regions are replicated later (**Figure 15.13**).

Eukaryotic cells contain **five** different polymerases, viz. α, β, γ, δ and ε. Among these α and δ are involved in DNA replication while β and ε are involved in DNA repair. All these are located within the nucleus excepting γ DNA polymerase found in cytoplasm which is responsible for replication of extrachromosomal (mitochondrial) DNA replication.

The basic scheme of DNA replication in eukaryotes is similar to that in prokaryotes. Here also, there is formation of lagging and leading strand. Lagging strand involves formation of small fragments of DNA or Okazaki fragments. The speed of DNA replication is much slower in eukaryotes than in prokaryotes and also the two strands are synthesized by two different polymerases. The DNA

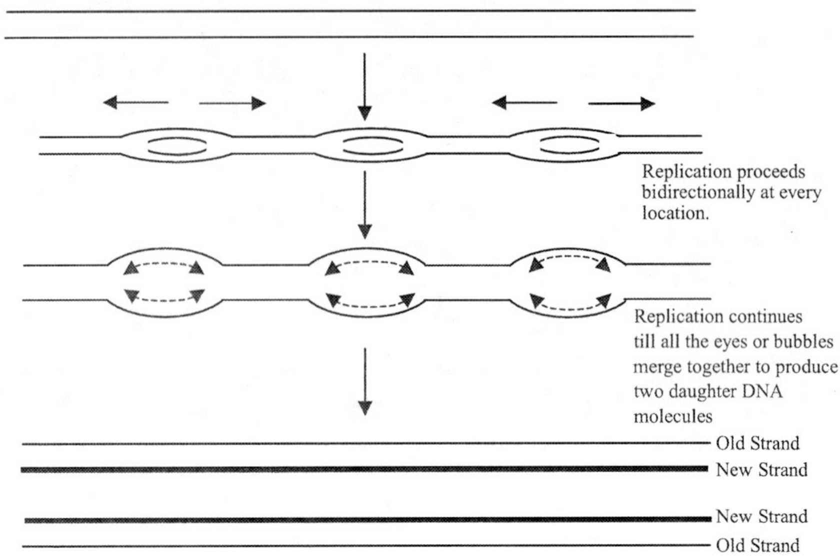

Figure 15.13: Replication of eukaryotic chromosomal DNA begins at many origins and proceeds bidirectionally at every location.

polymerase α catalyzes synthesis of Okazaki fragments (lagging strands) whereas DNA polymerase δ synthesizes continuous or the leading strand.

The rate of DNA replication has been found to depend upon the average replicon size and the multiplicity of replicons during replication. For example, in yeast *Saccharomyces cerevisiae* (genome size- 20,000 kb), the average replicon size is 40 kb and there is formation of 500 replicons during replicaton with a rate of DNA synthesis averaging 50–60 bp / seconds.

The size of human genome is $6x10^6$ Kb. Considering the average rate of replication to be 2.2 kb/minute, it will take 45,000 hrs for the complete replication of human genome. However, actually it takes only 9 hrs because of multiple origins or replicons during eukaryotic DNA replication.

TERMINATION OF DNA REPLICATION

There are specific sequences which cause the termination of replication at the point of juncture of two opposite replication forks. *E. coli* has circular genome and the replication forks originating from single origin may meet half the way down. Two characteristic sequences terD and terA carry out termination of *E. coli* DNA replication of forks. Each of these *ter* regions has a consensus sequence of 23 base pairs (**AATTAGTATGTTGTAACTAAAGT**). This sequence is shown to cause termination of DNA synthesis *in vitro* as well. Termination is mediated by a protein coded by *tus* gene which recognizes *ter* region and binds to it, and interferes with the functioning of DNA polymerase to halt the DNA replication.

SUGGESTED READING

Bloomfield V.A., Crothers D.M. and Tinoco I. 1974. *Physical Chemistry of Nucleic Acids*. Harper and Row, New York.

Fraenkel-Conrat H. 1969. *Physical and Chemical Properties of Nucleic Acids*. Ann. Rev. Biochem., 36: 407–448.

Henderson J.F. and Paterson A.R.P. 1973. *Nucleotide Metabolism: An Introduction*. Academic Press, Inc., New York.

Kornberg A. 1974. *DNA Synthesis*. W.H. Freeman and Company, New York.

Sanger W. 1984. *Principles of Nucleic Acid Structure*. Springer-Verlag, New York.

Watson, J.D. and Crick, F.H.C. 1953. *Molecular structure of nucleic acids: a structure for deoxyribose nucleic acid*, Nature, 4356.

REVIEW QUESTIONS

1. Describe the constituents of nucleic acids.
2. Draw the structure of deoxyribonucleic acid and explain following:
   a) Antiparallel nature of strands
   b) Complementarity among base pairs
   c) Hydrogen bonding among nitrogenous bases
3. Discuss the organization of prokaryotic and eukaryotic genome.
4. Describe replication of DNA and explain following:
   a) DNA polymerases
   b) Semi-conservative nature of DNA replication
   c) Okazaki fragments
   d) Leading strand
5. Give a comparative account of eukaryotic and prokaryotic DNA replication.

UNIT XVI

Types and Properties of RNA

RNA is a polymer of ribonucleotides linked together by 3′–5′ phosphodiester bonds. Ribonucleic acid is named so because of the ribose sugar group in the molecule's backbone and its occurrence in nucleolus. It is called acid because phosphate group present in the backbone ionizes into an anion giving it a negative charge. Although these are much similar, several important differences exist between RNA and DNA. RNA has a sugar-phosphate backbone with nitrogenous bases attached to it but RNA does not contain thymine, and bears uracil (U), a different pyrimidine, complementing with adenine (A). RNA is a single-stranded molecule produced by DNA by a process called **transcription** but does not remain confined in nucleus and carries out translation (protein synthesis) in the cytoplasm during which the information coded in DNA is finally translated into making of a protein molecule. RNA also serves to be a genetic material in some viruses (Reovirus, Retroviruses e.g. HIV). Mostly, RNAs are single-stranded but there are double-stranded regions because of intramolecular base complementarity. Such double-stranded loops are exhibited by secondary conformations of transfer RNA and ribosomal RNA.

During transcription a specific portion (template) of DNA is operated by specific RNA polymerases to create RNA. All RNA polymerases are dependent upon a DNA template in order to synthesize RNA. The resulting RNA is complementary to the template strand of the DNA duplex and identical to the non-template strand. The non-template strand is called the coding strand because its sequences are identical to those of the transcribed mRNA excepting that U is substituted for T.

Three different kinds of RNA (rRNA, mRNA, tRNA) having entirely different structure and specific functions related with expression of genetic information are found in biological systems.

RIBOSOMAL RNA (rRNA)

This class of RNA associates itself with ribosomal proteins to make a large ribonucleic protein complex called as ribosome. Ribosomes provide a platform or workstation to the mRNA where tRNAs charged with amino acids participte in formation of polypeptide. The proteins of the ribosome catalyze all the functions of protein synthesis. In prokaryotes, it is produced on a DNA template referred as ribosomal DNA. In eukaryotes, it is typically synthesized from nucleolar DNA. rRNA makes about 80% of the total RNA content of any cell. Depending upon physico-chemical conditions (pH, temperature), rRNA can exist in various forms such as small rod, coil or linear strand. Primarily, it is a single strand in which some double-stranded loops can be seen due to intramolecular hydrogen bonding among complementary nitrogenous bases. rRNAs provide a mechanism for interaction between mRNA in small ribosomal subunit and the tRNAs during translation by providing peptidyl transferase activity carried out in the large ribosomal subunit. rRNAs are always found associated with proteins and are collectively referred as **ribosomes**. Based on sedimentation coefficient, the prokaryotic ribosome is called as 70S ribosome with a mass of 2.5 MDa and 66% rRNA content. It consists of 30S and 50S subunits. The 30S subunit contains 16S rRNA (1542 bp +21 proteins) while the 50S subunit contains 23S (2904 bp) and 5S rRNA (120 bp) molecules along with 31 proteins. The eukaryotic ribosomes are heavier with a sedimentation coefficient of 80S. These are also made up of two subunits, 40S and 60S. The 40S subunit contains 18S rRNA (1874 bp +33 proteins) while the 60S subunit is made up of three rRNA molecules, viz. 28S, 5.8S and 5S rRNA containing 4718, 160, and 120 base pairs, respectively along with 49 protein molecules (**Table 16.1**).

The bacterial genome has the genes for 23S, 16S and 5S rRNA arranged in a string or continuous sequence. The transcription of these genes results in production of a 30S rRNA, called as p30S, containing all the three rRNA sequences and spacers in between. It undergoes post-transcriptional modification by action of ribonuclease (RNase) resulting in formation of 23S, 16S and 5s rRNAs. These molecules undergo some more modifications (e.g. methylation) during their assembly with ribosomal proteins. Contrastingly, in eukaryotes, the genes encoding rRNAs are dispersed as clusters all over the genome.

Table 16.1 Ribosomes and their RNAs

	Prokaryote		Eukaryote	
Ribosome	70S		80S	
Subunits	50S 23S rRNA 5S rRNA + 31 proteins	30S 16S rRNA + 21 proteins	60S 28S rRNA 5.8S rRNA 5S rRNA + 49 proteins	40S 18S rRNA + 33 proteins

MESSENGER RNA (mRNA)

These contain transcribed genetic code to be used by the translational machinery for determination of the correct order of amino acids to be added into an elongating polypeptide during translation. One strand of DNA duplex encoding a gene, referred as template strand or antisense strand, is transcribed into mRNA. It makes about 3–5% of total RNA content of the cell. The nucleotide sequence in an mRNA should be read as 'triplets or codons' where each nucleotide triplet in mRNA represents one specific amino acid that will be incorporated into the protein upon translation.

The size of the mRNA molecule varies greatly; however, the average molecular weight of mRNA molecule is 500,000 with an 8S sedimentation coefficient. mRNA molecules carrying 300 to 12,000 base pairs (forming proteins containing approximately 100 and 4,000 amino acid since 1 amino acid is coded by 3 base pairs) can be seen. The length of an *E. coli* mRNA ranges from 900–1500 bp.

Apart from A, T, U and C bases, mRNA may contain some unique and unusual bases. mRNA is always found as a linear molecule, without any intramolecular base pairing. Base pairing, if any, may render it useless. mRNAs are transcribed from DNA templates carrying genes. Depending upon the number of genes there can be a variety of mRNA types (1000 to 10,000) in a cell. These differ with respect to the nucleotide sequence and number of base pairs. An mRNA can be monocistronic, i.e. product of one gene (cistron) that singularly codes for one protein molecule. It can also be polycistronic, i.e. product of more than one adjacent cistrons that code for more than one protein molecule.

In prokaryotes, mRNA is transcribed and translated simultaneously in one cellular compartment. Many ribosomes attach on the bacterial mRNA (polysome) during the transcription while mRNA is still attached to DNA template. The life of bacterial mRNA is very short and it is translated within few minutes. Thus, the cells do not contain much of mRNA because it is rapidly degraded by riobnucleases. The average half-life of an *E. coli* mRNA is only two minutes.

mRNA is synthesized by transcription of a DNA template through the action of RNA polymerase in the 5′ to 3′ direction. The enzyme gets attached to the initiation site and uncoils a short section of double helix. As the enzyme moves down the template, there is addition of ribonucleotides (A, U, G and C) based on complementarity with DNA template which results in formation of mRNA strand. The enzyme gets stuck at the termination site and the nascent mRNA is removed from the DNA template. This newly constructed mRNA strand usually undergoes post-transcriptional modification that makes it functionally stable. The processing is very little in case of prokaryotes and as such the transcription and translation are almost simultaneous.

On the other hand, eukaryotic mRNA has to traverse from nucleus to cytoplasm due to cellular compartmentalization. It is therefore much more stable than prokaryotic mRNA. For these reasons, the eukaryotic mRNA molecule undergoes a great deal of post-transcriptional modification leading to significant alteration and reduction in size. The nascent or new mRNAs containing pre mRNA transcripts are referred as heterogenous nuclear RNA or hn RNA. These undergo capping at

5′end, adenylation at 3′end and methylation (**Table 16.2**). Firstly, its 5′end is modified by addition of a cap replacing the triphosphate of the original transcript with a nucleotide (guanine) in reverse orientation, thus sealing the end. This is referred as capping which is understood to be essential in the formation of mRNA-ribosome complex. Secondly, the 3′ end is modified by addition of many adenine nucleotides in presence of poly (A) synthetase enzyme, leading to formation of a **poly A** tail (**Figure 16.1**). Evidence suggest that polyadenylated mRNA are resistant for degradation than otherwise. Finally, some of these adenylate residues are methylated along with two or three nucleotides at 3′ end. Hence, any eukaryotic mRNA exhibits typical structural features with a methylated 5′ cap and polyadenylated 3′ tail. The synthesis and maturation of mRNA occurs in the nucleus and then it moves in to cytoplasm for translation. Methylated caps help mRNA bind to the ribosome and hence enhance the rate of protein synthesis. The cap is followed by a non-coding region (NC1, 10–100 nucleotide long) rich in A and U residues. NC1 is followed by the coding region starting with initiation codon AUG and averages up to 1500 nucleotides in length. The coding region is translated into protein molecule. This is flanked by termination or non-sense codons UAA, UAG or UGA. These codons do not code for any amino acid and bring translation to a halt. Termination codons are followed by another non-coding region (NC2, 50–150 nucleotides long) and a polyadenylated tail or poly (A) tail.

Table 16.2 Some prominent differences among prokaryotic and eukaryotic mRNA

S.No.	Prokaryotic mRNA	Eukaryotic mRNA
(i)	Mostly polycistroninc, i.e. contain transcripts for more than one gene (cistron) and hence contain many sites for initiation and termination of polypeptide synthesis.	Mostly monocistroninc, i.e. contain transcripts for one gene.
(ii)	In most bacteria the transcription and translation are simultaneous.	There are cellular compartments in the eukaryotes and after transcription in nucleus, mRNA moves into cytoplasm through pores of nuclear membrane, followed by its translation in cytoplasm.
(iii)	None or very little post-transcriptional modification.	Significantly altered by post-transcriptional modification that brings about reduction in size, capping, polyadenylation and methylation.
(iv)	Short-lived and metabolically unstable.	Have longer life and are metabolically stable.

TRANSFER RNA (tRNA)

These are relatively smaller and have two important functions, viz. (i) They code, carry and form covalent attachments to respective amino acids and (ii) their specific

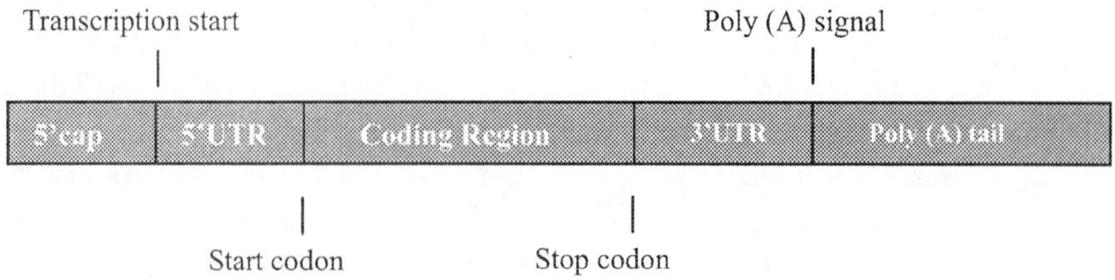

Figure 16.1: Structure of mRNA.

anticodon binds at the correct codon of the mRNA template, leading to correct insertion of amino acids into the elongating polypeptide chain. It is the second most abundant RNA of the cell after rRNA making about 10–20%. Also called as soluble RNA because of its smaller size (25,000–30,000; 3.8S, less than 100 nucleotides), it remains dissolved and cannot be precipitated out by centrifugation. Like other two RNAs, tRNA is synthesized in the nucleus on a DNA template with only 0.25% of DNA involved as a template for tRNA. tRNA molecules are transcribed, in eukaryotic cells, by RNA polymerase III, unlike mRNA which is transcribed by RNA polymerase II. A characteristic feature of tRNA is that three ribonucleotides, (–CCA) are added to its 3′ end after its separation from DNA template. The 5′ end of the tRNA always ends with either of the purines, G or C. The function of the tRNA is to carry amino acid from the cytoplasm to the protein synthesis station of ribosome and incorporate the correct amino acid coded by mRNA to the elongating polypeptide. Each amino acid is carried by a specific tRNA (as defined by genetic code) and the insertion of amino acid is based upon the pairing of codon of mRNA and anticodon of tRNA. Since there are 20 standard amino acids, there must be at least 20 specific tRNAs. However, there exist sixty types of tRNA since the genetic code is degenerate – that is, it contains more than one codon that specify the same amino acid indicating that multiple types of tRNA molecules bearing different anticodons may carry the same amino acid.

STRUCTURE OF tRNA

Holley and coworkers (1969) were the first to elucidate the primary structure of yeast alanine tRNA. Holley also proposed a secondary structure of tRNA based on intramolecular H-bonding. This clover leaf model is the most widely accepted model since it provides a most plausible explanation of tRNA functions (**Figure 16.2 a**). tRNA has primary structure (the linear sequence of nucleotides from 5' to 3'), secondary structure (usually visualized as the cloverleaf structure), and tertiary structure which is a L-shaped 3D structure that accommodates it into P and A sites of the ribosome. The secondary and tertiary structures have been elucidated on the basis of X-ray diffraction pattern of tRNA.

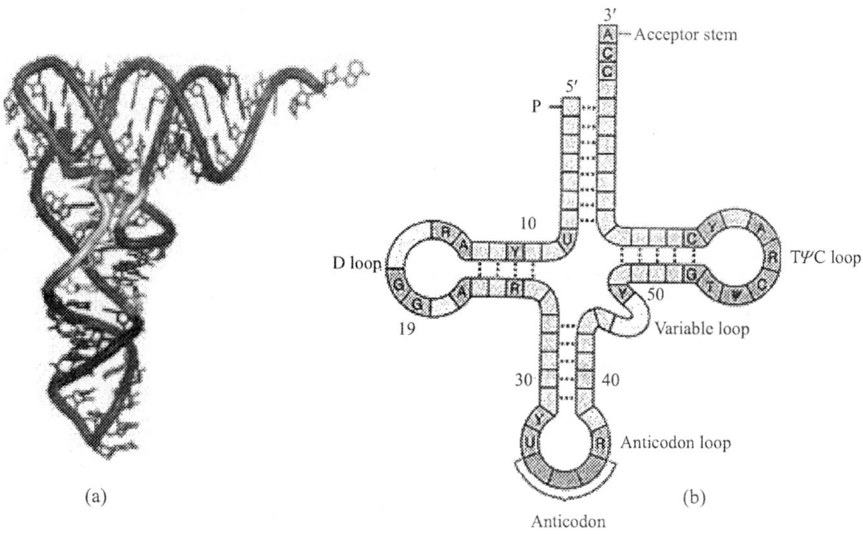

Figure 16.2: (**a**) Clover leave model of tRNA; (**b**) tRNA schematic showing various arms.

The ribonucleotide sequence enfolds and undergoes extensive intramolecular H-bonding to form a five armed structure, similar to the clover leaf. As a result the 3' and 5' ends of the chain are juxtaposed and five different arms (viz. acceptor arm, D arm, anticodon arm, variable arm and TΨC arm) stretch out in different directions. H-bonding among the complementary base pairs (A = U; G ≡ C) is seen throughout the different stems, excepting the loops. In addition to these bases, tRNA contains some 15–20% of unusual bases, e.g. methylcytosine, methylguanine, pseudouridine (Ψ), dihydroxyuridine (DHU), inosine, etc. It is believed that these unusual bases are modified after the tRNA molecule is transcribed. These modifications protect tRNA from ribonucleases in the cytoplasm.

The tRNA molecule enfolds to acquire a unique clover leaf shape. The two ends of the tRNA polypeptide come together to make an **amino acid acceptor arm**. It is a 7-bp stem that includes the 5′-terminal nucleotide with a 5′ terminal phosphate group and the 3′-terminal nucleotide with the 3′-terminal OH group which is used to attach the amino acid (**Figure 16.2 b**). The bases in the acceptor arm pair according to complementarity.

The 3′ end is longer with a CCA tail which is important for the recognition of tRNA during translation. The prokaryotic tRNA has the CCA sequence transcribed from parent DNA template itself while in eukaryotes, the CCA sequence is added during post-transcriptional modification. The D arm is a 4 bp stem ending in a loop containing dihydroxyuridine (D-loop). The anticodon arm is a 5-bp stem whose loop contains the anticodon coding for specific amino acid. The T arm is a 5 bp stem containing *TΨC* loop where *Ψ* is an unusual base, pseudouridine. tRNA is unique among all the three RNAs in having typical five arm structure and modified nucleotide bases occurring at several positions.

An anticodon on tRNA molecule is a base triplet unit made up of three nucleotides that are complementary to the three bases of the codon on the mRNA. Each tRNA contains specific anticodon triplet sequences that can base-pair to one or more codons for a specific amino acid. For example, a codon for glycine is GGU and accordingly the anticodon on mRNA must be CCA. However, some anticodons can pair with more than one codon due to a phenomenon known as **wobble base pairing**. According to the wobble hypotheis, the pairing of the third base of a codon is less stringent than for first two bases. This allows one type of tRNA to base pair with more than one codon. Particularly, inosine and pseudouridine bases of tRNA can hydrogen bond to more than one base in the corresponding codon position. In the genetic code, it is common for a single amino acid to occupy all four third-position possibilities; for example, the amino acid glycine is coded for by the codon sequences GGU, GGC, GGA, and GGG. To provide a one-to-one correspondence between tRNA molecules and codons that specify amino acids, 61 tRNA molecules would be required per cell. However, many cells contain fewer than 61 types of tRNAs because the wobble base is capable of binding to several, though not necessarily all, of the codons that specify a particular amino acid.

TRANSCRIPTION (RNA SYNTHESIS)

Transcription or RNA synthesis is the transfer of information in form of codons from a double-stranded template DNA molecule to single-stranded RNA molecule. Transcription is a key phenomenon of central dogma which enables the expression of genetic information coded in the DNA. Genes include sequence of base pairs that are transcribed and thus transcription process is also called as **gene expression**. Genes that code for an mRNA, and hence for a protein are called structural genes. RNA synthesis occurs by a mechanism that is similar to DNA synthesis except that (1) the precursors are **ribonucleotides** rather than deoxyribonucleotides, (2) only one strand of DNA is used as a template for the synthesis of the complementary RNA and (3) synthesis of RNA does not require any primer. The RNA molecule produced is complementary to the DNA **template strand** and identical (to the non-template strand), except that uracil replaces thymine. If the RNA molecule is an mRNA, it will specify amino acids in the form of triplet codons. They are also called **sense strands** because their nucleotide sequences "make sense" as they can be translated into sequences of amino acids or proteins. The synthesis of RNA chains, like DNA chains, occur in the 5' → 3' direction, with the addition of ribonucleotides to the 3'-hydroxyl group at the end of the chain. The reaction involves a nucleophilic attack by the 3'-OH on the phosphorus atom of the ribonucleoside triphosphate precursor with the elimination of pyrophosphate. This reaction is catalyzed by enzymes called RNA polymerases. RNA polymerases initiate transcription at specific nucleotide sequences called promoters. A single type of RNA polymerase catalyzes all the transcription in prokaryotes. Three different RNA polymerases are present in eukaryotes and different polymerase is responsible for the synthesis of a different class of RNAs. RNA synthesis takes place inside a locally unwound segment of DNA, or transcription bubble.

TRANSCRIPTION IN PROKARYOTES

As discussed earlier, the vital features of transcription are similar in both prokaryotes and eukaryotes, but many of the fine details such as the promoter sequences and other factors are different. A segment of DNA that is transcribed to produce one RNA molecule is referred as a transcript unit. Transcription units may be comparable to individual genes (monocistronic), or they may include several contiguous genes (polycistronic). Large transcripts that carry the coding sequences of several genes (polycistronic) are common in bacteria. RNA synthesis always occurs in the 5′ to 3′ direction. The process of transcription can be divided into three stages (**Figure 16.4**):

1. Initiation
2. Elongation
3. Termination of transcription and release of the nascent RNA molecule.

RNA Polymerases: The RNA polymerase that catalyzes transcription in prokaryotes is a complex, multimeric protein. The *E. coli* RNA polymerase has a molecular weight of an about 480,000 and consists of five peptides. Two of these are identical and thus the enzyme consists of four distinct polypeptides. The complete RNA polymerase molecule, the **holoenzyme,** has the composition $\alpha_2\ \beta\ \beta'\ \sigma$. The two α subunits play a role in the aggregation of the **tetrameric core** ($\alpha_2\beta\ \beta'$) of RNA polymerase. The β subunit contains the ribonucleoside triphosphate binding site and the β' subunit bears the DNA template binding region.

The **sigma (σ) factor** is involved only in the initiation of transcription; it plays no role afterwards and gets released after initiating transcription. Chain elongation is catalyzed by core enzyme ($\alpha_2\beta\ \beta'$). The function of sigma is to recognize the promoter sites in DNA and bind RNA polymerase to it. The core enzyme alone can also catalyze RNA synthesis but the transcription will occur at unspecified place.

1. Initiation: The initiation of transcription involves following three steps: (1) the RNA polymerase holoenzyme binds to a promoter region in DNA; (2) the localized uncoiling of the two strands of DNA, and (3) the formation of phoshodiester bonds between the first few nucleotides in the newly formed RNA chain. The holoenzyme remains bound at the promoter region during the synthesis of the first eight or nine bonds then the sigma factor is released. During initiation, only short chains of the two to nine ribonucleotides are synthesized and released. This abortive synthesis stops once chains of ten or more ribonucleotides have been synthesized and RNA polymerase has begun to move downstream from the promoter.

By convention, the nucleotide pairs or nucleotides within and adjacent to transcription units are numbered with reference to initiation site. The nucleotide pair corresponding to the first (5′) nucleotide of the RNA transcript is designated +1. Base pairs earlier to the initiation site (upstream sequences) are given minus (–) sign and those ahead of the initiation site (downstream sequences) are given plus (+) sign.

Many *E. coli* promoters have been sequenced and found to be much different. However, two small sequences of these promoters are sufficiently conserved to be recognized. The midpoints of these two conserved sequences occur at about 10 and 35 nucleotide pairs, respectively, ahead of the transcription-initiation site and hence they are called the **–10 sequence** and the **–35 sequence**, respectively. Such highly conserved sequences most often are called as **consensus sequences.** The –10 consensus sequence in the non-template strand is **TATAAT** and the –35 consensus sequence is **TTGACA**. At first, the sigma (σ) unit initially recognizes and binds to the –35 sequence, called as the **recognition sequence (Figure 16.3).** The A:T-rich –10 sequences help the localized unwinding of DNA, which is a primary requirement for the synthesis of a new RNA chain. The distance between the –35 and –10 sequences is also found to be highly conserved in *E. coli* promoters, and is in between 15 and 20 pairs in length.

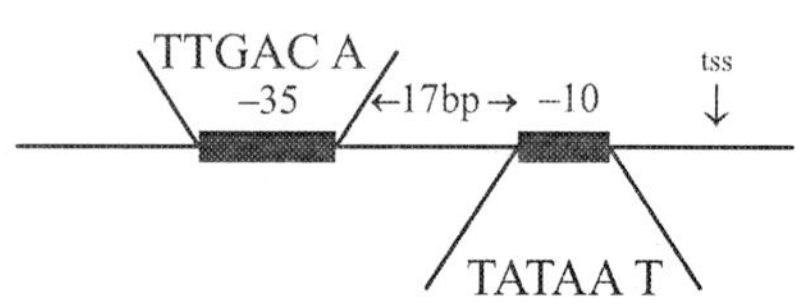

Figure 16.3: Recognition sequences in *E. coli* promoters.

2. Elongation: After the release of the σ subunit elongation of RNA is catalyzed by RNA polymerase core enzyme. The extension of RNA chain takes place within the transcription bubble. The RNA polymerase molecule contains both DNA coiling and DNA rewinding activity and unbinds the DNA double helix ahead of the polymerization site and reassociates the complementary DNA strands behind the polymerization site. In prokaryotes, the average length of a transcription bubble is 18 nucleotide base pairs, and about 40 ribonucleotides are added to the elongating RNA chain every second. The nascent RNA chain is displaced from the DNA template strand as RNA polymerase traverses along the DNA.

3. Termination: Termination of transcription occurs when RNA polymerase gets stuck at **termination sequences** and the transcription complex dissociates, releasing the nascent RNA molecule. Two types of transcription terminators are found in *E. coli*. Those work in presence of a protein factor called **rho (ρ)** are

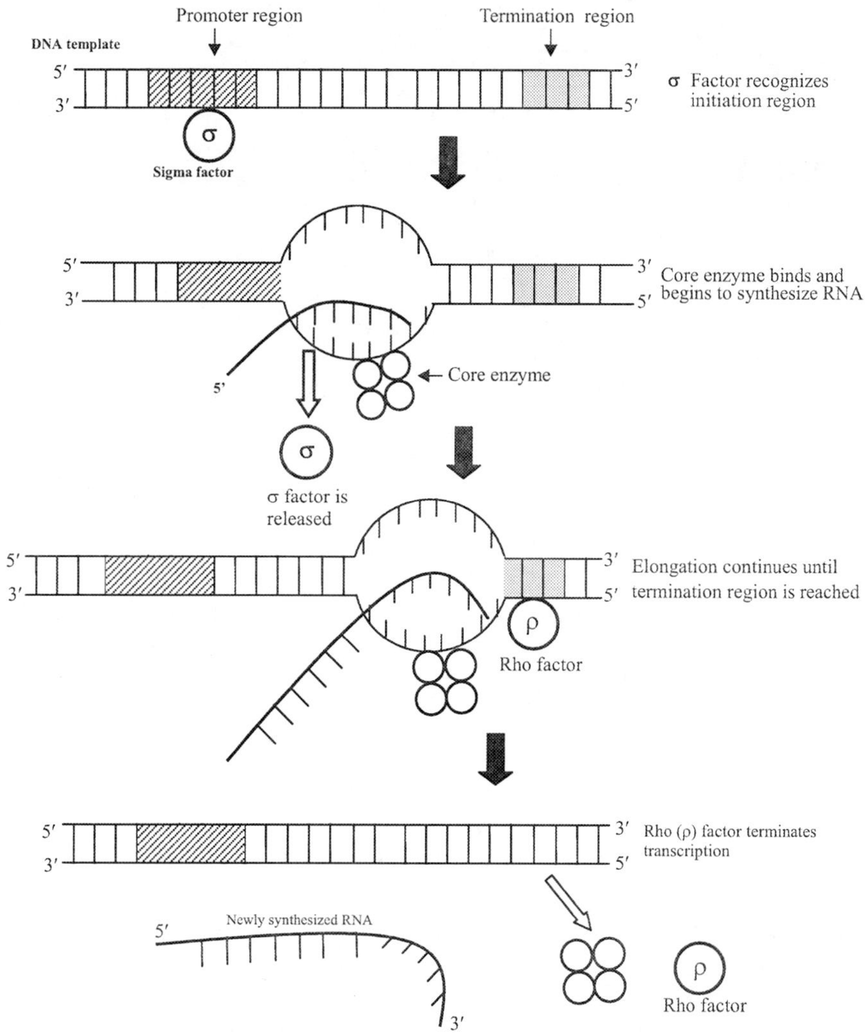

Figure 16.4: General transcription in prokaryotes.

called as **rho-dependent terminators**. The other types of sequences result in the termination of transcription without any involvement of rho; such sequences are called **rho-independent terminators**.

Rho-independent terminators are marked by presence of G:C-rich region followed by six or more A:T base-pairs, with the A's present in the template strand. The nucleotide sequence of the G:C-rich region is such that regions of the single-stranded RNA can base-pair and acquire a hairpin like structure. The RNA hairpin structure is formed immediately after the synthesis of the participating regions of the RNA chains. This structure retards the movement of RNA polymerase along

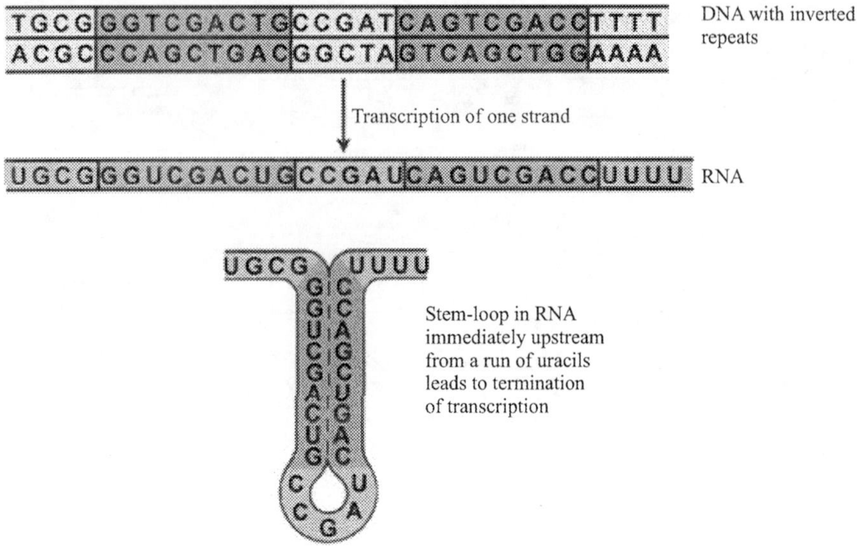

Figure 16.5: Sequence of a rho-independent transcription terminator. The sequence contains G:C rich regions followed by A: T base pairs. Transcription of such DNA segment leads to formation of an RNA with high G:C rich RNA chain, which base pair to form a hairpin structure. This structure stops movement of RNA polymerase along the DNA molecule resulting in termination of transcription and release of nascent RNA molecule.

the DNA, causing halt in chain elongation. A series of U's flanking the hairpin region help in the release of the newly synthesized RNA chains from the DNA template since U:A bonding is much weaker (**Figure 16.5**).

Rho-dependent termination sequences are 50 to 90 base pair long and specify RNA transcript that are rich in C residue and largely devoid of G's. The rho protein binds to the elongating RNA chain and moves from 5′ to 3′ terminus along the mRNA, following the RNA polymerase molecule. As and when RNA polymerase halts at the rho-dependent termination sequence, **rho** binds up with the polymerase and pulls the nascent RNA chains from the transcription bubble. However, the complete mechanism of rho-dependent termination is still unknown.

TRANSCRIPTION IN EUKARYOTES

Although the overall process of transcription seems to be alike in prokaryotes and eukaryotes, it is highly complex in eukaryotes. In eukaryotes, transcription occurs in the nucleus, and the RNAs that encode proteins are transported to the cytoplasm for translation on ribosomes. Prokaryotic mRNAs are often polycistronic, i.e. they contain the coding regions of two or more genes in tandem. Most eukaryotic mRNAs that have been characterized contain the coding region of a single gene (monogenic) but some multigenic mRNAs are also known to occur. Three different RNA polymerases are present in eukaryotes and each catalyzes the transcription of a specific class of genes. Moreover, eukaryotic, primary transcripts of genes that encode polypeptides undergo three major post-transcriptional modifications before they are transported to the cytoplasm for translation.

(i) Methylation or capping: 7-Methyl guanosine caps are added to the 5'ends of the primary transcripts.

(ii) Addition of poly 'A' tail: Poly (A) tails are added to the 3' ends of the transcripts.

(iii) Splicing: introns if present are spliced out of transcripts.

7-Methyl guanosine cap is added to the initial nucleotide of the transcript by a 5′–5′ phosphate linkage. The 3' poly (A) tail is a polyadenosine series of 20 to 200 nucleotides.

The nascent RNA primary transcripts found in the eukaryotic nucleus are called heterogeneous nuclear RNA (hn RNA). Major parts of these hnRNAs are non-coding intron sequences, which are excised from the primary transcripts and degraded in the nucleus during post-transcriptional modification. Most RNA transcripts get covered with RNA-binding proteins during or immediately after their synthesis in the nucleus for their protection from degradation by ribonucleases. This enhances the average half-life of RNA in eukaryotes to about five hours, in contrast to an average half-life of less than five minutes in *E. coli*.

Eukaryotic RNA Polymerases: We have seen that a single RNA polymerase catalyzes the transcription in *E. coli.* However, in eukaryotes, three different RNA polymerases are present. These are complex proteins, with 10 or more subunits and are designated as RNA polymerase I, II, and III. All these RNA polymerases need the presence of other proteins called **'transcription factors'** in order to initiate the synthesis of RNA chains.

RNA polymerase I is located in the nucleolus, a distinct region of the nucleus where rRNA are synthesized and combined with ribosomal proteins. RNA polymerase I catalyzes the synthesis of all ribosomal RNA except the small 5S rRNA. RNA polymerase II transcribes nuclear genes that encode proteins and possibly other genes specifying hnRNA leading to formation of mRNA. RNA polymerase III catalyzes the synthesis of the transfer RNA molecules, the 5S rRNA molecules, and small nuclear RNAs (sn RNAs).

Three RNA polymerases can also be distinguished by their different sensitivities to inhibition by α-amanitin, a poison produced by the mushroom *Amanita phalloides*. Among these, RNA polymerase I is insensitive to α- amanitin, but RNA polymerase II is inhibited completely at low concentrations of α- amanitin, while RNA polymerase III shows an intermediate level of sensitivity to this drug. α-amanitin has been used to determine which RNA polymerase catalyzes the transcription of a particular gene.

Mechanism of Eukaryotic Transcription

The eukaryotic transcription is also marked by three similar events, i.e. initiation of transcription, elongation of nascent RNA chain and termination of transcription. However, as compared to prokaryotes a great deal of complexity is involved here.

Initiation of RNA Chains: Unlike prokaryotic RNA polymerases, eukaryotic RNA polymerases cannot initiate transcription on their own. All three eukaryotic RNA polymerases require the support of transcription factors to initiate the RNA synthesis. These transcription factors bind to a promoter region in DNA to start initiation of transcription. Different promoters and transcription factors are specified by the RNA polymerases I, II, and III. The initiation of mRNA synthesis by RNA polymerase II, which transcribes the eukaryotic enzymes and functional proteins is described here.

Initiation of transcription involves uncoiling of a segment of DNA, allowing the DNA strand to function as a template for the synthesis of RNA. This uncoiling is brought about by transcription factors with specific sequences in the promoter for the transcription unit. The promoter recognized by RNA polymerase II consists of short conserved sequences, located upstream from the transcription start point. The conserved sequence nearby the transcription start point (position +1) is called the **TATA box**; it has the consensus sequence 5′ TATAAAA 3′ (on the nontemplate strand) and is located at about position -30. The TATA box plays an important role in localizing the transcription start site. Another conserved element is called the **CAAT box**; which usually occurs near position –80 and has the consensus sequence GGCCAATCT. Two other conserved elements, the **GC box,** consensus GGGCGG, and the **octamer box,** are also present in RNA polymerase II promoters. The G:C and octamer box enhance the efficiency of a promoter in initiating transcription.

Initiation of transcription by RNA polymerase II requires several basal **transcription factors**. These must interact with promoters in the correct sequence for successful initiation of transcription. These factors are denoted as **TFIIX** (Transcription factor for polymerase II, where X is a letter identifying the individual factor.

Among these transcription factors, TFIID is the first to interact with the promoter; it contains a TATA-binding protein (TBP) and several small TBN-associated proteins. Following this, TFIIA joins the complex. This is followed by TFIIF and RNA polymerase II joining the transcription initiation complex together. TFIIF contains two subunits, one of which has DNA-unwinding activity. TFIIF catalyzes the localized unwinding of the DNA double helix. TFIIE then joins the initiation complex, binding to the DNA downstream from the transcription start point. The role of two other factors, TFIIH and TFIIJ, which join the complex after TFIIE, is still unknown (**Figure 16.6**).

In case of RNA polymerase I and III, similar transcription factors and promoter complexes are required to initiate transcription. However, the promoters of genes

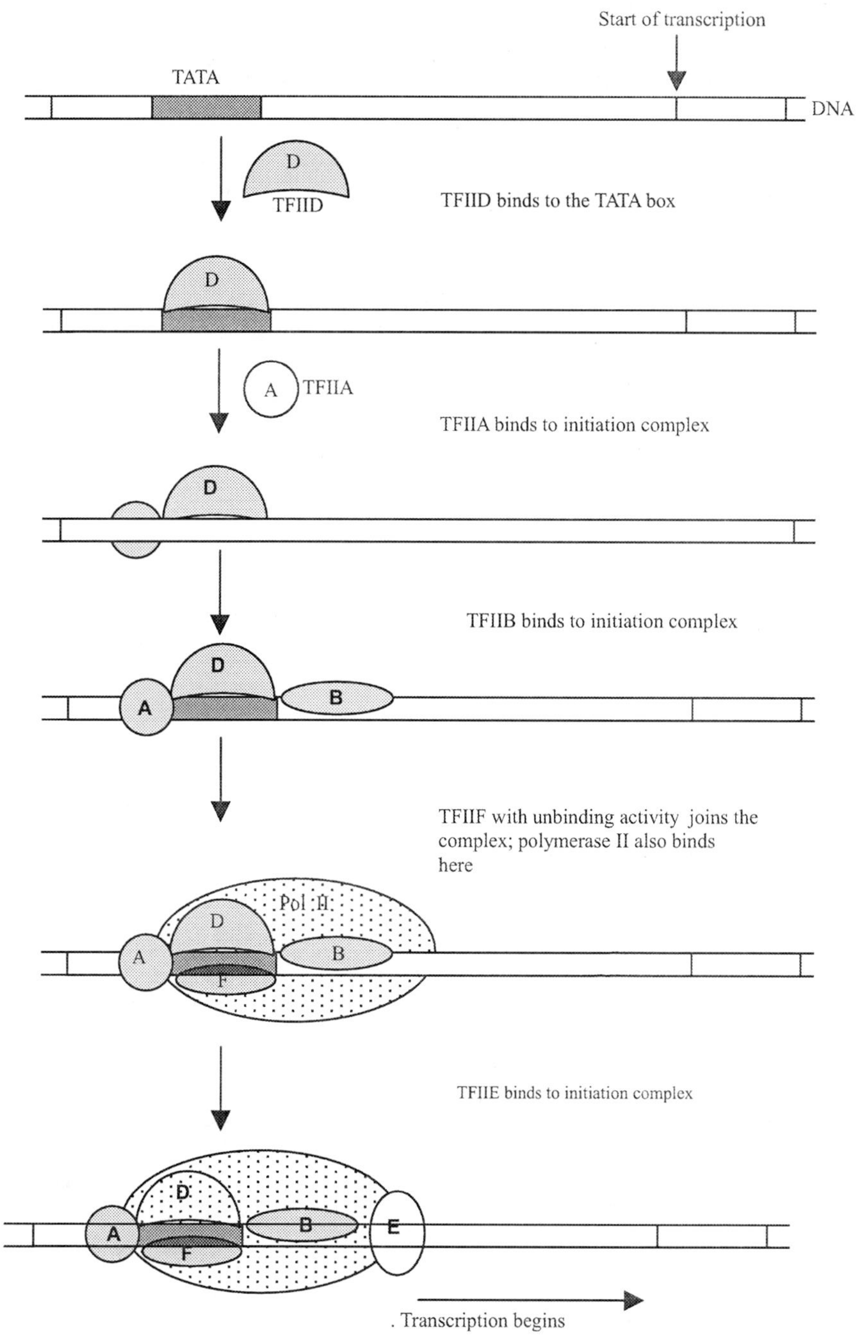

Figure 16.6: Initiation of transcription in eukaryotes.

transcribed by RNA polymerase I and III are different from the promoters utilized by RNA polymerase II.

RNA Chain Elongation and 5' Methyl Guanosine Capping: Once eukaryotic RNA polymerases move ahead from their initiation complex; they catalyze RNA chain elongation by the same mechanism as the RNA polymerase of prokaryotes.

Initially, early in the elongation process, the 5' ends of eukaryotic pre-mRNAs are modified by the addition of 7-methyl guanosine (7-MG) caps. These 7-MG caps are added when the growing RNA chains are only about 30 nucleotides long. The 7-MG cap contains an unusual 5'–5' triphosphate linkage and two or more methyl groups. The 7-MG caps are recognized by protein factors involved in the initiation and these also protect the growing RNA chains from degradation by RNAases (ribonucleases).

Termination and Addition of 3'Poly (A) Tail: The 3' end of nascent RNA transcribed by RNA polymerase II is produced by cleavage of the primary transcripts by endonucleases rather than by the termination of transcription. The actual termination event often occurs at multiple sites that are located 1000 to 2000 nucleotides downstream from the site that will become the 3' end of the mature transcript. Hence, transcription proceeds beyond the site that will become the 3'

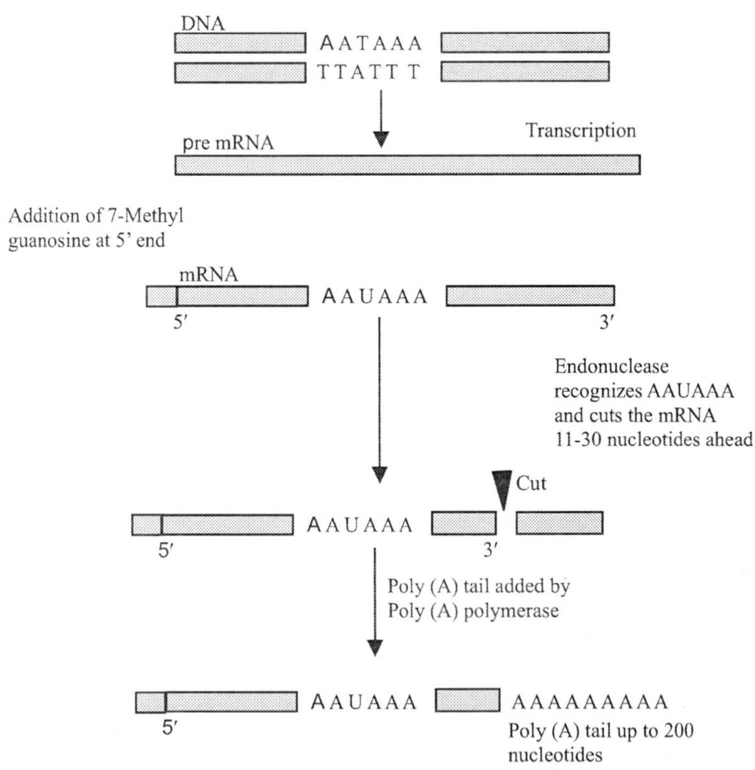

Figure 16.7: Termination of eukaryotic transcription and addition of poly (A) tail.

terminus and extra distal segment is removed by endonucleolytic cleavage. The cleavage event that produces the 3' end of a mRNA transcript occurs at site 11 to 30 nucleotides downstream from a sequence, AAUAAA. After cleavage, the enzyme **poly (A) polymerase** carries out addition of poly (A) tail about 200 nucleotide lengths, to the 3′ ends of the transcripts. The addition of a poly (A) tail to eukaryotic mRNA is called **polyadenylation.** This poly (A) tail of eukaryotic mRNA enhances its stability and plays an important role in its transport from the nucleus to the cytoplasm (**Figure 16.7**).

SUGGESTED READING

Brawerman G. 1974. *Eukaryotic messenger RNA.* Ann. Rev. Biochem., 43: 621.

Darnel J.E. Jr 1985. **RNA.** Scientific American 197 (3).

Rich A. and Kim H. 1978. *The three-dimensional structure of transfer RNA.* Scientific American .

Schimmel P.R.D., Soll D. and Abelson J.N. 1979. *Transfer RNA: Strucuture, properties and Recognition.* Cold Spring Harbour, New York.

Singer M.F. and Leder P. 1966. *Messenger RNA: An evolution.* Ann. Rev. Biochem., 35: 195–230.

REVIEW QUESTIONS

1. Give a detailed account of structure and importance of RNA.
2. Discuss types and structure of RNA.
3. Discuss the organization of prokaryotic and eukaryotic ribosomal RNA (rRNA).
4. Compare between prokaryotic and eukaryotic mRNA.
5. Give the structure, properties and role of tRNA in transcription.
6. What is transcription? Discuss the mechanism of transcription in prokaryotes and eukaryotes.
7. Describe the post-transcriptional modification of RNA.

UNIT XVII

Genetic Code and Protein Synthesis

The DNA sequence of a gene is co-linear with the amino acid sequence of the polypeptide it encodes or the nucleotide sequence of the coding DNA strand in the 5'–3' direction specifies the same order of the amino acids in the polypeptide from N-terminal to C-terminal. The sequence of DNA is transcribed into mRNA which is then translated into specific protein as per the sequence. The relationship between the nucleotide sequence of mRNA and the amino acid sequence of the polypeptide is dictated by the code, called as Genetic code **(Table 17. 1)**. mRNA sequences are read in triplets or as groups of three nucleotides each specifying one amino acid. Such triplets are called as codons. Three codons UAG, UGA and UAA do not specify any amino acid and are called as nonsense or terminator or stop codons. Whenever any of these codons is encountered by a RNA polymerase, it leads to termination of the protein synthesis. The initiation of translation in prokaryotes happens with a modified or formylated methionine, i.e. N-formylated methionine. Similarly, the eukaryotic translation begins with methionine. These amino acids are coded by initiator codon AUG which happens to be the first codon to be read by polymerase.

DEGENERACY AND WOBBLE HYPOTHESIS

RNA is made up of four types of nucleotides and therefore there can be $4^3 = 64$ possible different triplets. However, there are only 20 amino acids reported so far in natural proteins. Thus, one amino acid is coded by more than one or several different codons. This is why the genetic code is called to be degenerate. Only two amino acids, methionine and tryptophan are coded by single codons, AUG and UGG. Codons specifying same amino acid are called as synonyms. Most of such synonyms differ in only their third nucleotide of the triplet, e.g. GUU, GUC, GUA and GUG all code for valine. During translation, each codon is bound by a complementary anticodon on the tRNA carrying a specific amino acid. The pairing at the third base pair is less stringent than for the first two bases allowing a single kind of tRNA to pair with more than one anticodon, e.g. phenylalanine tRNA having

Table 17. 1 The genetic code

1st base at 5' end	2nd base: U	C	A	G	3rd base at 3' end
U	Phe	Ser	Tyr	Cys	U
	Phe	Ser	Tyr	Cys	C
	Leu	Ser	**Stop**	**Stop**	A
	Leu	Ser	**Stop**	Trp	G
C	Leu	Pro	His	Arg	U
	Leu	Pro	His	Arg	C
	Leu	Pro	Gln	Arg	A
	Leu	Pro	Gln	Arg	G
A	Ile	Thr	Asn	Ser	U
	Ile	Thr	Asn	Ser	C
	Ile	Thr	Lys	Arg	A
	Met	Thr	Lys	Arg	G
G	Val	Ala	Asp	Gly	U
	Val	Ala	Asp	Gly	C
	Val	Ala	Glu	Gly	A
	Val	Ala	Glu	Gly	G

codon GAA recognizes both UUU and UUC anticodons on mRNA. The genetic code throughout the living kingdom shares several common features but there are some differences too. This is particularly true in case of extra chromosomal genetics, e.g. in mitochondrial mRNAs some codons code for different amino acids. In yeasts and mammals, AUA specifies methionine (Met) rather than isolucine (Ile). However, such cases are limited to translation of few extrachromosomal proteins and hence do not affect much.

The sequence of mRNA molecule is read in form of triplets from the 5' end to 3' end. Depending upon the fact that which nucleotide is considered first nucleotide or the first base by the polymerase, there can be three reading frames possible. The correct reading frame is set by recognition of the initiation codon, AUG, by the ribosome. Conversely, in case of small genomes of some bacteriophages the genes overlap and each has a different reading frame.

PROTEIN SYNTHESIS

Protein synthesis or translation stands for meaningful deciphering of the information transcribed in mRNA molecules into the sequences of amino acids or proteins.

Translation occurs on ribosomes, which are complexes of rRNA and ribosomal proteins scattered in the cytoplasm. Translation involves three types of RNA, namely, mRNA, tRNA and rRNA. rRNA molecules are present as part of the structure of each ribosome, and about 60 small soluble tRNA molecules mediate the integration of the required amino acids into polypeptides in response to specific nucleotide sequences in mRNAs. The amino acids are attached to the correct tRNA molecules as specified by anticodon by a set of activating enzymes called **aminoacyl-tRNA synthetases.**

The nucleotide sequence of an mRNA molecule is translated into amino acid sequence according to the language of the genetic code. Each amino acid is coded by one or more codons, and each codon contains three nucleotides. Of the 64 possible nucleotide triplet, 61 specify amino acids—two of which also function as polypeptide chain initiation signals; and three specify polypeptide chain termination. The tRNA molecules contain one specific nucleotide triplet **anticodons**, which base-pair with the codons in mRNA during the translation process.

Ribosomes are made up protein and rRNA in almost equal proportions. They are composed of two subunits, one large and one small, which separate when the translation of an mRNA molecule is completed and reassociate during the initiation of translation. Ribosome size is expressed as their rates of sedimentation during centrifugation or **Svedberg units**. Francis Crick (1958) proposed that some kind of an adapter molecule must dictate the specification of amino acids by codons in mRNA during protein synthesis. tRNA molecules were first called soluble RNA (sRNA) molecules and subsequently transfer RNA (tRNA) molecules. tRNAs molecules contain a triplet nucleotide sequence, the anticodon, which is complementary to and base-pairs with the specific codon mRNA during translation. There are one to four tRNA for each of the 20 amino acids. The amino acids are attached to the tRNAs by high energy bonds between the carboxyl group of the amino acid and the 3'-hydroxyl termini of the tRNAs by the ATP consuming process called as tRNA activation or charging. The tRNA is activated or charged with amino acid in a two-step process in the presence of aminoacyl-tRNA synthase. There is at least one aminoacyl-tRNA synthase for each of the 20 amino acids. The first step in protein synthesis involves the activation of the amino acid driven by ATP energy:

$$\text{Amino acid} + \text{ATP} \xrightarrow{\text{Aminoacyl-tRNA synthase}} \text{Amino acid-AMP} + \text{P} - \text{P}$$

The amino acid-AMP intermediate then reacts with the appropriate tRNA:

$$\text{Amino acid-AMP} + \text{tRNA} \xrightarrow{\text{Aminoacyl-tRNA synthase}} \text{Amino acyl-tRNA} + \text{AMP}$$

Such aminoacyl-tRNAs are substrates for polypeptide synthesis on ribosomes. Each aminoacyl tRNA recognizes the correct mRNA codon and binds to it with its anticodon and presents the amino acid in a suitable steric configuration that facilitates peptide bond formation. Ribosome has three distinct tRNA binding sites. The ***A*** or **aminoacyl site** binds the new incoming aminoacyl-tRNA. The ***P*** or **peptidyl site** binds the tRNA whose amino acid is being used for elongation and the ***E*** or **exit site** bears the departing uncharged tRNA.

The aminoacyl-tRNA binding sites are located on the larger subunit (50S) and the mRNA molecule is bound by the 30S subunit. The specificity for aminoacyl-tRNA binding in each site is dictated by the mRNA codon that makes up part of the binding site. As the ribosome moves along an mRNA, the specificity for the aminoacyl-tRNA binding A, P, and E sites changes as different mRNA codons move through the ribosome.

The translation of the sequence of nucleotides in an mRNA molecule into the sequence of amino acids in its polypeptide product can be divided into three steps:

1. Chain initiation
2. Chain elongation
3. Chain termination

Chain Initiation

In prokaryotes, the initiation of protein synthesis involves several factors including 30S subunit of the ribosome, an initiator tRNA, an mRNA molecule, three soluble protein **initiation factors: IF-I, IF-2** and **IF-3**, and one molecule of GTP. Translation occurs on a complete unit of 70S ribosome. In the first stage of the initiation, a 30S subunit binds with an mRNA molecule and three initiation factors. This complex is joined by the 50S subunit to form 70S ribosome. The synthesis of polypeptides is initiated by a special tRNA that codes for methionine (**$tRNA_f^{Met}$**) in response to a translation initiation codon (AUG). All prokaryotic translation is initiated with methionine. Methionyl-$tRNA_f^{Met}$ binds to the ribosome in response to AUG initiation codons in mRNAs. Sometimes it may also bind to another initiation codon, GUG which otherwise codes for valine.

Initiation starts with the simultaneous formation of two complexes: one contains initiation factor IF-2 and charged methionyl-$tRNA_f^{Met}$ and the other contains mRNA, a 30S ribosomal subunit and initiation factor IF-3. The initiation factor IF-3 is very crucial as 30S subunit / mRNA complex will form only in its presence. Prokaryotic mRNA contains a conserved polypurine segment, consensus AGGAGG, the **Shine-Dalgarno sequence** located about seven nucleotides upstream from the AUG initiation codon.

The IF-2-methionyl-$tRNA_f^{Met}$ complex and the mRNA-30S subunit-IF-3 complex ultimately combine together and initiation factor IF-1 and one molecule of GTP join to complete 30S initiation complex. Finally, the 50S subunit binds to the 30S initiation complex to produce the complete 70S ribosome. Initiation factor IF-3 must be released from the complex before the 50S subunit can join the complex.

The addition of the 50S subunit is coupled with consumption of energy from GTP and the release of the two initiation factors, IF-1 and IF-2 **(Figure 17.1)**.

The addition of the 50S ribosomal subunit to the complex puts the methionyl-tRNA$_f^{Met}$ in the peptidyl (P) site with the anticodon of the tRNA aligned with the AUG initiation codon of the mRNA. Methionyl-tRNA$_f^{Met}$ is the only aminoacyl-tRNA that can enter the P site directly, without passing through the aminoacyl (A) site. The initiator AUG is at P site and the next codon of the mRNA is in queue at the **A** site. Next aminoacyl-tRNA will bind at this site accordingly.

The initiation is more complex in eukaryotes and involves several soluble initiation factors. The overall process has two distinguishing features: firstly, the amino group of the methionine on the initiator tRNA is not formylated as in the prokaryotes. Also the initiation complex in eukaryotes is formed at the 5' terminus of the mRNA contrasting to the binding at Shine-Dalgarno sequence in prokaryotes. In eukaryotes, the initiation complex moves through the mRNA, starting at the 5' end, searching for an AUG initiation codon. Therefore, translation begins at the AUG nearest to the 5' terminus of the mRNA molecule. However, eukaryotes have a special initiator tRNA, **tRNA$_i^{Met}$** ('i' for initiator) contrasting to the tRNA$_f^{met}$ of prokaryotes. The initiator methionyl-tRNA$_i^{Met}$ cooperates with a soluble initiation factor and enters the P site directly through the initiation process.

After this, a cap-binding protein (CBP) binds to the 7-methyl guanosine cap at the 5' end of the mRNA. This is followed by binding of other initiation factors to the CBP-mRNA complex. Finally, small (40S) subunit joins the complex. The entire initiation complex moves 5'→3' along the mRNA molecule, searching for an AUG codon. When an AUG triplet is found, the initiation factors dissociate from the complex, and the large (60S) subunit binds to the tRNAi^{met} – mRNA – 40S subunit complex, forming the complete (80S) ribosome. This complex now takes up polypeptide chain elongation.

Chain Elongation

The process of polypeptide chain elongation is almost similar in both prokaryotes and eukaryotes. The addition of new amino acids to the growing polypeptide occurs in the three steps: (i) binding of the specified aminoacyl-tRNA to the A site of the ribosome, (ii) shifting of nascent polypeptide from tRNA in the *P* site to the tRNA in the *A* site by the formation of a new peptide bond, and (iii) movement of the ribosome along the mRNA to put the next mRNA codon in the A site. The polypeptide-tRNA and the discharged tRNA are shifted from the A and P sites to the P and *E* site, respectively. The above three steps are repeated throughout the polypeptide elongation.

At the onset of elongation, an aminoacyl-tRNA as specified by mRNA binds to the *A* site of the ribosome. The three nucleotides in the anticodon of the incoming aminoacyl-tRNA pair with the nucleotides of the mRNA codon present at the A site. This binding requires **elongation factor Tu** carrying a molecule of GTP (**EF-**

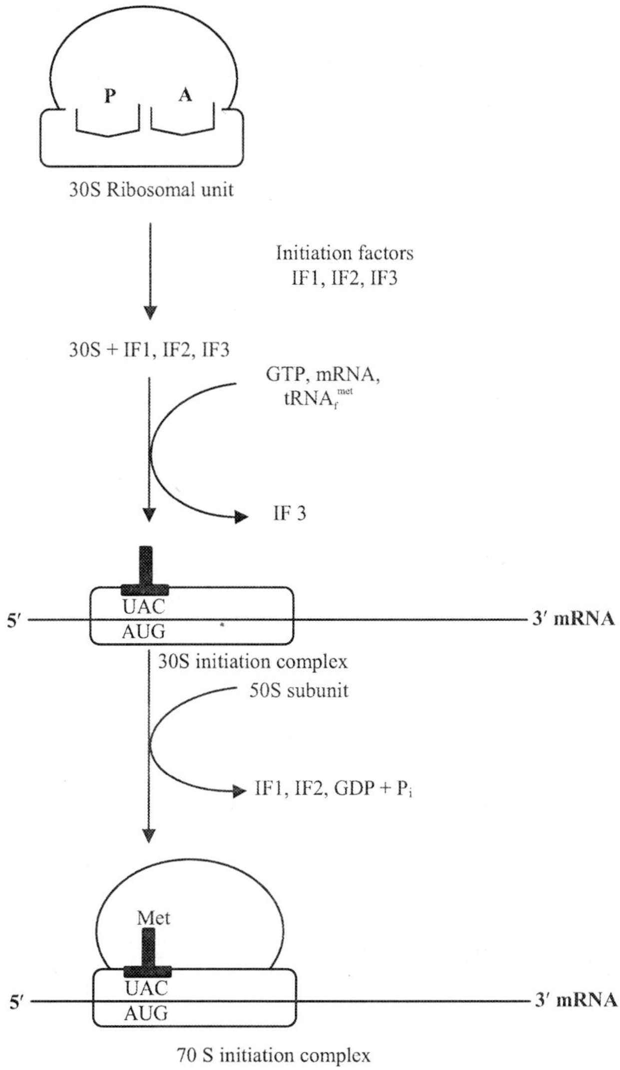

Figure 17.1: Initiation of protein synthesis in prokaryotes.

Tu-GTP). This is followed by the formation of a peptide bond between the amino group of the aminoacyl-tRNA at the *A* site and carboxyl terminus of the growing polypeptide chain attached to the tRNA at the *P* site. This removes the growing chain from the tRNA in the P site and covalently binds the chain to the tRNA at the *A* site. This reaction occurs in the presence of an enzyme, **peptidyl transferase**, found associated with the 50S subunit of the ribosome.

In the third and the last step of chain elongation, the peptidyl-tRNA present in the *A* site of the ribosome is moved to the *P* site, and the uncharged tRNA in the P

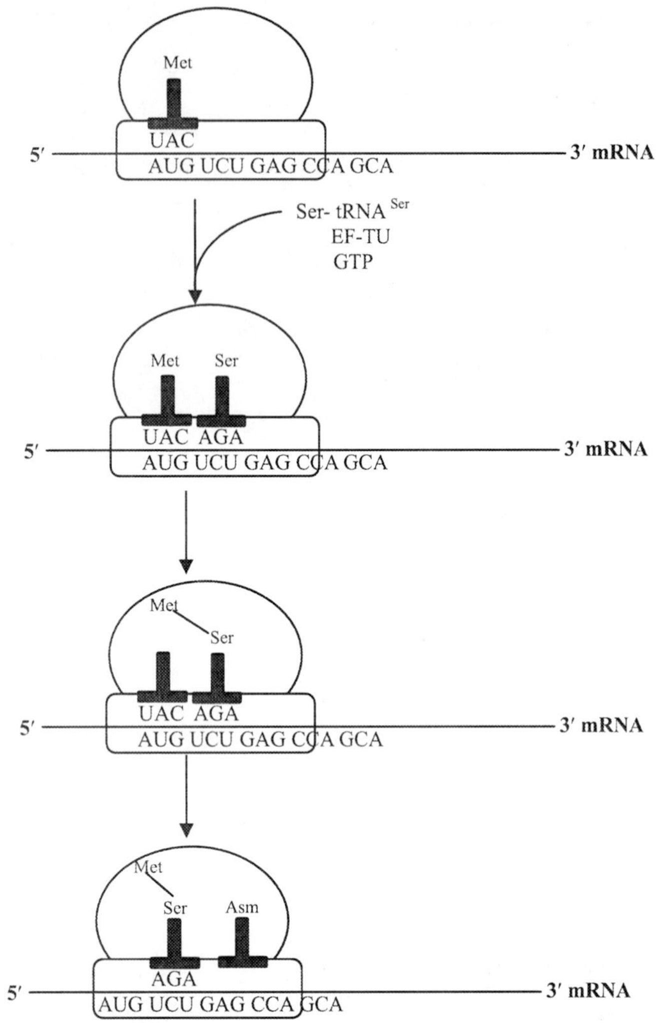

Figure 17.2: Elongation of peptide during protein synthesis.

site is moved further. This happens because the ribosome moves three nucleotides toward the 3' end of the mRNA molecule. The movement requires GTP energy and **elongation factor G (EF-G)**. This movement of the peptidyl-tRNA from the A site to the P site leaves the A site empty thus allowing the ribosomal ribosome platform to take up new charged tRNA and start another cycle of elongation (**Figure 17.2**).

Chain Termination

Polypeptide chain elongation is arrested when the ribosome encounters any of three chain-termination or non-sense codons (UAA, UAG or UGA). These three stop codons are recognized by soluble proteins, also called **release factors (RFs)**. Among the two release factors, RF-1 recognizes termination codon UAA and UAG, where as RF-2 recognizes UAA and UGA. Contrastingly, in eukaryotes, only one release factor **(eRF)** can identify all the three termination codons. The presence of a release factor in the A site alters the activity of peptidyl transferase. The enzyme then adds a water molecule to the carboxyl terminus of the nascent polypeptide. This leads to the release of the polypeptide from the tRNA molecule in the P site **(Figure 17.3)**.

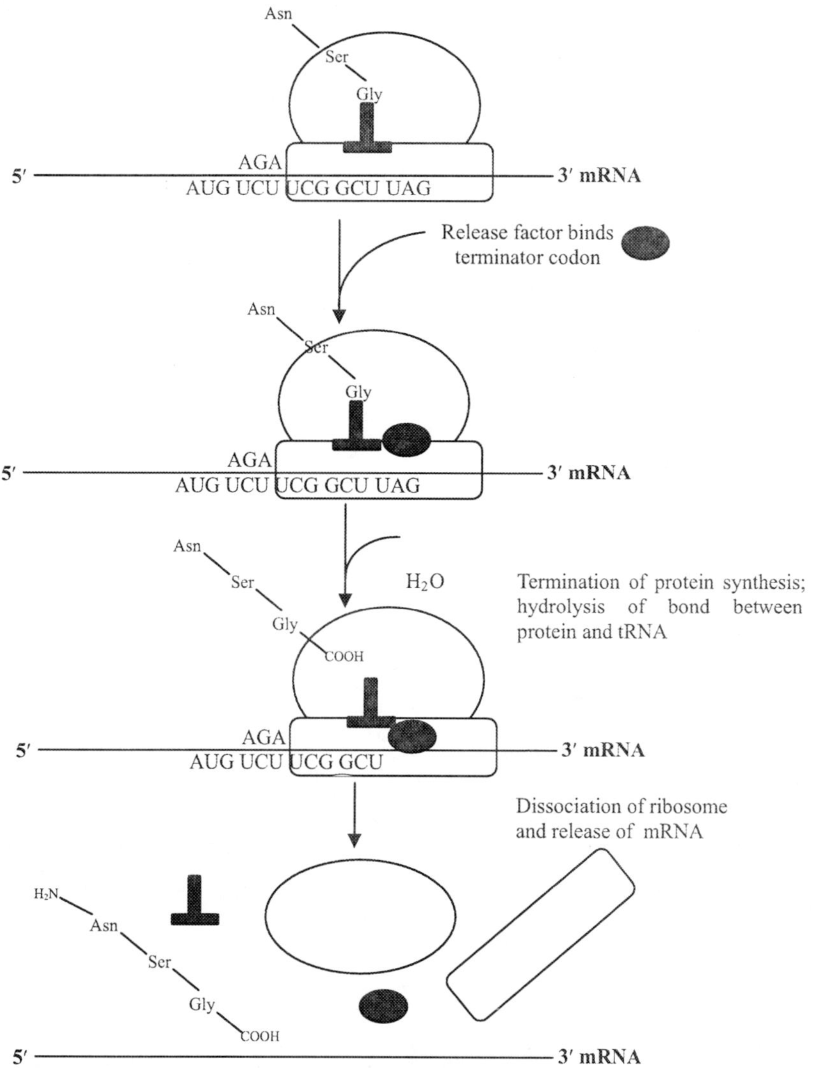

Figure 17.3: Termination of protein synthesis.

Chain termination should also be understood to be an equally important step in protein synthesis as any undesirable addition to the polypeptide will also lead to production of abnormal protein.

SUGGESTED READING

Chambliss G. 1980. *Ribosomes: Structure, Function and Genetics*. University Park Press, Baltimore.

Losick R. and Chamberlin M. 1976. *RNA Polymerases*. Cold Spring Harbour, New York.

Merrick W.C. 1992. *Mechanism and regulation of eukaryotic protein synthesis*. Enzyme. 44 (1–4): 7–16.

Zamecnik P. 1984. *The machinery of protein synthesis*. Trends Biochem. Sci. 9:464–466.

REVIEW QUESTIONS

1. What is meant by genetic code? Explain wobble hypothesis and degeneracy of the genetic code.
2. Discuss the important features of genetic code and give details of initiation and stop codons.
3. Describe various steps (initiation, elongation and termination) involved in protein synthesis.
4. Write a note on following:
   a) Triplet codons
   b) Aminoacyl-tRNAs
   c) Role of ribosome in protein synthesis
   d) RNA polymerase

UNIT XVIII

Gene Regulation

The operon model was proposed in 1961 by Jacob and Monod to elucidate the regulation of genes compulsory for lactose utilization in *E. coli*. Later in 1965, these two scientists were awarded Nobel Prize for discovering that the transcription of a set of contiguous (arranged one after the other) structural genes is regulated by two controlling elements. One of the elements, the repressor gene, encodes a repressor, which binds to the operator. The operator is always contiguous with the structural genes whose expression it regulates.

Transcription is initiated at promoters located just upstream (5') from the coding regions of structural genes. When repressor is bound to the operator, RNA polymerase cannot bind the operator due to physical hindrance and thus prevents transcription of the structural genes in operon. Operator is usually located between the promoter and the structural genes that they regulate. Such a complete unit, including the operator, the promoter and structural genes is referred as an operon.

Whether the repressor will bind to the operator and turn off the transcription of the structural genes in an operon is determined by the presence or absence of the effector molecule. Inducible operons and repressible operons can be distinguished from each other on this basis. In case of an inducible operon, the free repressor binds to the operator, turning off the transcription. In case of a repressible operon, the position is reversed. The free repressor cannot bind to the operator. Only the repressor-effector molecule (co-repressor) complex is effective in binding to the operator.

The mRNAs consisting of more than one structural gene are polycistronic or multigenic. For example, the tryptophan operon (trp operon) in *E. coli* contains the coding sequences of five different genes and the resulting mRNA is polycistronic. The concentration of gene products may differ depending upon the relative efficiency of initiation of translation. In some cases, the level of transcription signals can alter the amounts of gene product synthesized.

A cell has to spend a lot of energy in producing structural or functional proteins from amino acids for its normal functioning. The expenditure of energy and basic raw materials (monomers) is done in such a way so that only proteins and macromolecules desirable at that point of time are built. This is simply to avoid any loss of energy and chemical resources. This is true for most bioactive molecules including hormones, enzymes and structural components. Most such regulatory mechanisms take place at the transcriptional level. Enzyme synthesis, particularly that of (inducible and repressible enzymes) presents a good example of expression and regulation of genes as per need of the cell. We know that formation of protein is directed by the central dogma (DNA $\rightarrow$ RNA $\rightarrow$ Protein). We will see in this chapter how this regulation is governed by presence or absence of inducer / repressor. The understanding of this phenomenon is necessary from point of view of industrial microbiology so as to study the effect of various factors on production of any microbial product and than use this knowledge to achieve maximum productivity.

POSITIVE AND NEGATIVE CONTROL OF GENE EXPRESSION

The regulation of any gene is achieved either by *induction*, or turning genes on, and *repression*, or turning genes off, which represent positive control mechanisms and negative control mechanisms, respectively. This is achieved through functions of regulator genes that encode for products that bind to and regulate the expression of other genes. In positive control mechanisms, the product of the regulator gene is required to turn on the expression of one or more genes (gene specifying the amino acid sequences of enzyme or structural protein), while in negative control mechanisms, the product of the regulator gene is necessary to put off the expression of genes. Positive and negative regulation is demonstrated for both inducible and repressible systems in **Figures 18.1 and 18.2**.

We know that a gene is expressed when RNA polymerase binds to its promoter and transcribes the DNA to an RNA transcript that includes the transcribed code of desired gene. The product of the regulator gene acts by binding to a site called the regulator-protein-binding-site (RPBS) ahead of the promoter region of the structural gene(s).

The binding of the product of the regulator gene at the regulator affects the transcription of the structural gene(s) either positively, i.e. the gene is turned on or in a negative manner, i.e. the expression is halted. In case of positive control, the regulator gene products are called **activators**—because they activate gene expression. In case of negative control such products are called as repressors because they repress the gene expression. The binding of the activator or repressor molecule to the operator may depend upon the presence or absence some **effector** molecules in the cell. These are generally molecules such as amino acids, sugars, and similar metabolites. The effector molecules involved in induction (positive control) of gene regulation are called **inducers;** those involved in repression (negative control) of gene expression are called **co-repressors**.

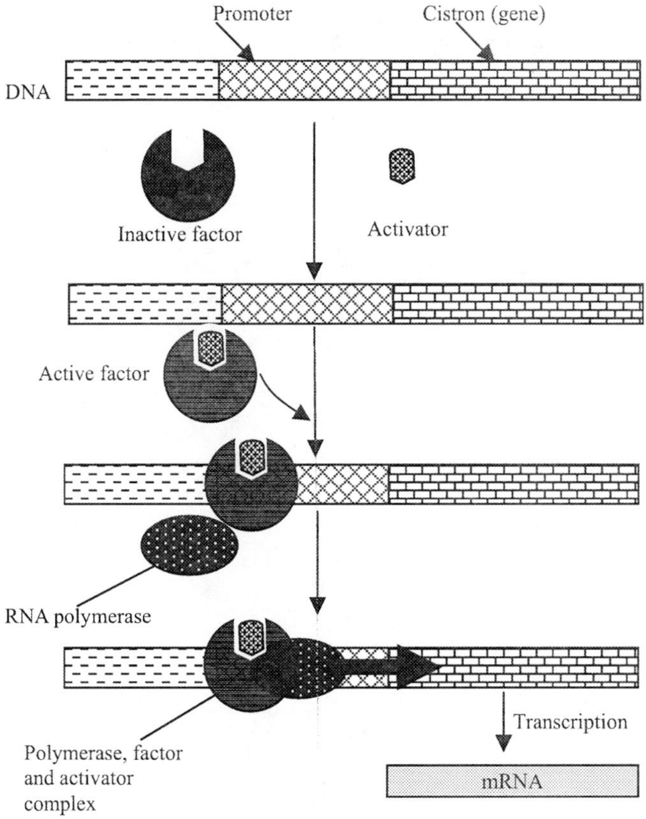

Figure 18.1: Schematic showing mechanism of positive gene regulation.

The effector molecules (inducers and co-repressors) when present inside the cell, bind tightly with the regulator gene products (activators and repressors) and bring about some conformational changes in these proteins. This change is also called as allosteric transition which alters the binding of the effector molecule to the regulator sites neighbouring to the promoters of the structural genes they control.

In a **negative inducible control mechanism**, the free repressor binds the *RPBS* and prevents the transcription of the structural gene(s) in the absence of inducer. When inducer is present, it binds with the repressor protein and does not allow repressor to bind at regulator. With no repressor bound to the regulator, RNA polymerase binds to the promoter and moves on to transcribe the structural gene (s). In positive **inducible control mechanisms**, the activator cannot bind to the regulator unless inducer is present, and RNA polymerase cannot transcribe the structural gene(s) unless the activator/ inducer complex is bound to the repressor. Thus, transcription of the structural genes is turned on only in the presence of inducer (**Figure 18.1**).

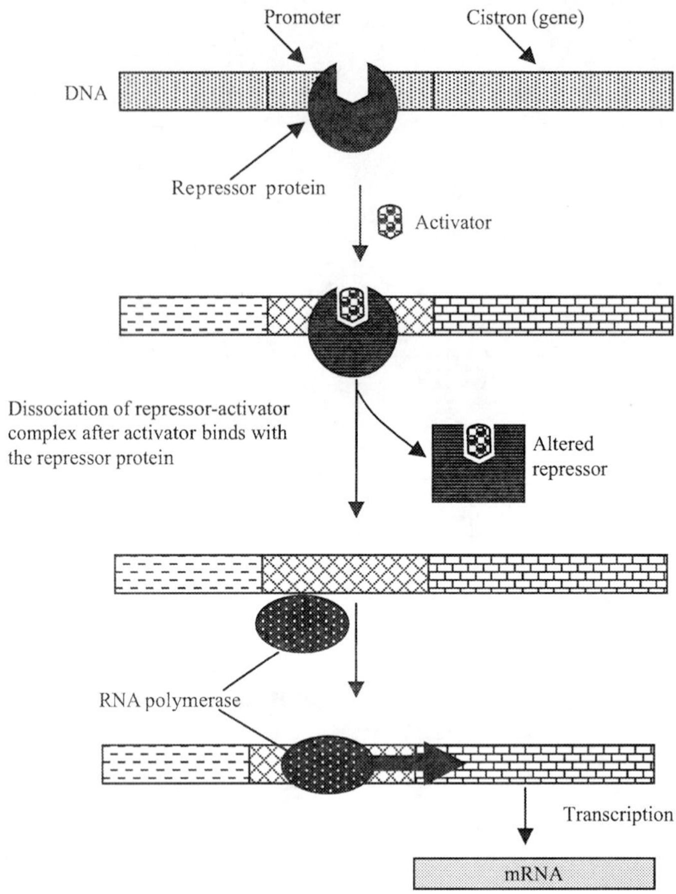

Figure 18.2: Schematic showing mechanism of negative gene regulation.

In a **negative repressible regulatory mechanism**, transcription of the structural gene(s) occurs in the absence of the co-repressor only. When the repressor / co-repressor complex is bound to the regulator, it prevents RNA polymerase from transcribing the structural genes. In the absence of co-repressor, repressor alone cannot bind to the regulator and thus RNA polymerase cannot bind to the promoter and transcribe the structural genes. In a **positive repressible control mechanism**, the product of regulator gene, the activator, must be bound to the regulator in order for RNA polymerase to bind the promoter and transcribe the structural gene (s). When co-repressor is present, it forms a complex with the activator protein, and the activator/ co-repressor complex is unable to the bind to the regulator. Consequently, RNA polymerase cannot bind to the promoter and transcribe the structural gene (s).

It is important to understand the major differences in the mechanism of these four methods of gene regulation in prokaryotes. In the positive control mechanism, the activator participates in turning on gene expression, whereas in a negative control mechanism, the regulator gene product, the repressor, turns off the gene expression by binding to the regulator. In both positive and negative control mechanisms, whether gene expression is inducible or repressible depends on whether the free regulator protein alone or the regulator protein/ effector molecule complex binds to the regulator protein binding site (*RPBS*).

Gene Regulation in Eukaryotes

As discussed earlier in the case of prokaryotic cells, the regulation of gene expression is applied to the majority of cases at the level of transcription. Similar to prokaryotes, gene regulation in eukaryotes also involves both activation and repression, because at any given point of time, genes may have to be switched on or off as per the presence or absence of inducer / repressor. Such DNA sequences involved in eukaryotic regulation have also been identified by DNA footprinting and sequencing.

Activation of Transcription

The structure and function relationship of DNA has been studied in great detail for last few decades and with the advent of PCR, automated DNA sequencer and many more techniques in proteomics and genomics there has been a greater understanding of otherwise intricate eukaryotic system. Initially, these studies involved formulation of theoretical models based on the occurrence of an enormous proportion of repetitive DNA in eukaryotes. Experiments in chromatin reconstitution showed existence of a positive control exercised by non-histone chromosomal proteins.

Britten-Davidson Model for Eukaryotic Gene Regulation

Britten and Davidson in 1969 proposed a model for regulation of protein synthesis in eukaryotes known as Britten-Davidson model or the gene-battery model. It was subsequently elaborated by them in 1973. The model assumes the presence of four classes of genes scattered in the operon:

(i) Producer gene – Similar to structural gene of prokaryotic operon.
(ii) Receptor gene- A receptor site is comparable to operator to gene in bacterial operon.
(iii) Integrator gene- It is comparable to regulator gene and is responsible for synthesis of an activator RNA that may or may not give rise to proteins before it activates the receptor site.
(iv) Sensor site - A sensor site regulates the activity of integrator gene, which can be transcribed only when the sensor site is activated. The sensor sites are recognized by agents such as hormones and proteins that play a crucial role in altering the pattern of gene-expression. For, instance, hormone-

protein complex or a transcription factor may bind to a sensor site and cause the transcription of integrator.

Among these sites (receptor and sensor sites) are those sequences which help only in recognition without taking part in RNA synthesis while the genes (producer gene and integrator gene) are the sequences involved in RNA synthesis. This model also states that receptor sites and integrator genes may be repeated a number of times, so as to control the activity of still larger number of genes scattered in the same cell. Repetition of receptor ensures that same activator recognizes all of them at a given point of time and several enzymes of one pathway are simultaneously synthesized as per the need of the cell. The multiplicity of receptor sites and integrator genes also allows the transcription of same gene in different stages of development. Each structural gene may have several receptor sites, each responding to one activator thus different activators may activate the same gene at different times. An integrator gene may also fall in cluster with same sensor site.

In Britten-Davidson model a set of structural genes controlled by one sensor site is termed as a gene battery. Such a configuration is suitable to the need and performance of multiple functions in a eukaryotic cell. Sometimes, when major changes are needed, it is necessary to activate several sets of genes. If one sensor site is associated with several integrators, it may cause transcription of all integrators at the same time thus causing transcription of several producer genes through receptor sites. The repetition of integrator and receptor sequence is consistent with the knowledge that sufficient repeated DNA occurs in eukaryotes.

The Lactose Operon

The lactose (lac) operon was the first set of genes that revealed the genetic control mechanism of protein synthesis. The study was initiated when Francois Jacob and Jacques Monod observed that the expression of certain genes in mutants of *Escherichia coli* was induced by the presence of lactose in the medium. As described earlier, an **operon** is a segment of DNA that has a cluster of contiguous genes which are transcribed as a single polycistronic mRNA. A cistron is a functional unit of gene and the lac operon involves expression of three functional genes leading to production of three different enzymes of lactose metabolism. The lac operon is governed by a promoter (P) and an operator (O). The three genes and their respective enzymes are: (i) ***lac Z***- This gene codes for the enzyme β-galactosidase which breaks the β-galactoside linkage in the lactose molecule to generate glucose and galactose, (ii) ***lac Y***- This gene codes for the enzyme β-galactoside permease which transports lactose into the cell and (iii) ***lac A***-This gene codes for the enzyme β-galactoside transacetylase. The role of this enzyme is uncertain yet.

The **promoter** and **operator** of the lac operon are correspondingly located instantly upstream from these genes (**Figure 18.3**). The promoter is the area in the DNA where the RNA polymerase binds to initiate transcription while the operator

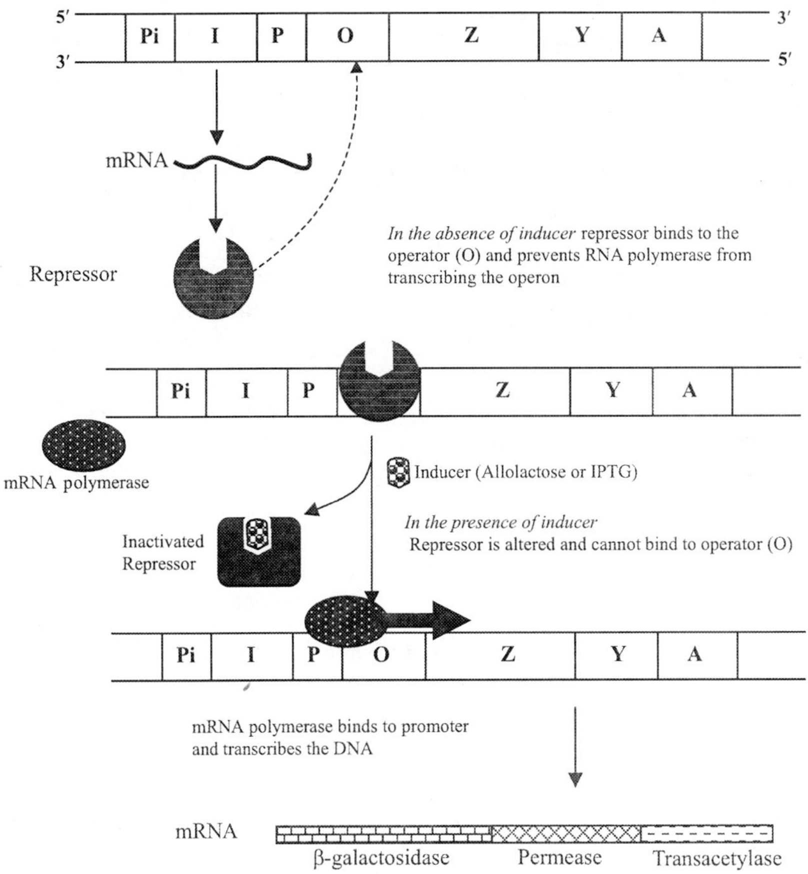

Figure 18.3: Schematic showing mechanism of *lac* operon.

is the area of DNA where the **repressor** binds to halt the transcription. The ***lac I*** operon is located instantly upstream from the *lac* operon. The *lac I* operon consists of the *lac I* promoter and the *lac I* regulator gene. The lac *I* regulator gene codes for the repressor protein.

As discussed earlier, the gene regulation can be negative or positive. In **negative regulation**, a **repressor** molecule binds to the operator of an operon and does not allow RNA polymerase to transcribe genes. In **positive regulation**, an **activator** interacts with the RNA polymerase in the promoter region to initiate transcription. The **lactose operon** is an example of **negative** regulation. When *Escherichia coli* cells are growing in a medium containing lactose as the only carbon source, some of the lactose is converted to allolactose. Allolactose acts as an inducer and turns on the *lac* operon. Both, in the presence of glucose and in the absence of lactose, allolactose is not produced and thus the *lac* operon is turned off. This is because glucose is a readily available energy source and cell machinery will not need to metabolize lactose till any glucose is present. However, this putting off of operon is also a subject to catabolite repression.

The study of the gene expression in the *lac* operon proceeds by induced mutagenesis and selection of the desired mutants. Mutants can be grouped on the basis of type and location of mutation as evidenced by expression (auxotrophy). The mutation can also be localized using the ***cis-trans test*** which is also called as complementation test. When a gene acts in *cis* mutation, it affects only the expression of genes in the same DNA molecule whereas when a gene acts in *trans* mutation, it affects the expression of genes on other DNA molecules not containing the mutation. Some mutations can make the host *E. coli* constitutive for synthesis of the *lac* operon products where the transcription of structural *lac* genes is not affected either by presence or absence of lactose in the medium. Two constitutive mutants of the *lac* operon are designated as ***lacI*** and ***lacOc***. Another type of mutant in which the *lac* operon remains inactive in the presence of lactose is called as super-repressor mutant ***lacIs***.

CATABOLITE REPRESSION

The presence of glucose has long been known to put off the induction of the *lac* operon, as well as other operons controlling enzymes involved in carbohydrate catabolism (glucose effect). This **catabolite repression** guarantees that only glucose is metabolized in preference to the, less efficient, energy sources such as disaccharides (lactose, maltose) and polysaccharides (starch, cellulose), etc.

The catabolite repression of the *lac* operon and several other operons is mediated by a regulatory protein called **CAP** (catabolite activator protein) and a small effector mononucleotide molecule called **cyclic AMP** (cAMP, adenosine-3', 5'-monophosphate). Since CAP binds cAMP, it is also called as the cyclic AMP receptor protein.

The *lac* promoter contains two separate binding sites, one for RNA polymerase and one for the CAP /cAMP complex. This complex must be present at its binding site in the *lac* promoter for the induction of the operon. The CAP/ cAMP complex consequently exerts positive control over transcription of the *lac* operon. It has an effect accurately opposite to that of repressor binding to the operator. Though the defined mechanism by which CAP/cAMP stimulates RNA polymerase binding to the promoter is still uncertain, its positive control of *lac* operon transcription is confirmed by *in vivo* and *in vitro* experiments.

Only the CAP/ cAMP complex binds to the *lac* promoter; in the absence of cAMP, CAP does not bind to the promoter. Thus, cAMP acts as the effector molecule, influencing the effect of CAP on *lac* operon transcription. The intracellular cAMP concentration is sensitive to the presence or absence of glucose. High concentration of glucose brings about decrease in the intracellular concentrations of cAMP. Glucose prevents the activator of adenylcylase, the enzyme that catalyzes the formation of cAMP from ATP. Consequently, the presence of glucose results in a decrease in the intracellular concentration of cAMP. In the presence of a low concentration of cAMP, CAP cannot bind to the *lac* operon promoter. Consecutively, RNA polymerase cannot bind efficiently to the *lac* promoter in the absence of bound CAP/ cAMP complex. Thus, in the presence of glucose, *lac* operon transcription never exceeds two per cent of the induced rate observed in the absence of glucose. By related mechanisms, CAP and cAMP keep the arabinose (*ara*) and galactose (*gal*) operons of *E. coli* from being induced in the presence of glucose.

THE TRYPTOPHAN OPERON

The lactose operon invokes negative regulation of gene expression underlying the mechanism of production of regulatory proteins that prevent expression. As seen earlier that in the absence of lactose there is no need of producing lactose metabolizing proteins and hence the regulatory protein—the repressor blocks the expression of genes. However, it is important to note that this is a case of catabolism and one would wonder will the idea of operon suits to the anabolism or synthesis of any particular compound.

In such cases, the repressor does not bind to the operator alone but needs to form a complex with the effector molecule. Thus, the structural genes producing enzymes involved in synthesis are expressed at a level that is determined by the strength of the promoter and as and when, the ample amount of that molecule is synthesized, an effector molecule will bind with the repressor and this complex will block the operon from being transcribed.

The tryptophan (**trp**) operon is an example of a biosynthetic operon whose expression is regulated by an effector. The operon consists of 5 structural genes that code for the three enzymes required to convert the precursor, chorismic acid into tryptophan and an operator and promoter. Here the operator lies within the promoter region. In the absence of tryptophan, the repressor is made normally but it is inactive and not able to bind to the operator of the operon. Thus, the genes of the operon are transcribed and the enzymes produced are able to synthesize tryptophan. After a particular or sufficient level of the tryptophan is reached in the cell, the amino acid binds with the inactive repressor and changes its conformation. This makes the repressor active and allows it to bind at the operator. This stops transcription of the structural genes. Thus, the tryptophan in this case acts as a co-repressor. This is also an example of negative control as it involves repressor to stop transcription. The R gene that encodes the **trp** repressor is not located away from the operon. The operator (O) region of the **trp** operon lies within the promoter (P). Two transcription termination sequences are also located at the end of the operon downstream the gene A. The L region carries a 162 nucleotide long mRNA leader sequence (**Figure 18.4**). The rate of transcription in the absence of co-repressor, tryptophan, is 70 times more than that under repression (in the presence of tryptophan). Addition of the tryptophan in *E. coli* culture brings about

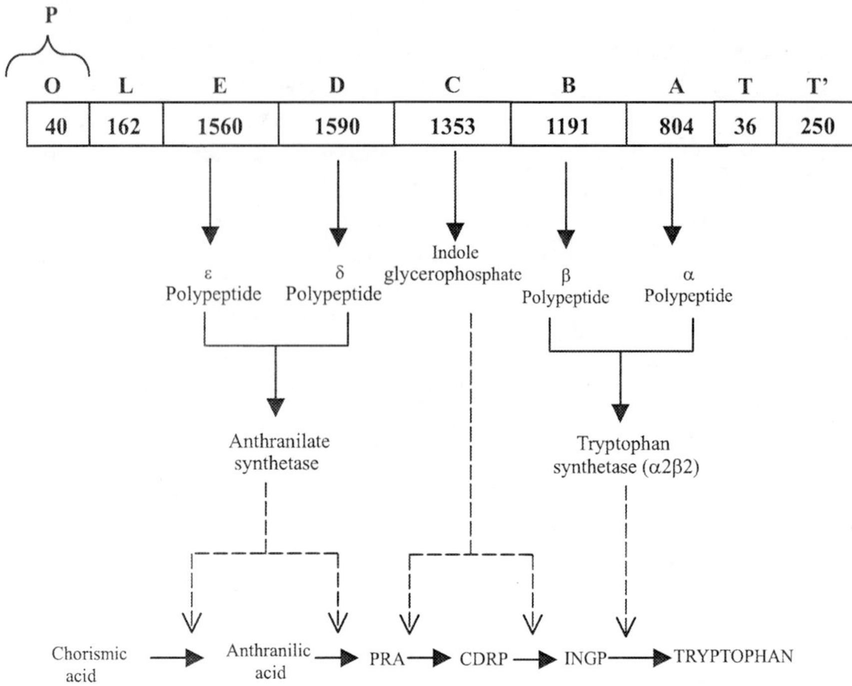

Figure 18.4: The trp (tryptophan) operon in *E. coli* contains five structural genes that encode enzymes involved in the biosynthesis of tryptophan from its precursor chorismic acid. Numbers given below genes specify the length of the genes (base pairs). PRA: phosphoribosyl anthranilate; CDRP: carboxyphenylamino-deoxyribulose phosphate; INGP: indole-glycerol phosphate.

a sharp decrease in the reduction of tryptophan synthesizing enzymes, however, this is attributed to a separate phenomenon, attenuation. Attenuation is a different mechanism controlled by sequences called as attenuatror. This occurs by the premature termination of the transcription just after the formation of leader sequence.

	Enzyme	**Encoding gene**
(i)	Anthranilate Synthetase	
	Component I (ε)	Trpe
	Component II (δ)	Trpd
(II)	N-(5′-Phosphoribosyl)-Anthranilate Isomerase	
	and Indole-3-Glycerol Phosphate Synthase	Trpc
(III)	Tryptophan Synthase	
	α-Subunit	Trpb
	β-Subunit	Trpa

CATABOLITE REPRESSION: ROLE OF CYCLIC AMP

cAMP is an abbreviation for 3'–5' adenosine monophosphate given by Sutherland in 1965. Cyclic AMP is a nucleotide and it regulates the speed of several biochemical reactions in the living system. It is also referred as the second messenger (first messenger being hormones). Presence of cyclic AMP has been noticed in all bacteria and animal tissues. Majority of cAMP is found associated with the cell membrane.

Structure of cAMP: It is a nucleotide having adenosine as nitrogenous base and a phosphoric acid and ribose sugar. The cAMP differs from AMP in possessing the phosphate group in a cyclic manner and involving both 3' and 5' carbons of ribose sugar. As per nomenclature it is 3', 5' cyclic adenylic acid or adenosine 3'5'cyclic phosphoric acid (**Figure 18.5**).

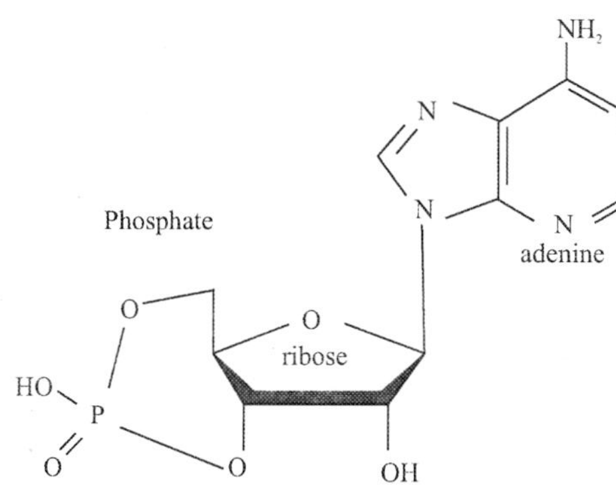

Figure 18.5: Structure of cAMP.

Synthesis of cAMP: It is synthesized in the cell membrane of a variety of cells from ATP under the influence of the enzyme adenyl cyclase in the presence of Mg^{+2} ions.

$$\text{ATP} \xrightarrow[\text{Adenyl cyclase}]{Mg^{+2}} \text{cAMP} + \text{PPi}$$

Degradation of cAMP: Hydrolysis of cAMP leads to formation of AMP. This is done by an enzyme called phophodiesterase as follows:

$$\text{cAMP} + H_2O \xrightarrow[\text{phosphodiesterase}]{Mg^{+2}} \text{AMP}$$

Biological Functions of cAMP

cAMP regulates cellular metabolism and called as secondary messenger since it transmits the primary hormonal message from the cell surface to the cell interior.

1. **Role of cAMP in glycogenesis:** Glycogenesis is synthesis of glycogen by the linking of glucose units. The last step in glycogenesis is catalyzed by the enzyme glycogen synthetase and this step is activated by cyclic AMP.
2. **Role of cAMP in lipolysis:** Lipolysis is controlled by cyclic AMP molecules present in the adipose tissues. cAMP is also called as lipolytic agent. cAMP activates some of the lipases which leads to hydrolysis of lipids.
3. **Role cAMP as hunger signal:** cAMP gives a hunger signal in bacterium. As long as glucose is present in the medium the bacterium does not give any hunger singal. As soon as the glucose concentration is lowered, the cAMP increases and sends a hunger signal to synthesize the enzyme responsible for using other sugars available.
4. **Role of cAMP in controlling transcription:** It plays a major role in controlling transcription in *Lac* operon. The cAMP binds itself to a catabolic gene activator protein to form a complex called cAMP-CAP, which binds itself to the promoter site of the *Lac* operon. The initiation of transcription is stimulated in response to this binding.
5. **Role of cAMP in glycogenolysis:** Gycogenolysis is a process in which liver glycogen is converted to blood glucose. The signal for this process is given by glycogen. The message of glycogen is transmitted by cAMP to interior of the cell. The cAMP initiates a chain of reactions for the conversion of glycogen to glucose.

Catabolite Gene Activator Protein (CAP)

CAP is a regulatory protein referred to as catabolite gene activator protein. It is a homodimer (subunit Mr. 22,000) with binding sites for DNA and cyclic AMP. When glucose is absent CAP binds to a specific site near lac promoter and stimulates RNA transcription 50 folds. CAP is therefore, a positive regulatory element in response to glucose level.

SUGGESTED READING

Alberts B., Bray, D., Lewis, J., Raff, M., Roberts, K. and Watson, J.D. 1994. *Molecular Biology of the Cell*, Garland Publ. Inc., N.Y.

Cooper, G.M.1997. *The Cell: A Molecular Approach*, ASM Press, USA.

Dale, J.W. 1998. *Molecular Genetics of Bacteria*, John Wiley & Sons.

Darnell, J., Lodish, H. and Baltimore, D. 1990. *Molecular Cell Biology*, Scientific American Books Inc., NY.

Freifelder, D. and Malacinski, G.M. 1993. *Essentials of Molecular Biology*, Jones and Bartlett Publishers Inc.

Glick, B.R. and Pasternak, J.J. 1994. *Molecular Biotechnology*, ASM Press, USA.

Hartl, D.L., Freifelder, D. and Snydder, L.A. 1988. *Basic Genetics*, Jones and Bartlett Publishers Inc.

Hawkins, J.D., 1996. *Gene Structure and Expression*, Cambridge Univ. Press.

Karp, G. 1996. *Cell and Molecular Biology: Concepts and Experiments*, Jhon Wiley & Sons Inc. N.Y.

REVIEW QUESTIONS

1. Explain the mechanism of positive and negative regulation of gene expression.
2. What is an operon? Explain giving suitable examples.
3. Explain the mechanism of gene regulation in eukaryotic cells.
4. Write note on the following:
   a) Lac operon
   b) Britten Davidson model
5. What is meant by catabolite repression? Explain the role of cyclic AMP in catabolite repression.

UNIT XIX

Mutation and Repair

Mutation can be defined as a sudden heritable change in DNA or genetic material or a process by which DNA base-pair change or a chromosome change occurs is called Mutation. Mutations occur spontaneously *in vivo* or may be induced experimentally *in vitro* by exposing living cells to mutagens. Mutations occur naturally and can also be induced with chemical or physical mutagenic agents. Apart from these, they can also be induced at specific sites by genetic engineering techniques. The rate of spontaneous mutation depends on the growth conditions and surrounding environment of the organism and is estimated to be between 10^{-10} and 10^{-5} per generation per gene. Although dominated by deletions, all kinds of mutations are observed in nature. The underlying causes include errors in the functioning of the enzymes such as DNA polymerases, integration and excision of jumping elements (transposons), erroneous recombination and natural exposure to mutagens. The use of mutagenic agents enhances the frequency of mutation up to 10^{-5} to 10^{-3} for improved isolation of mutants producing high level of secondary metabolites and even higher frequencies 10^{-2} and 10^{-1} for isolation of auxotrophic mutants.

Mutations can be classified on the basis of their size in terms of structural effects:

(i) **Genome mutation** causes change in the number of chromosomes;
(ii) **Chromosome mutation** causes change in the order of genes within the chromosome such as deficiency, deletion, inversion and translocation
(iii) **Gene or point mutations** change in the base sequence in a gene.

Genome and chromosome mutations are more relevant in case of higher eukaryotic organisms such as plant and animals while mutations used for strain improvement in industrial microbiology are usually point mutations.

THE MOLECULAR BASIS OF MUTATION

The double-helical structure of DNA undergoes semi-conservative replication based on complementary and specific base-pairing. This leads to accurate transmission of genetic information from parents to offsprings. However, the structures of the bases in DNA are not static, e.g. hydrogen atoms can shift from one position in a purine or pyrimidine to another position, for example, from an amino group to ring nitrogen. This chemical fluctuation is called **tautomeric shift (Figure 19.1).** Such shifts may have immense significance in DNA metabolism because the pairing potential of the bases may be changed. The more stable forms of thymine and guanine and the amino forms of adenine and cytosine may infrequently undergo tautomeric shifts to less stable enol and imino forms, respectively. If a base acquires its rare tautomeric form at the point of being incorporated into a nascent DNA chain, a mutation would result. When the bases are present in their rare imino or

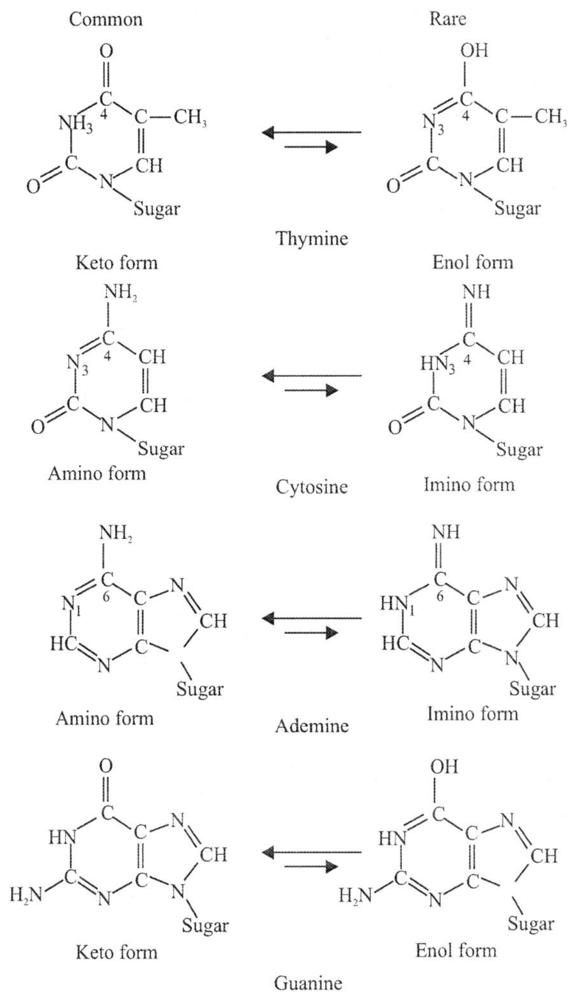

Figure 19.1: Tautomeric forms of the four common bases in DNA. The shifts of hydrogen atoms between the number 3 and number 4 positions of the pyrimidines and between the number 1 and number 6-positions of the purines change their base-pairing potential.

enol states, they can form adenine-cytosine and guanine-thymine base pairs resulting in a **mismatch** base pair.

Mismatch mutations due to tautomeric shifts in the nitrogenous bases of DNA involve the replacement of a purine in one strand of DNA with the other purine and the replacement of pyrimidine in the complementary strand with the other pyrimidine. Such base-pair substitutions are called **transitions**. Substitutions involving the replacement of a purine with a pyrimidine and vice versa are called **transversions**. Mathematically, four different transitions and eight different transversions are possible (**Figure 19.2**). Another type of point mutation involves the addition or deletions of one or a few base pairs. Base-pair additions and deletions are collectively referred to as **frameshift mutations** because they alter the reading frame of all triplet codons of transcribed mRNA. The addition of one extra base pair in DNA shifts the complete reading frame and results in incorporation of wrong amino acids, leading to formation of an altered or useless protein. All three types of point mutations, **transitions, transversions and frameshift mutations**, are spontaneously or naturally occurring mutations and have a slow rate.

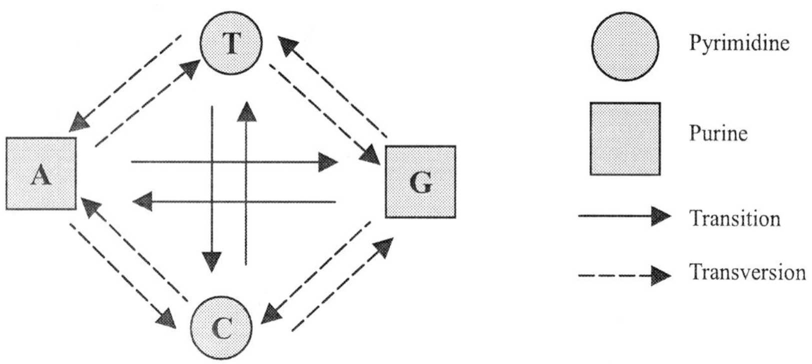

Figure 19.2: Twelve different base substitutions can occur in DNA.

Induced Mutation

Mutations can also be induced artificially at will by introducing any mutagen around any living organism. Several such physical and chemical mutagens are known that alter the DNA upon exposure. These mutagens interfere with the regular structure of DNA by forming dimers or by intercalating in between the DNA ladder and thus leading to either incorporation or deletion of base pairs. Traditionally, these methods have been used for development of industrial strains in industrial microbiology where a population of microorganism is given sub-lethal dose of a mutagen and the resulting colony forming units (cfu's) are screened for induction of desired trait.

TYPES OF MUTATIONS

Mutations can be interpreted and described in various ways. Primarily, these can be described as any inheritable change in DNA sequence. On the basis of number and type of base pair change and its effects, mutations have been classified into following categories.

(i) Point mutation: When a single base-pair is altered, the mutation is called point mutation. Further, they are classified into four types

(a) Base substitution
(b) Base insertion
(c) Base deletion
(d) Base inversion

(ii) Missense mutation: A base-pair change in the DNA causes a change in mRNA codon, so that a different amino acid is inserted into the polypeptide in place of specified codon, resulting in altered phenotype.

For example, an AT to GC transition mutation changes the DNA from

5′-AAA-3′ to 5′GAA-3′

3′-TTT-5′ to 3′-CTT5′

This will result in change in the mRNA codon from 5′-AAA-3′ (lysine) to 5′-GAA-3′ (glutamic acid) resulting in incorporation of glutamic acid in place of lysine in the polypeptide.

(iii) Nonsense mutation: A base-pair change in the DNA that results in the change of an mRNA codon from one that specifies an amino acid to a chain-terminating (nonsense) codon (UAG, UAA or UGA). For example, an AT-to –TA transversion mutation changes the DNA from 5′-AAA-3′ to 5′-TAA-3′ and 3′-TTT-5′ to 3′-ATT-5′. This changes the mRNA codon mRNA 5'-AAA-3' (lysine) to 5'-UAA-3' which is a nonsense codon leading to abrupt termination of polypeptide elongation. This may result in formation of a non-functional truncated protein.

(iv) Neutral mutation: A base-pair change in a gene that changes a codon in the mRNA such that the resulting amino acid substitution produces no detectable change in the formation and function of transcribed protein.

For example, an AT- to –GC transition mutation changes the codon from 5′-AAA-3′ to 5′-AGA-3′ which substitutes arginine for lysine. Both are basic amino acid and possibly the resulting protein is fully functional.

(v) Silent mutation: A base-pair change in a gene that alters a codon in mRNA such that same amino acid is inserted in the protein and hence there is no change.

(vi) Back mutation: Base pair change that reverts the mutated gene into normal gene (wild) is called back mutation or reverse mutation.

(vii) Polar mutation: Segments of DNA that jump from one position to another are called transposons or jumping genes. The mutations resulting due to insertion of these jumping elements are called as polar mutations.

MUTAGENS

Mutagens are agents or substances that can bring about a permanent alteration to the physical composition of a DNA such that the genetic message is altered. These are of two types

1. Chemical mutagens
2. Physical mutagens

I. Chemical Mutagens

Chemical mutagens can be classified in to two types.

(i) Which are mutagenic to both replicating and non-replicating DNA, e.g. alkylating agents and nitrous acid.

(ii) Those are mutagenic only to replicating DNA.

There are many chemical agents that can cause mutation:

(A) Base Analogues: The mutagenic base analogues have structures similar to the normal bases and are incorporated into DNA during replication. However, their structures are sufficiently different from the normal bases in DNA resulting in increase in the frequency of mispairing, and the mutation, during replication. The two most commonly used base analogues are 5-bromouracil and 2-aminopurine. The pyrimidine 5-bromouracil is a thymine analogue; the bromine at the 5 position is similar in several respects to the methyl ($-CH_3$) group at the position in thymine (**Figure 19.3**). However, the bromine at this position changes the charge distribution that results in increased frequency of tautomeric shift. In its more stable keto form, 5-bromouracil pairs with adenine. Where as after tautomeric shift to its enol form, it pairs with guanine. The mutagenic effect of 5-bromouracil is the same as that observed for tautometic shifts in normal bases, namely, **transitions**. Quite similary, 2-aminopurine gets incorporated in place of adenine or guanine base and leads to transition.

(B) Nitrous Acid (HNO_2): Nitrous acid is a potent mutagen that acts on both replicating and non-replicating DNA. Nitrous acid causes **oxidative deamination** of the amino groups in adenine, guanine, and cytosine. The reaction converts the amino groups to keto groups and changes the hydrogen bonding potential of the modified bases. Adenine is deaminated to hypoxanthine, which base-pairs with cytosine rather than thymine. Cytosine is converted to uracil, which base-pairs with adenine instead of guanine. The deamination of guanine is not mutagenic since it results in formation of xanthine, and xanthine-just like guanine, base-pairs with cytosine. Because the deamination of adenine results in A:T $\rightarrow$ G:C transitions,

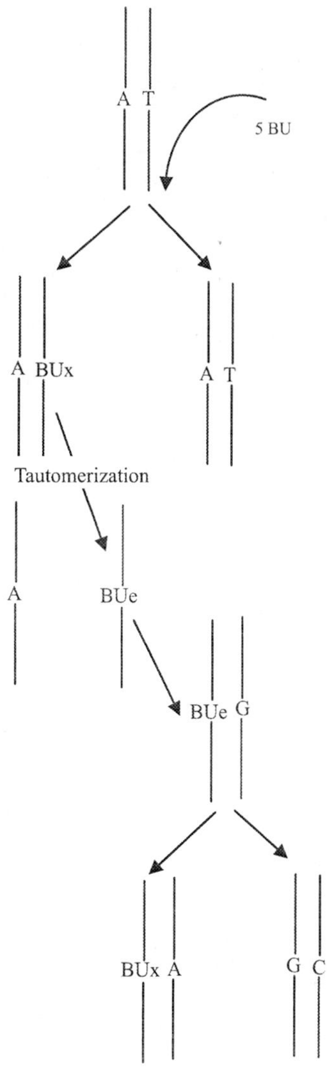

Figure 19.3: Mutation via addition of base analogue 5' Bromo-uracil (BU) leading to AT → GC transition.

and the deamination of cytosine produces G:C → A:T transitions, nitrous acid induces transitions in both direction A:T ↔ G:C. As a result, nitrous acid-induced mutations may also revert back to wild-type genotype for similar reasons (**Figure 19.4**).

(C) Acridine Dye: Acridine dyes such as acridine orange, proflavin and a series of derivatives that intercalate between DNA ladder causing slight bends or kinks in the molecule are potent mutagens. When DNA molecules containing intercalated

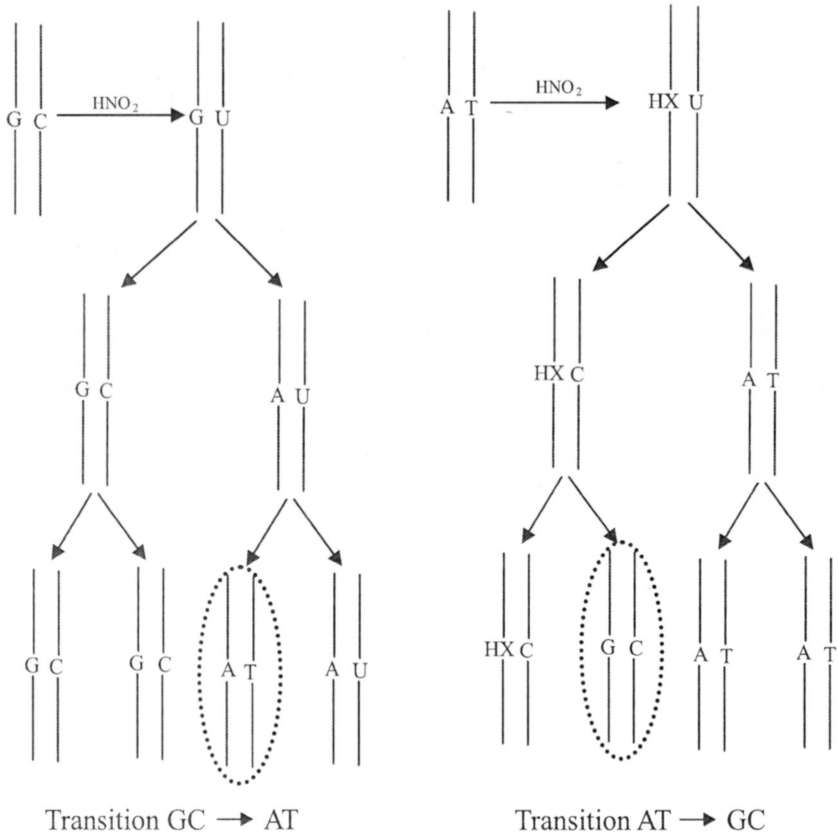

Figure 19.4: Base transition due to a chemical agent- Nitrous acid. (HX-thypoxanthine)

acridines replicate, additions and deletions of one to a few base-pair occur. These additions and deletions, usually of a single base pair, result in frame shift mutation leading to incorporation of wrong amino acids in the protein resulting in formation of non-functional proteins.

(D) Alkylating Agents: Alkylating agents are chemicals that donate alkyl group to the nitrogenous bases. They include chemicals that have multiple effects on DNA including nitrogen mustard, and methyl methane sulphonate, ethyl methyl sulphonate (MMS and EMS). Alkylating agents induce all types of mutations, including transitions, transversions, frameshifts, and even chromosome aberrations. Alkylating agents involve the transfer of methyl or ethyl groups to the nitrogenous bases, resulting in their altered base-pairing potentials, e.g. ethyl methyl sulphonate (EMS) causes ethylation of the bases in DNA at the 7-N and the 6-O positions. The resulting base, 7-ethylguanine, base-pairs with thymine to cause a G:C→A:T transition.

Hydroxylamine (NH_2OH) a hydroxylating agent has a typical mutagenic effect causing G:C → A:T transitions. Hydroxylamine hydroxylates the amino group of cytosine leading to formation of hrdroxylaminocytosine. The altered base-pairs with adenine, leading to G:C → A:T transitions.

Physical Mutagens

The electromagnetic spectrum with shorter wavelengths and higher energy than visible light can be subdivided into **ionizing radiations** (X-rays, gamma rays, and cosmic rays) and **non-ionizing radiations** (ultraviolet light).

Ionizing Radiations: On behalf of their high energy, ionizing radiations induce gross changes in DNA and chromosome structure including, deletions, duplications, inversions, translocations and also chromosome aberration or breakage due to radiation-induced breaks in chromosomes. X-rays and other forms of ionizing radiations are quantified in Roentgen (r) units, which is a measure of the number of ionizations per unit volume under a standard set of conditions. For most practical purposes, X-rays are widely used as an ionizing radiation to induce mutation.

Non-ionizing Radiation: The nitrogenous bases of DNA readily absorb ultraviolet radiations and become more reactive. Non-ionizing radiations such as ultraviolet rays do not ionize molecules because they have less energy than ionizing radiations

Thus if given exposure, ultraviolet light acts as a power mutagen for microorganisms. The maximum absorption of UV by DNA occurs at a wavelength of 254 nm and correspondingly maximum mutagenicity also occurs at 254 nm. These suggest that UV-induced mutagenesis process is mediated by the absorption of UV rays by purines and pyrimidines. Pyrimidines absorb strongly at 254 nm and become very reactive. Two major products of UV absorption by pyrimidine are pyrimidine hydrate and pyrimidine dimers. Thymine dimers affect mutations in two ways. Firstly, dimers distort the structure of DNA double helices and interfere with accurate DNA replication; secondly, wrong bases are added during repair of such defects (**Figure 19.5**).

MUTATION REPAIR MECHANISM

In nature organisms are exposed to a variety of chemicals, radiations, stress, etc. Spontaneous mutations are a routine process in the development and evolution of any organism. Corresponding to these changes, organisms have evolved very meticulous mechanisms of damage control. These mechanisms are called as repair systems.

The variety and efficiency of repair mechanisms witnessed in organisms indicate the importance of keeping check at mutation. *E. coli* alone exhibits mechanisms for the repair of mutations and defects in DNA including, light-dependent repair or photoreactivation, excision repair, mismatch repair, post replication repair, and error-prone repair system (SOS response).

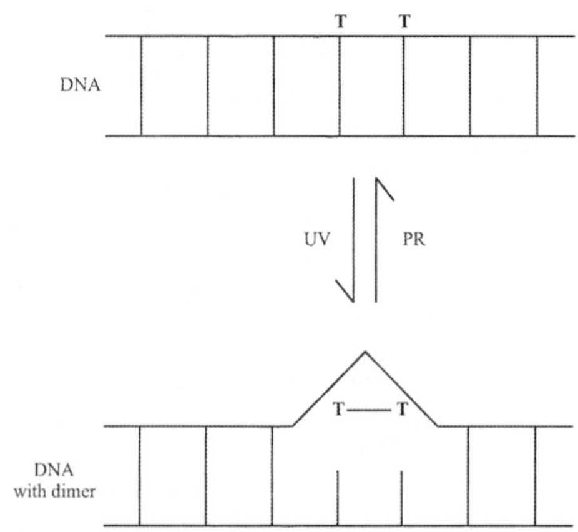

Figure 19.5: Formation of thymine dimer upon exposure to UV and its reversal by photoreactivation (PR).

Photoreactivation

DNA absorbs ultraviolet light and adjacent thymine molecules get covalently linked to make thiamine dimers. These dimers create hindrance in the replication of DNA. However, this mutation is reverted upon exposure to normal visible light. This reversal is due to photoreactivation by photolyase enzyme which splits the thymine dimers. This process of DNA repair in bacteria is carried out by a light dependent enzyme called **DNA photolyase**. This enzyme identifies and binds with thymine dimers in DNA and using visible light (300–400 nm) cleaves the covalent cross-links. The enzyme can also split cytosine dimers and cytosine-thymine dimers. Thus, whenever ultraviolet light is used to induce mutations in bacteria, the culture is grown in the dark for some length of time to achieve maximum mutation frequency and avoid reversion (**Figure 19.5**).

Excision Repair

Exonucleases are able to recognize the damages or misincorporations in the DNA molecules such as alkylation, deamination, pyrimidine dimer formation. Such DNA damages are repaired by excision repair which involves cutting out and replacement of wrongs nucleotides. The DNA exonuclesase and DNA polymerase I split phosphodiester bond of the backbone and remove a 30 nucleotide long segment. This void is filled by a newly synthesized segment by DNA polymerase I and ligated by DNA ligase.

Excision repair of modified (alkylated or deaminated) DNA employs special enzymes, e.g. in case of alkylated DNA of *Micrococcus luteus,* 3-methyladenine formed is specically cleaved off by an N-glycosylase. Similarly, AP endonucleases (AP = apurinic or apyrimidinic bases) are also known to remove alkylated bases

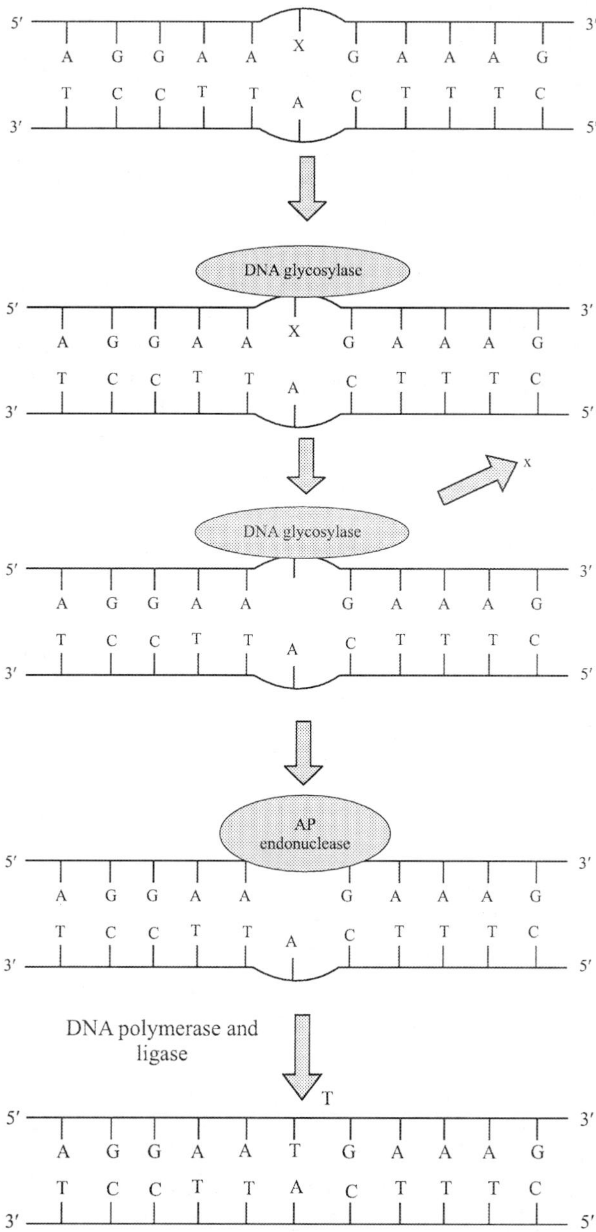

Figure 19.6: Mechanism of base excision repair. Damage in the the DNA (X) is localized and removed by DNA glycosylases. The removal leads to formation of a AP (Apurinic or Apyrimidinic site). AP endonuclease hydrolyses the phosphodiester bond between each AP site removing the phoshphate backbone. The resulting void is finally filled up by the DNA polymerase and the nick is sealed by ligase.

while Uracil DNA glycosylase, hypoxanthine DNA glycosylase are involved in the repair of deaminated DNA (**Figure 19.6**).

Excision repair of damaged DNA precedes through the DNA endonuclease. Firstly, a DNA endonuclease identifies and binds to the damaged base or bases in DNA and removes them. This is followed by DNA polymerase activity that fills in the gap by synthesizing the undamaged complementary strand of DNA as template. Finally, the enzyme DNA ligase seals the nick between new segment and the old DNA strand.

There are two basic types of excision repair: **(i) base excision repair** system that remove abnormal or chemically modified bases from DNA, and **(ii) nucleotide excision repair** pathways remove larger defects like thymine dimers. Both these pathways are light independent and are known to occur in a variety of organisms including humans.

(i) Base excision repair

Base excision repair is mediated by DNA glycosylase enzymes which recognize modified or abnormal bases in DNA. Glycosylases recognize a specified altered base, such as deaminated bases, oxidized bases, methylated bases, etc. and cleave the glycosidic bond between the abnormal base and 2-deoxyribose sugar, leading to formation of apurinic or apyrimidinic sites (AP sites). AP sites are recognized specialized endonucleases called AP endonucleases, which act together with phosphodiesterases to excise the sugar-phosphate groups at AP sites. DNA polymerase then replaces the missing nucleotide according to the specifications of the complementary strand, and the nick is sealed by DNA ligase.

(ii) Nucleotide excision repair

Nucleotide excision repair removes larger DNA segments like thymine dimers and abnormal bases with large side-groups. In nucleotide excision repair, a unique excision nuclease activity produces cuts on either side of the damaged nucleotide (s) and excises an oligonucleotide containing the damaged base (s). This nuclease is also referred to as **excinuclease** to distinguish it from routine nucleases.

The *E. coli* nucleotide excision repair pathway is shown in **Figure 19.6**. In *E.coli*, excinuclease activity requires the products of three genes, *uvr*A, *uvr*B, and *uvr*C (where *uvr* means UV repair). A trimeric protein containing a UvrA dimer and one UvrB polypeptide recognizes the defect in DNA, binds to it, and uses energy from ATP to bend the DNA at the damaged site. The UvrA dimer is then released, and the UvrC protein binds to the UvrB/DNA complex. The UvrB protein cleaves the fifth phosphodiester linkage from the damaged nucleotide(s) on the 3'side, and the UvrC protein hydrolyzes the eightth phosphodiester linkage from damaged on the 5' side. The *uvr*D gene product; i.e. DNA helicase II, releases the excised dodecamer. Finally, DNA polymerase I fill in the gap, and ligase seals the nick.

Mismatch Repair

Mismatch repair, helps in replicative proofreading by correcting mismatched nucleotides inducted in DNA after replication. Mismatches often involve the usual four bases in DNA. A thymine may be mispaired with a guanine and since both these are usual bases of DNA, mismatch repair systems need some specialized way to determine whether the T or the G is the wrong base at the given site. The repair systems distinguish it by identifying the template or sense strand, which contains the original nucleotide sequence, and strand synthesized *de novo*, which contains the wrong base. This distinction between template and newly synthesized strand can also be made on the basis of the pattern of methylation in newly replicated DNA. The mismatch repair system recognizes and uses the difference in methylation pattern to excise the mismatched nucleotide from newly formed strand and replaces it with the correct nucleotide according to complementarity of template strand.

In *E. coli*, mismatch repair requires the products of four genes, *mut*H, *mut*L, *mut*S, and *mut*U (*mut*D). Initially, the MutS protein recognize mismatches and binds there. This is followed by joining of MutH and MutL proteins at the site. MutH contains a **GATC-specific endonuclease activity** that cleaves the unmethylated strand at GATC. The incision sites may exceed 1000 exonucleotide pairs. The subsequent excision process requires MutS, MutL, DNA helicase II (MutU), and a specific exonuclease. If the incision occurs at a GATC sequence 5' to mismatch, a 5' $\rightarrow$ 3' exonuclease similar to *E. coli* exonuclease VII is required. If the incision occurs 3' to the mismatch, a 3' $\rightarrow$ 5' exonuclease activity like that of *E. coli* exonuclease I is needed. After the excision process has removed the mismatched nucleotide from the unmethylated strand, DNA polymerase fills in the gap as per complementarity rule, and DNA ligase seals the nick.

Post-replication Repair

If a damaged DNA with hindered base pairing (e.g. due to intercalation of dyes) undergoes replication, a gap of approximately 1000 nucleotides is formed during replication. These gaps are filled by recombination with the parent strands by the action of *recA* gene product.

When DNA polymerase III finds a thymine dimer in the template strand, its movement is blocked. The enzyme restarts DNA synthesis at some position past the dimer, leaving a gap in a newly produced strand opposite the dimer. This gap is repaired by a recombination-dependent repair process mediated by the *E. coli rec*A gene product. The RecA protein, which is required for homologous recombination, stimulates the exchange of single strand of DNA at the gap and mediates pairing with the homologous double helices. During postreplication repair, the RecA protein binds to the single strand of DNA at the gap and mediates pairing with the parent homologous DNA segment of the sister double helix. The gap opposite the dimer is filled with the homologous DNA strand from the sister DNA molecule. The resulting gap in the sister double helix is then filled in by DNA polymerase, and the nick is sealed by DNA ligase. The thymine dimer remains in the template strand of the other progeny DNA molecule, but the complementary strand is now intact. The thymine dimer may be removed by photoreactivation by photolyase.

The Error-prone Repair System (SOS or Save Our Soul Response)

When the DNA of *E. coli* cells is heavily damaged by mutagenic agents such as UV light, the cells take some major steps in their attempt to survive. Such an action is called as **SOS response.** During this a whole battery of DNA repair, recombination, and replication proteins are synthesized. Two of these proteins, encoded by *umu*C and *umu*D genes, allow DNA replication to proceed across damaged segments of template strands. This mechanism is referred as **error-prone repair system** as it eliminates gaps in the newly synthesized strands but, increases the frequency of replication error. The SOS response appears to be a rapid, risky and inaccurate attempt to evade the fatality. This system also leads to increase in the rate of mutation.

SUGGESTED READING

Lewin B. 2000. *Genes*. Oxford Univ. Press N.Y.

Moloy, S.R., Cronan, J.E., and Freifelder, D. 1994. *Microbial Genetics*. Jones & Bartlett.

Maxson, L.R., and Daugherty, C.H. 1992. *Genetics*, Wm. C. Brown Publishers.

Prescott, L.M., Harley, J.P. and Klein, D.A. 2003. *Microbiology* Wm. C. Brown Publishers.

Sandhya Mitra. 1994. *Genetics-A Blueprint of Life*, Tata McGraw-Hill Publishing Company.

Stanier, R.Y., Ingraham, J.L., Wheelis, M.L. and Painter, P.R. 1992. *The Microbial World*, 5th ed. Prentice-Hall Publishers.

REVIEW QUESTIONS

1. What do you understand by mutation? Describe various types of mutations.
2. List important chemical mutagens and explain their mode of action.
3. List important physical mutagens and explain their mode of action.
4. Write notes on the following:
   (a) Transition
   (b) Transversion
   (c) Frameshift mutation
   (d) Nonsense mutation
5. Explain how base transition is brought about by nitrous acid base analogues.
6. Describe various mechanisms of mutation repair seen in bacteria.

UNIT XX

Genetic Recombination in Bacteria

Much of the biological diversity among living organisms could be accounted for the genetic recombination and sexual reproduction. Microorganisms also exhibit a great range of diversity including strain level genetic diversity. Although elaborate sexual mechanisms are not seen among bacteria, at least three mechanisms of genetic recombination are seen.

Three major types of genetic transfers are seen in bacteria that include **transformation**, in which a DNA molecule is taken up from the external environment and incorporated into the **genome**; **conjugation**, in which donor DNA is transferred from one bacterial cell to another by direct contact and **transduction**, in which DNA is transferred from one bacterial cell to another by a bacterial virus (bacteriophage).

TRANSFORMATION

Transformation is the manifestation of change brought about by the transfer of genetic material in organisms by small extracellular piece of DNA. It is the unidirectional transfer of extracellular DNA into any cell, resulting in a change in phenotype of recipient. In this process, donor DNA is taken up from the external environment and incorporated into the genome of a recipient cell and expressed. Griffith, a British bacteriologist, while working with *Streptococcus pneumoniae*, a bacterium that causes pneumonia in both humans and mice observed this phenomenon for the first time. Ultimately, his observations lead to the identification of DNA as the physical basis of heredity or genetic material.

Microorganisms get naturally transformed by taking up foreign DNA elements from their natural surrounding. Alternatively, these can also be transformed *in vitro* where we provide appropriate conditions for the uptake of desired DNA by making cells competent.

Transfer of DNA molecules across the cell membrane and into the cytoplasm of recipient bacteria is an energy requiring process and hence transformation is not observed in all bacteria, but only in those species that possess enzymatic machinery required to bind free DNA molecules in the surrounding medium and transport them to the cytoplasm, e.g. *S. pneumoniae, Bacillus subtilis* and *Haemophilius influenzae.* However, only *competent cells*, which secrete *competence factor*, a small protein that induces the synthesis of 8-10 new proteins required for transformation are capable of serving as recipient cells during transformation. Thus, as mentioned above very few species of some genera are able to transform naturally.

Griffith (1928) showed that an avirulent, rough, non-encapsulated strain of *S. pneumoniae* gets converted into a virulent, smooth, capsulated strain by addition of killed virulent cells in its medium.

(1)	Living S cells (smooth)	→	Mouse contracts pneumonia	→	S colonies isolated from tissue of dead mouse
(2)	Living R cells	→	Mouse remains healthy	→	R colonies isolated from tissue of mouse
(3)	Heat killed S cells	→	Mouse remains healthy	→	No colonies isolated from tissue
(4)	Living R cells plus heat killed S cells	→	Mouse contracts pneumonia	→	R and S colonies from tissue of dead

The S (smooth) type of *S. pneumoniae* synthesizes a gelatinous capsule compound of complex carbohydrates. The R (rough) strain is unable to synthesize the capsular polysaccharide, they form small colonies that have a rough surface. This strain does not cause pneumonia because without the capsule the bacteria are unable to cause pneumonia as they get inactivated by the immune system of the mouse.

Oswald Avery, Colin McLeod, and Maclyn McCarty (1944) demonstrated that mechanism of transformation proceeds through DNA transfer. Transformation in *Streptococcus* was originally discovered in 1923, but it was not until 1944 that the chemical substance responsible for changing the R cells into S cells was identified. They showed that the substance causing the transformation of R cells into S cells was DNA.

They noted following points and drew inferences from these:

(1) The transforming activity in S cells is not destroyed by heat.

(2) The transforming activity is not destroyed by either protease or RNase thus the transforming principle is neither protein nor RNA.

(3) The transforming activity is destroyed by DNase. Transforming activity is therefore most likely due to DNA.

The bacterial cell can be transformed only in its competent state where DNA receptors and other transformation-related proteins are present on the cell surface. Competence depends upon growth stage and conditions and may last for a few minutes to several hours. In some species such as *Neisseria gonorrhoeae*, competence is marked by the presence of external appendages such as pili.

For practical reasons, competence can be induced in otherwise not naturally transformable species, e.g. in *E. coli* by chemical treatment with calcium chloride and a brief heat shock. Permeation of foreign DNA can also be facilitated by creating pores on cell surface under electric pulse (electroporation). During artificially induced transformation, usually plasmids are used to transform cells and they remain independent of the bacterial chromosome. *Bacillus subtilis* cells become competent only in the presence of high concentration of competence pheromone, which is a small peptide, excreted by growing bacteria.

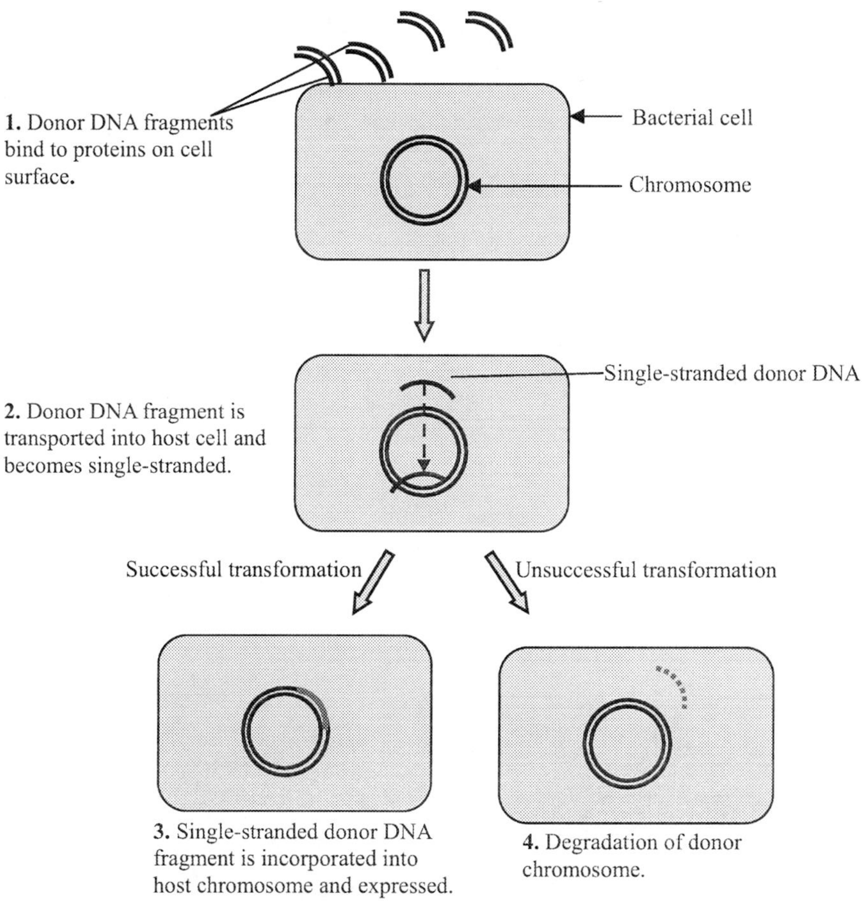

Figure 20.1: Bacterial transformation.

In common mechanism of transformation as seen in gram +ve *Bacillus* and *Streptococcus*, a competent cell binds a large double-stranded fragment of DNA at specific receptor sites on the surface of recipient bacterium following which,

DNA is moved across the cell membrane by an energy consuming process. One of the DNA strands is hydrolyzed by a membrane-bound exonuclease, providing energy to help fuel DNA transport across the membrane. The undegraded DNA strand associates with membrane proteins that were induced by the competence factor and is transported across the membrane. The single-stranded DNA fragment is coated with DNA-binding protein, and recombination (*rec* A) protein that search for the same or very similar DNA sequences in the chromosome of recipient cell (**Figure 20.1**). Other recombination enzymes catalyze the integration of single strand of donor, replacing the equivalent strand of the chromosome of the recipient cell. However, the nature of DNA to be taken up by natural transformation and its mechanism varies among different species, e.g. *Haemophilus*, a Gram-negative bacterium takes up only those DNA having a 10 bp specific sequence and takes up double-stranded DNA as such. Transformation is an important and routine technique in molecular biology and genetic engineering. Genetically engineered DNA always requires a suitable host for its expression. Recombinant plasmids carrying gene of interest are routinely transformed in *E. coli* for studying expression (**Figure 20.2**).

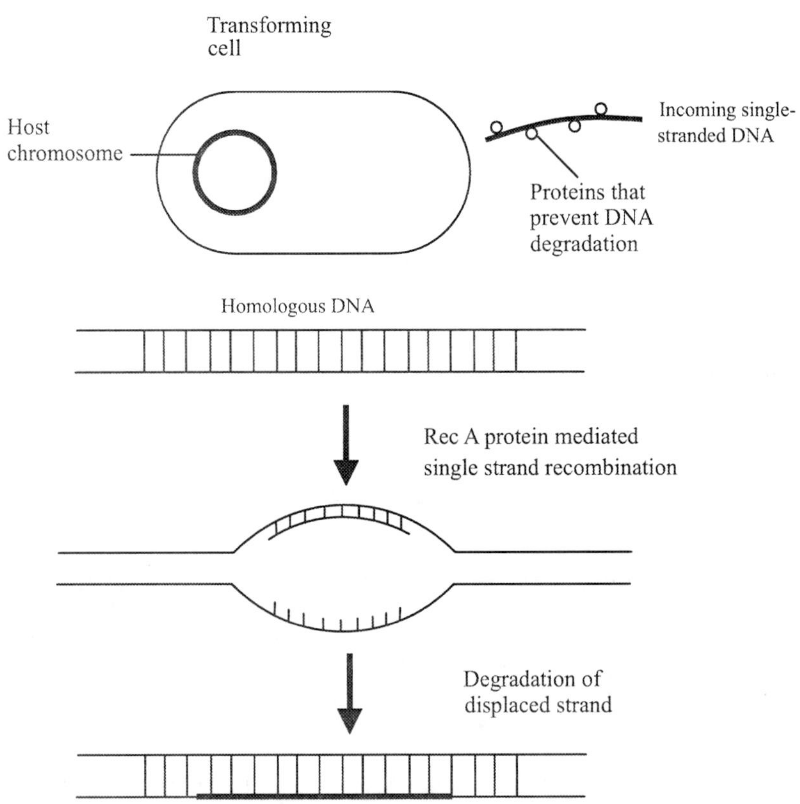

Figure 20.2: Transformation-recombination of the foreign DNA in bacterial chromosome.

Some strains (e.g. *E. coli* DH5 α) are made competent by incubating in $CaCl_2$ solution and then transformed.

CONJUGATION

Conjugation is a process in which transfer of genetic material is accomplished by direct cell-cell contact between donor and recipient bacterial cell. The contact is followed by a physical bridge between both cells. However, since genetic material is transferred unidirectionally and does not involve gamete formation or gamete fusion, conjugation is referred as a para-sexual process. It was discovered by Lederberg and Tatum in 1946 in *E. coli* while studying the genetic control of metabolism in *E. coli*. In an experiment involving two auxotrophic mutant strains of *E. coli*, where auxotrophic strain (A) required methionine and biotin ($met^{-1}bio^{-1}$) and other strain (B) required threonine and leucine ($thr^{-1}leu^{-1}$) for growth, a prototroph was produced out of their mating.

The prototroph ($met^{+1}bio^{+1}thr^{+1}leu^{+1}$) was able grow in the minimum nutrient medium. Thus, it was believed that when the auxotroph A and B come in proximity or contact there occurs genetic recombination and the genes responsible for auxotrophy are complemented in the deficient strain.

Two variants of this para-sexual mechanism are found in nature:

1. $F^+ \times F^-$ mating (involving transfer of plasmid)
2. Hfr conjugation (involving transfer of chromosomal DNA)

(1) $F^+ \times F^-$ Mating

Donor cell (F^+) carry a plasmid called the fertility factor (**F factor** or **F plasmid**) that is a circular extra chromosomal molecule containing about 94,000 base pairs. Genes of this factor are responsible for making cell-cell contact and F^- factor transfer from donor to recipient cell. Many F^- Plasmid genes direct the synthesis of **Sex pili** on the bacterial surface of F^+ bacteria that attach the F^+ cell to F- cell. During $F^+ \times F^-$ conjugation a contact tube is made between F^+ cell and F^- cell. The F factor replicates by the rolling circle mechanism and its single-stranded copy is transferred in the recipient cell. Later, this single-stranded factor is replicated in the recipient (**Figure 20.3**).

(2) Hfr Conjugation

Some F factors are also capable of integrating into the bacterial chromosome. This ability of integration in based upon presence of homologous sequences on F factor and the bacterial chromosome. At first F factor is integrated into bacterial chromosome at several locations by recombination between homologous insertion sequences present on both the F factor and bacterial chromosome. The integrated F factor carries out synthesis of sex pili, undergoes rolling-circle replication, and finally transfer of genetic material to a F^- cell (recipient cell) occurs. DNA transfer

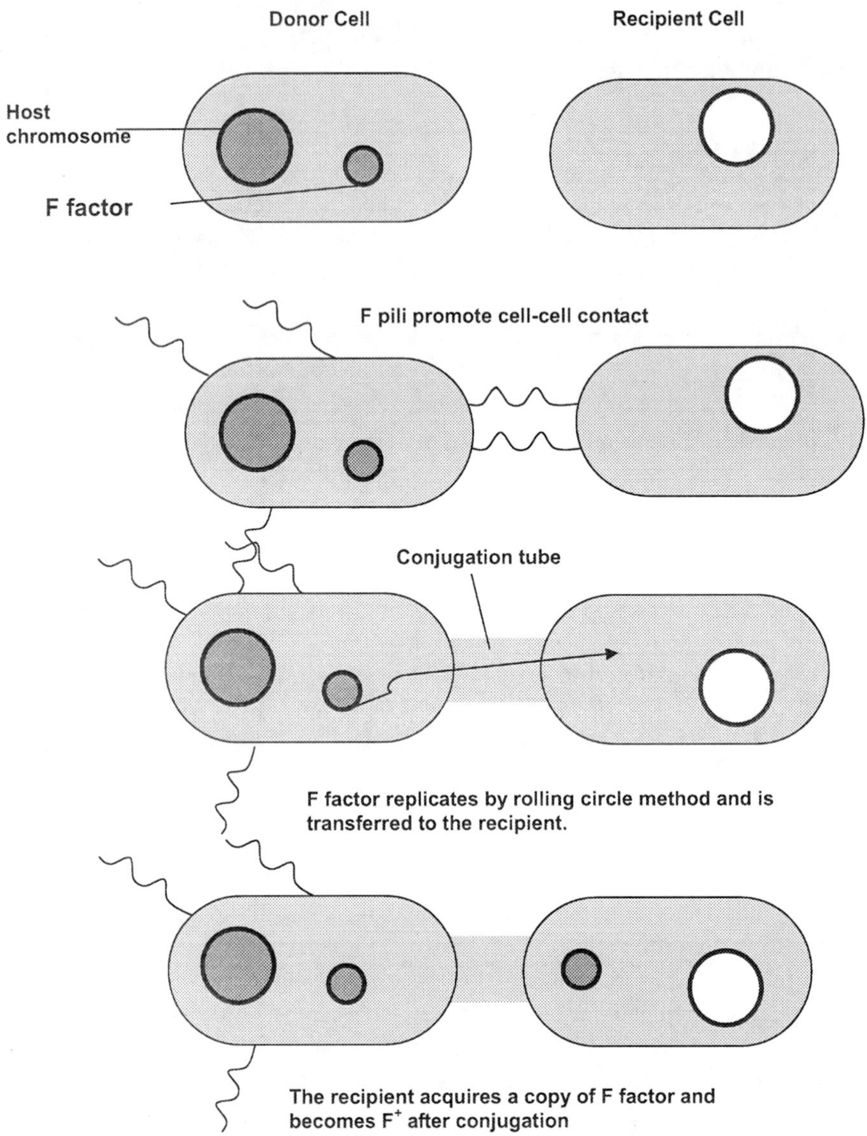

Figure 20.3: Conjugation in bacteria proceeds through mating, DNA transfer and separation.

is initiated at the origin of replication on the F factor, just as in $F^+ \times F^-$ mating. Rolling-circle replication transfers the donor strand from the main chromosome into the recipient cell where it becomes double-stranded. Usually, only a small part of F factor is transferred to the recipient, and the recipient remains F^-. It takes approximately one and a half hour for the entire bacterial chromosome to be transferred from Hfr strain to recipient, and the connection between conjugating bacteria is usually severed before the process is finished.

Ordinarily in Hfr strains the F factor after conjugation dissociates itself from the chromosomal DNA. Sometimes there is an error during excision of F factor and the incompletely separated F' factor contains a segment of bacterial chromosome. Mating between F'and F^- strain is very similar to $F^+ \times F^-$ excepting that bacterial genes of F'are transferred with a very high frequency into recipient cell. The recipient thus acquires a partially diploid set of the genes transferred with F'factor. This phenomenon is referred as a sexduction.

TRANSDUCTION

It is a process in which the transfer of bacterial DNA from donor bacterium to recipient bacterium is mediated by bacteriophages. In this process, bacterial genes get integrated with the phage capsid. The bacteriophage then injects these genes into another bacterium, transferring genetic material from one bacterium to another bacterium. Transduction is of two types (i) Generalized Transduction and (ii) Specialized Transduction.

In **generalized transduction** the phage transfers any segment of the bacterium genes to another bacterium (**Figure 20.4**). At the end of the phage life cycle, when phage DNA is being packaged inside phage head, fragments of the bacterial chromosome are erroneously packaged into the phage head. Due to this packaging mistake the phage particle carries bacterial DNA fragment in its capsid. The bacterial DNA fragment packaged into the phage varies with the size of the phage head, and approximately 30-60 genes may be packaged into head of a transducing phage.

Donor DNA must be integrated into the host (recipient) DNA for an effective transformation. The fragment of donor DNA is expressed into the recipient cell. However, the donor DNA is not replicated because it does not have any origin of replication, thus when the bacterial cell divides, only one of the two daughter cells receives the transduced DNA fragment. Other daughter cell having no donor fragment reverts back to its original phenotype. This is referred to as **abortive transduction** with respect to the reverted daughter cell. **Specialized transduction** denotes transduction of specific genes from donor bacterium to the recipient bacterium via. a specific bacteriophage that undergoes lysogeny. For example, λ-phage DNA proceeds into lysogeny by a site-specific recombination event. The λ-phage DNA gets integrated in bacterial genome and is referred to as **prophage**. At the end of lysogeny the circular phage chromosome loops out and is released. The excised λ-DNA may carry bacterial genes that flank its insertion sites. Such DNA is packaged into a phage coat and becomes a complete virion. Whenever this virion infects another bacterium, these bacterial genes get transduced or transmitted into their new host genome. However, the defective chromosome of this viron is unable to direct the synthesis of progeny phages but the bacterial genes carried by

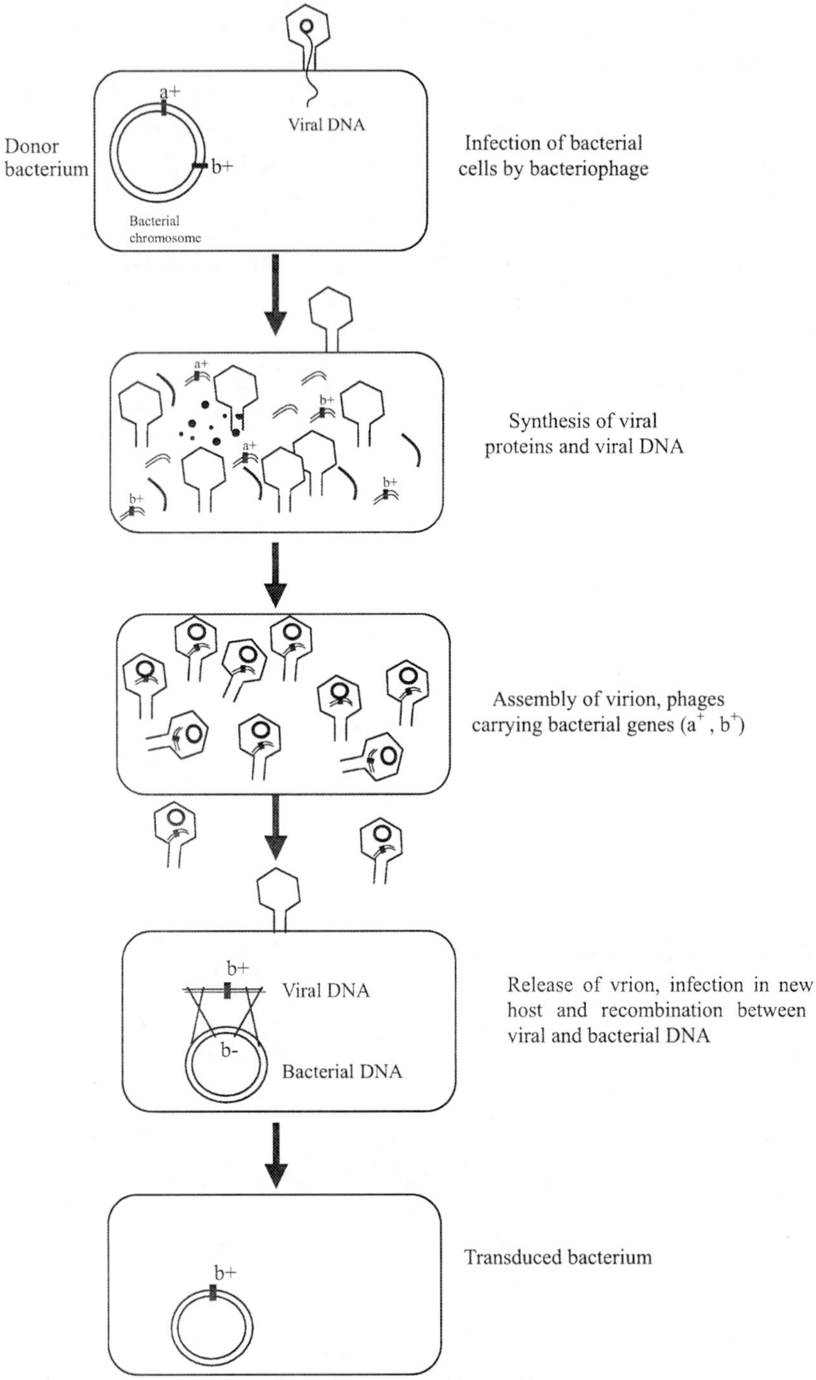

Figure 20.4: Generalized transduction in bacteria through bacteriophage

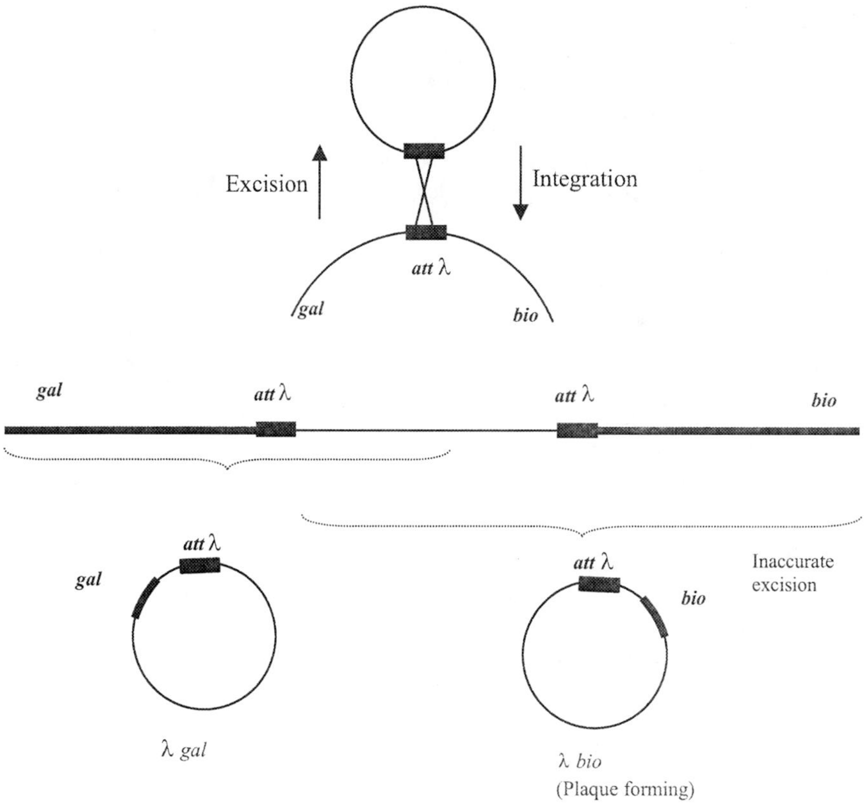

Figure 20.5: Specialized transduction in bacteria through bacteriophage. The λ DNA attaches in *E. coli* chromosome through site *att λ* at a locus between gal and bio genes. Termination of lysogeny results in erroneous detachment of λ phage DNA with bacterial DNA. Thus, the λ phage DNA may contain gal or bio genes of *E. coli*. Whenever, such a phage infects other host, it may tranduce these elements into it.

transducing particle recombine with the genes of the newly infected bacterium and become included into the chromosome of the recipient cell (**Figure 20.5**).

GENETIC MAPPING

In the post human genome era, there are many organisms whose complete genome has been sequenced. However, it is essential for us to localize various genes on chromosomal or plasmid DNA so as to manipulate them at our will using restriction endonucleases. The genetic map of *E.coli* genome showing location of various genes is shown in **Figure 20.6**. Gene map of a bacterial chromosome or plasmid can be obtained as described in the following text.

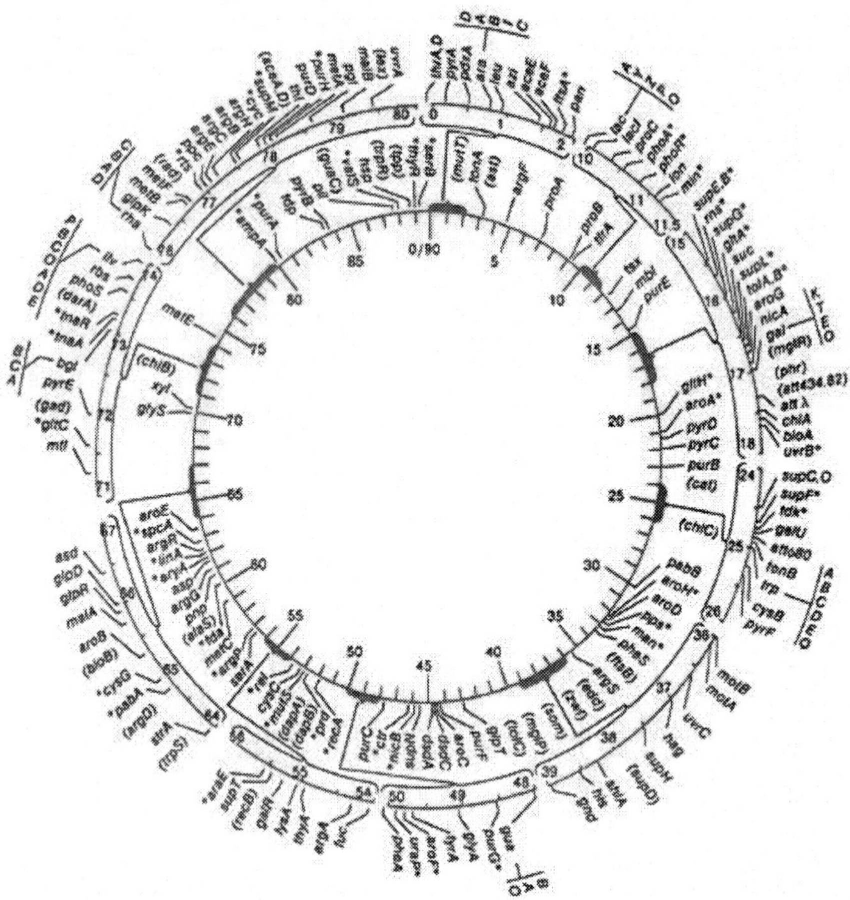

Figure 20.6: Circular genetic map of *E. coli*
Map distances are given in minutes and the total map length is 100 minutes.

Gene Mapping by Transformation

A schematic showing presence of genes at different loci on chromosome is referred as gene map. Experiments based on transformation and its abrupt intermittent stoppage has been extensively used in mapping of the bacterial genome. In transformation based gene mapping, two genes that are closely linked will be found frequently on the same DNA fragment. Such genes have a good chance of being co-transformed; On the other hand, if they are *far* apart they will never be found on the same fragment. To conclude about the position of any two genes we need three different types of donor DNA molecule, viz. $a^+ b^-$, $a^- b^+$ and $a^+ b^+$. The recessive cells are transformed with gradually decreasing concentration of all the three DNA molecules. Predictably, the frequency of single transformation should decrease linearly as per the decreasing concentration of respective DNA molecule. This is followed by determination of double transformation frequency of $a^- b^-$ cells by "gradually decreasing concentration of" $a^+ b^+$ DNA molecules of the two genes located close to each other. Similar kinetics of transformation is expected in both the experiments. However, if the genes are far apart, double transformation will occur only if two DNA fragment are taken up by recipient cells and two recombination occur. Thus, if DNA concentration is reduced ten times, the frequency of double transformation will fall hundred folds. Through a comparison of the frequencies of single and double transformations one can very easily conclude the relative position of two genes.

Gene Mapping using Conjugation

In a mating between Hfr × F^- bacteria, the chromosomal DNA moves from donor to recipient at constant rate. If the mating is stopped intermittently in an intermittent mating experiment by lysing cells open in a blender, the recipient cells will receive only some part of the DNA (**Figure 20.7**).

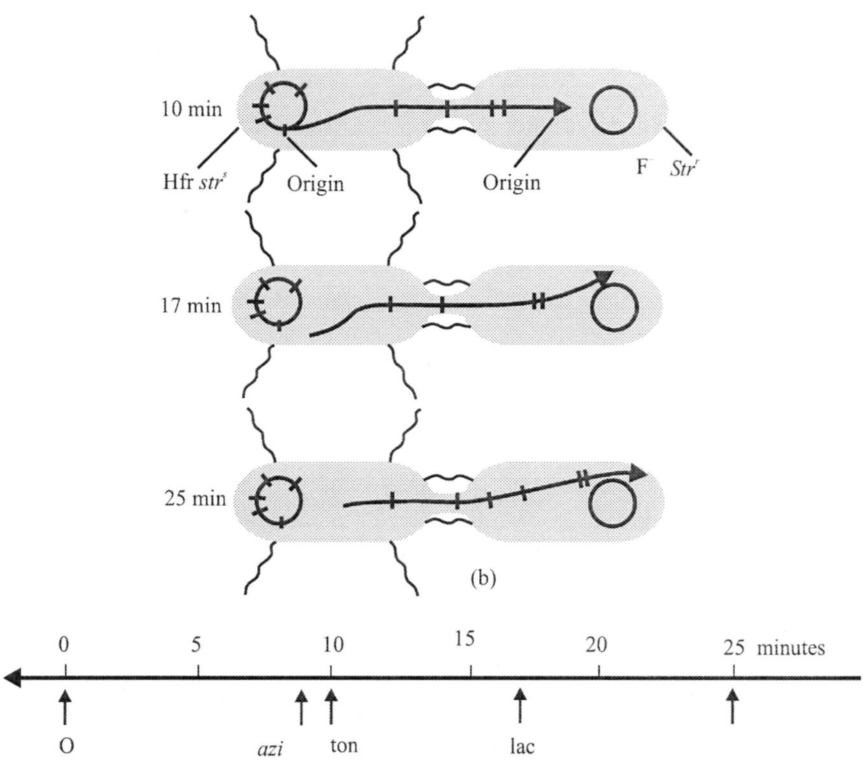

Figure 20.7: Genetic mapping by conjugation.

Based on time, the order of gene transfer can be deduced because first gene to enter the recipient cell is closest to the origin of transfer while the gene farthest from the origin will enter in the last. The map is generated by marking the genes on a circle in terms of time units.

PLASMIDS

Plasmids are self-replicating circular, double-stranded, supercoiled extra chromosomal DNA molecules. They make about 0.2-4% of the size of bacterial DNA content. Plasmids contain some special genes which are important for survival of bacterium in particular environment. Often such genes are used as indicator / marker of the plasmid. They depend on the metabolic state of the host cell for their replication; since they utilize most of the replication machinery of the host.

Types of Plasmids:

1. F plasmid (fertility or sex plasmid): These plasmids enable bacterium to conjugate and transfer chromosomal genes from a cell containing F plasmids to one that lacks it (F^-). Sometimes the F plasmid itself can also be transferred to a cell lacking the plasmid without conjugation (**Figure 20.8**).

2. R plasmid: R-plasmids denote drug resistance plasmids. They make the host cell resistant to one or more antibiotics, and many R plasmids can transfer the resistance to cells lacking R plasmid. These plasmids have gained more importance as they are responsible for evolution of multi drug resistance in pathogenic microbial forms (**Figure 20.9**).

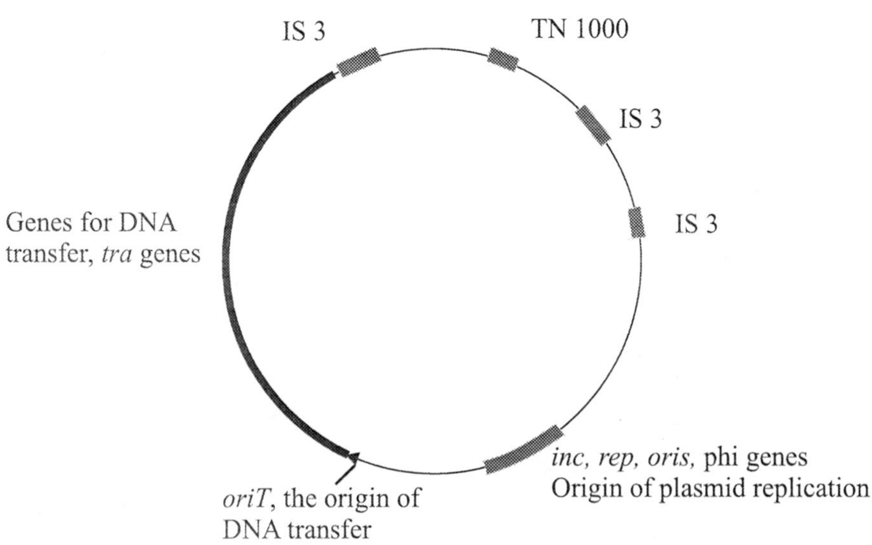

Figure 20.8: Map of F plasmid in *E. coli.*

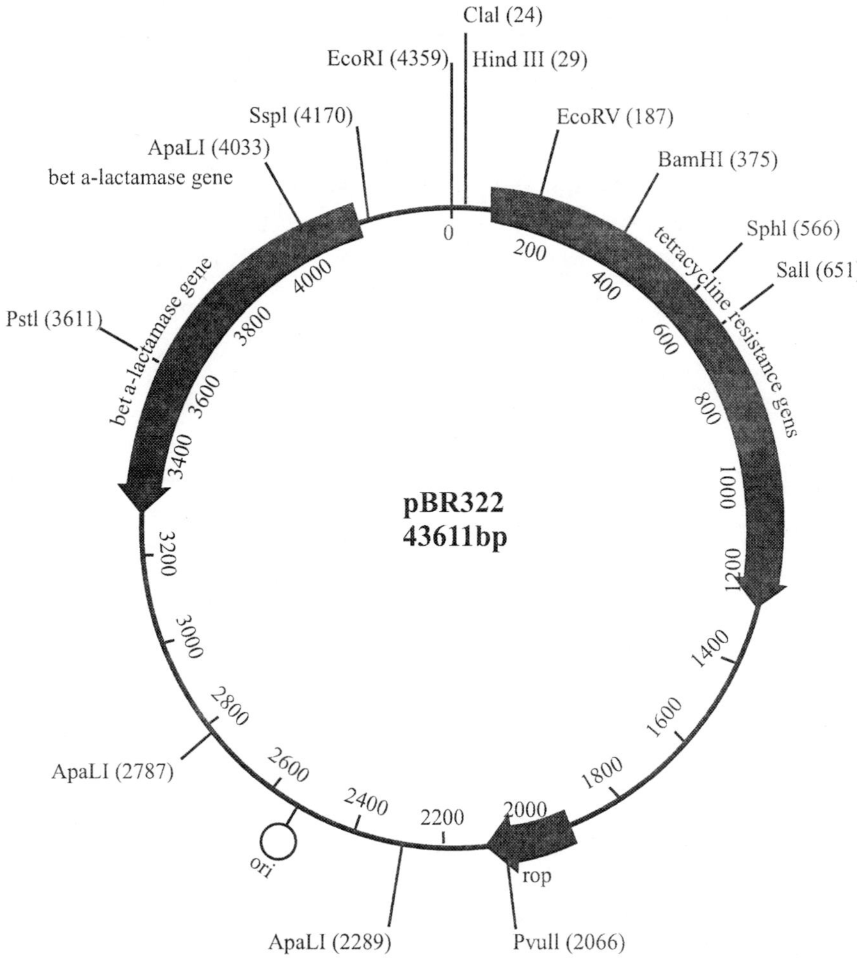

Figure 20.9: Map of pBR 322.

3. Col plasmid: These plasmids synthesize proteins, collectively called as Colicins (bacteriocins), which can kill closely related bacterial strains that lack a Col plasmid of the same type. Colicin producing plasmids can be classified into two types:

(a) Conjugative plasmids – 1-3 copies per cell

(b) Non-conjugative plasmids – 10-30 copies per cell

4. Degradative plasmid: These plasmids code for some special mechanism of degradation of some specific compounds. Some fermentative reactions are also specified by these plasmids, for example, the ability to ferment lactose (lac plasmid). Degradation plasmids are used to play an important role in bioremediation of xenobiotics and recalcitrant chemicals.

5. Mercury-resistance plasmid: Some strains of enteric bacteria, *Pseudomonas* and *Staphylococcus aureus* have plasmids that enable the host to survive under toxic concentrations of heavy metal ions. For example, *Pseudomonas* – plasmids code for resistance to mercury and *E. coli* and *Salmonella*- code for resistance to nickel, cobalt, mercury and arsenate.

6. Tumour-inducing plasmid: A large plasmid in *Agrobacterium tumefaciens* is found to cause crown gall disease in dicotyledonous plants. In this case, infection of plant cells by bacterial plasmid leads to development of tumour. These plasmids are used for development of transgenic plants.

7. Cryptic plasmid: These are obscure low molecular weight DNA molecules found in many bacteria and do not have any known biological role, therefore, they are called as cryptic plasmids.

COSMIDS

Plasmids carrying the phage λ 'cos' site are referred as cosmids. Cosmids are hybrids of a phage DNA molecule and a bacterial plasmid. Thus, cosmids are essentially plasmids that contain a minimum of 2500 bp of λ DNA which include-

(i) The cos site (the sequence yielding cohesive ends), and
(ii) Sequences needed for bringing about cleavage by terminases.

These vectors combine the features of a plasmid and phage λ to increase the probability of selecting a recombinant plasmid carrying foreign DNA. Cosmids characteristically have enzymes that pack the DNA molecule into the viral head on the cos site. The packaging reaction for any molecule that carries 'cos sites' separated by 37-53 kb of DNA can also be achieved. Cosmids need selectable marker such as amp^r or tet^r gene and an origin of replication of plasmid. Cosmids do not produce plaques, instead colonies are formed on selective media containing specific antibiotic as per the selectable marker of cosmid. Cosmids constructed as a hybrid molecule between a plasmid and a phage are also referred as **phagemids**. These have origin of replication from a plasmid (usually from pBR322) and the cos site from a λ phage. This gives the name cosmid (cos from cos site and mid from plasmid) to these elements. The cosmid infect the bacteria just like a phage. They are relatively small in size, but can accept the foreign DNA of very large size, often many times larger than length of cosmid DNA itself. The size of the cosmid should be appropriate enough to fit in the head space of the phage. These vectors are very convenient for making genomic libraries with large insert.

TRANSPOSONS

Transposons are found in both prokaryotes and eukaryotes. These were first described as controlling elements, because of their control on the expression of genes. R.W. Hedges and A.E. Jacob of London in 1974 used the term transposon for a DNA segment or genetic element which could move from one molecule to another and carried resistance for ampicillin. They noticed that transfer of antibiotic resistance from one plasmid to another is accompanied by an increase in size of the recipient plasmid (**Table 20.1**). Transposon can occupy different sites in the genome and can transponse itself between two bacterial chromosomes or two plasmids or between a plasmid and a bacterial chromosome. Transposition was also observed in those bacteria which were mutant for the gene *rec* A responsible for recombination, thus suggesting that these transfers do not involve normal recombination process. The two ends of each of the two strands in transposon consist of nucleotide sequences that are complementary to each other, but in reverse order. For instance, if one end is CCCAGAC, the other end will have GTCTGGG. These inverted repeats provide stem and loop structure characteristic to such sequences. These repeated sequences may range from short segment of nucleotides to as many as 1400 bp. Insertion of any gene between transposons makes the transfer of gene possible without any need of recombination process. Some transposons have their ends consisting of insertion sequences (IS) which resemble a transposon in having inverted repeats (**Figure 20.10**). Two identical transposons with an intervening sequence may give rise to a new but longer transposon.

Table 20.1 Properties of some well-known prokaryotic transposons

Transposon	Resistance against	Size (bp)	Size of inverted repeat (bp)
Tn1, Tn2, Tn3,Tn4	Ampicillin	4,957	38
	Ampicillin, Streptomycin, sulphonamide	20,500	Short
Tn5	Kanamycin	20,500	1500
Tn6	Kanamycin	4,200	Unknown
Tn7	Trimetroprin, Streptomycin	14,000	Unknown
Tn9	Chloramphenicol	2,638	18/38
Tn10	Tetracycline	9,300	1400
Tn204	Chloramphenicol, Fusidic acid	2,457	18/38
Tn402	Trimethoprim	7,500	Unknown
Tn551	Erythromycin	5,200	35
Tn903	Kanamycin	3,100	1050
Tn917	Erythromycin	5,100	Small
Tn1721	Tetracycline	10,900	Small

Transposable Elements in Yeasts

An example of transposable element in yeast is Ty1, which is present in 35 copies in yeast genome. At the ends are present direct repeats (not inverted repeats or IR as found in bacteria) called delta (d), of which 100 copies per genome are found.

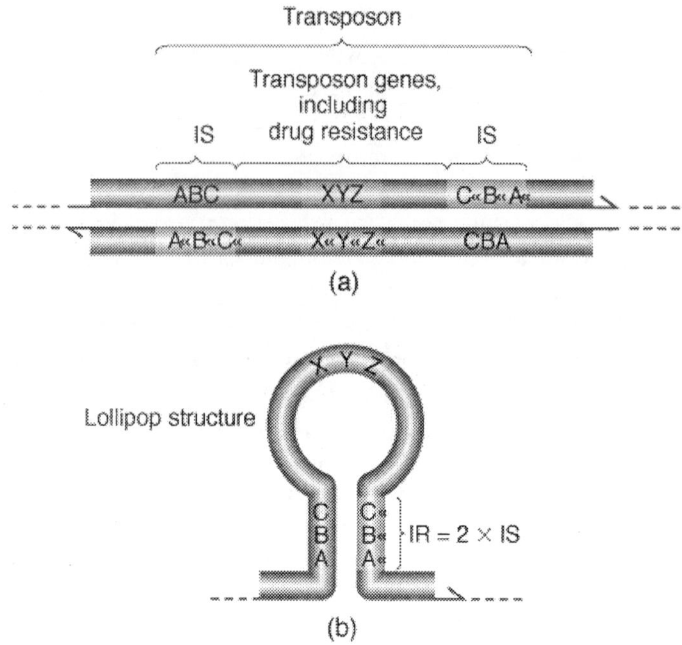

Figure 20.10: Mechanism of transposition **(a)** XYZ genes flanked by insertion sequences (IS), **(b)** looping out of the transposon.

Ty elements generate a repeated sequence (5bp long of the target DNA) during transposition and cause mutations due to insertion. Transposition in this case is caused through an RNA intermediate (class I transposons).

OVERLAPPING GENES

Any two genes are referred as overlapping genes when the signal to start transcription of the one gene resides inside the second gene. Also, the "stop" signal of the second gene must not be read by the ribosome during its **translation**. This is possible as RNA is read in triplets (codons) and hence there can always be three separate sequences (reading frames) that can be read by the ribosomes. Such reading frames are different in the RNA transcripts of the overlapping genes. Overlapping of genes allow the synthesis of more proteins from a given region of DNA than is possible if the genes were arranged in tandem or sequence. This hold true for most viral genomes, which are much smaller in comparision to the amount of information they code for. For example, the DNA present in the circular, single-stranded DNA genome of bacteriophage ϕX174 would not be large enough to encode all the eleven phage proteins if the genes were arranged in linear sequence. The overlapping genes provide an effective mechanism of genome economy. The genome project estimate says that the human genome contains around some 30000-70000 genes while it encodes for 100,000 to 200,000 proteins. Thus, there is a fair chance of several obscure overlapping genes, although several proteins come from alternative splicing of exons in a single gene. The nucleus (also called as nucleomorph owing to its ancient origin) of an algae, *Guillardia*, has only about 500,000 base pairs in genome but produces almost 500 proteins. This is because the genome contains 44 overlapping genes. Overlapping of genes enhances chances of fatality due to mutation, since a single mutation disturbs more than one protein.

SUGGESTED READING

Dale, J.W. 1998. *Molecular Genetics of Bacteria.* John Wiley & Sons.

Darnell, J., Lodish, H. and Baltimore, D. 1990. *Molecular Cell Biology*, Scientific American Books Inc., NY.

Freifelder, D. and Malacinski, G.M. 1993. *Essentials of Molecular Biology*, Jones and Bartlett Publishers Inc.

Glick, B.R. and Pasternak, J.J. 1994. *Molecular Biotechnology*, ASM Press, USA.

Hartl, D.L. Freifelder, D. and Snydder, L.A. 1988. *Basic Genetics*, Jones and Bartlett Publishers Inc.

Hawkins, J.D., 1996. *Gene Structure and Expression*, Cambridge Univ. Press.

Karp, G. 1996. *Cell and Molecular Biology: Concepts and Experiments*, John Wiley & Sons Inc. N.Y.

Lewin B. 2007. *Genes IX.* Jones and Bartlett Publishers, Sudbury, Massachusetts, USA.

REVIEW QUESTIONS

1. Describe various methods of genetic recombination in bacteria and their importance.
2. What is meant by transformation? Explain the process of transformation in bacteria.
3. Give details of types of conjugation seen in bacteria.
4. What is transduction? Differentiate between general and specialized transduction.
5. What is genetic mapping? Explain how gene mapping is done by conjugation?
6. Write notes on the following:
   (a) Plasmids
   (b) Cosmids
   (c) Transposons
   (d) Overlapping genes

Index